Ten Applications of Graph Theory

Mathematics and Its Applications *(East European Series)*

Hansjoachim Walther

Institute of Technology, Ilmenau, G.D.R.

Ten Applications of Graph Theory

D. Reidel Publishing Company

A MEMBER OF THE KLUWER ACADEMIC PUBLISHERS GROUP

Dordrecht / Boston / Lancaster

Library of Congress Cataloging in Publication Data

CIP

Walther, Hansjoachim.
Ten applications of graph theory.

(Mathematics and its applications. East European series ; v.)
Translation of: Anwendungen der Graphentheorie.
Bibliography: p.
Includes index.
1. Graph theory. I. Title. II. Series: Mathematics and its applications (D. Reidel Publishing Company). East European series ; v.
QA166.W3213 1984 511'.5 83-13854
ISBN 90-277-1599-8 (D. Reidel Pub. Co.)

Translated from the German by Ursula Nixdorf

Distributors for the Socialist Countries
VEB Deutscher Verlag der Wissenschaften, Berlin

Distributors for the U.S.A. and Canada
Kluwer Academic Publishers,
190 Old Derby Street, Hingham, MA 02043, U.S.A.

Distributors for all remaining countries
Kluwer Academic Publishers Group,
P. O. Box 322, 3300 AH Dordrecht, Holland

storage and retrieval systems, without written permission from the copyright owner.

Contents

Editor's Preface

Growing specialization and diversification have brought a host of monographs and textbooks on increasingly specialized topics. However, the "tree" of knowledge of mathematics and related fields does not grow only by putting forth new branches. It also happens, quite often in fact, that branches which were thought to be completely disparate are suddenly seen to be related.

Further, the kind and level of sophistication of mathematics applied in various sciences has changed drastically in recent years: measure theory is used (non-trivially) in regional and theoretical economics; algebraic geometry interacts with physics; the Minkowsky lemma, coding theory and the structure of water meet one another in packing and covering theory; quantum fields, crystal defects and mathematical programming profit from homotopy theory; Lie algebras are relevant to filtering; and prediction and electrical engineering can use Stein spaces. And in addition to this there are such new emerging subdisciplines as "completely integrable systems", "chaos, synergetics and large-scale order", which are almost impossible to fit into the existing classification schemes. They draw upon widely different sections of mathematics.

This program, Mathematics and Its Applications, is devoted to such (new) interrelations as exempla gratia:

— a central concept which plays an important role in several different mathematical and/or scientific specialized areas;
— new applications of the results and ideas from one area of scientific endeavor into another;
— influences which the results, problems and concepts of one field of enquiry have and have had on the development of another.

The Mathematics and Its Applications programme tries to make available a careful selection of books which fit the philosophy outlined above. With such books, which are stimulating rather than definitive, intriguing rather than encyclopaedic, we hope to contribute something towards better communication among the practitioners in diversified fields.

Because of the wealth of scholarly research being undertaken in the Soviet

Union, Eastern Europe, and Japan, it was decided to devote special attention to the work emanating from these particular regions.

Thus it was decided to start three regional series under the umbrella of the main MIA programme.

Graph theory, the topic of the present volume in the MIA (East European Series), is a fully-established subdiscipline in itself and related to many parts of (pure) mathematics. It is also eminently and directly applicable in a large number of concrete situations. This is what this book is about. For people who want to know how to apply graph theory (or who need examples for lectures on the topic) it has an almost ideal structure in that it first sets out the original problem, then proceeds to discuss the graph theory involved, and finally presents the algorithm(s) which exist to solve them.

The unreasonable effectiveness of mathematics in science ...

Eugene Wigner

Well, if you knows of a better 'ole, go to it.

Bruce Bairnsfather

What is now proved was once only imagined.

William Blake

As long as algebra and geometry proceeded along separate paths, their advance was slow and their applications limited.

But when these sciences joined company they drew from each other fresh vitality and thenceforward marched on at a rapid pace towards perfection.

Joseph Louis Lagrange

Amsterdam, March 1983

Michiel Hazewinkel

Preface

The present book is a translation of the textbook *Anwendungen der Graphentheorie* which was published by Deutscher Verlag der Wissenschaften in 1978. It is meant for students studying all branches of operations research, and for graduates and practical men to give them a means for modelling and solving organization and optimization problems, in particular with a combinatorial component.

The application of graph theory implies two aspects. On the one hand, it is applied graph theory with attention being given to the numerical ascertainment of the characteristic values of a given graph (e.g. the question arises of how to find a minimal set of arcs in a graph after the removal of which the graph is circuit-free; cf. Chapter 9). On the other hand, it implies the application of theorems and algorithms of graph theory in other scientific domains (when determining an optimal sequence of computation in algorithm, a decisive role is played, for example, by loops, and the question arises how many feedback arcs have to be cut to make the running of the algorithm loop-free; cf. Chapter 9). Both aspects are connected with each other and are discussed in the book.

The short introduction contains the most necessary concepts of graph theory which will be used in the text, and those concepts, which are required for one chapter only, are defined. Chapter 1 forms the basis for all other chapters dealing with flow problems, while the remaining chapters are, in essence, independent.

On the basis of well-known theorems of graph and network theory which will be discussed in more detail in the first chapter, a number of theorems will be formulated and proved, resulting in practicable algorithms. These will then be given at great length and elucidated by means of examples. Particular attention has been attached to the choice of examples used. They are by no means simple, but have been chosen, where possible, such that all difficulties, and also all subtleties of the algorithm, become apparent. Although the algorithms given are constructed such that they can be directly executed by computer engineering means, we had to desist from further preparation for those with practical experience, since the character of a textbook should be preserved.

In some chapters (Chapter 5 and 10) the bounds of current applications research are touched upon but, in general, mention is made in the bibliography of more or less dispersed methods which give an insight into the variety of spheres of application. Of course, completeness has not been aimed at and numerous examples of application (e.g. graph spectra in chemistry) could not be included. Much attention has been given to flow and tension and supply and transportation problems, and to planarity studies.

All theorems mentioned are proved and the reader who is primarily interested in the approaches and algorithms may skip the proofs for the time being. The more mathematically inclined reader, however, has the chance of checking his own capabilities in attempting to solve the numerous exercices contained in the text. The large number of figures are not only helpful for understanding, but they also point out the advantages offered by the possibility of representing graph and network problems in an illustrative way. The reader should thus become acquainted with standard problems and procedures and should also be enabled to recognize the combinatorial core of many other problems and, finally, to solve them on the basis of graph- and network-theoretical approaches.

In the framework of the translation the problematic nature of the complexity of problems (NPC problems), which has come to the fore at an even higher degree in recent years, has been indicated only briefly (cf. Chapter 6 "Assignment and travelling salesman problems").

Many thanks to all my colleagues who helped me to elaborate on the German original version and I wish to thank Mrs. Brigitte Schönefeld for having conscientiously typed the manuscript. My special thanks are due to Mrs. Ursula Nixdorf, who made the translation, and to my wife Ute who showed a lot of understanding for this work.

Ilmenau, February 1983 **Hansjoachim Walther**

Chapter 0

Introduction

Many problems encountered in various spheres of life involve graphs, directly or indirectly. Who would not have tried, as a schoolboy, to draw the "House of Santa Claus" with a single stroke (cf. Fig. 0.1) or to solve the problem of the three houses and the three factories (cf. Fig. 0.2) requiring each house to be connected with each factory by strokes which do not cross? Though less obviously, some of them may be formulated as problems of graph theory like, for example, the following:

A group of seven chess-players wants to find the best lightning-chess-player among them. Each one of them has to play twice for five minutes against each of the others. How can the match be organized so that after seven rounds each of the players has played one game against each of the others?

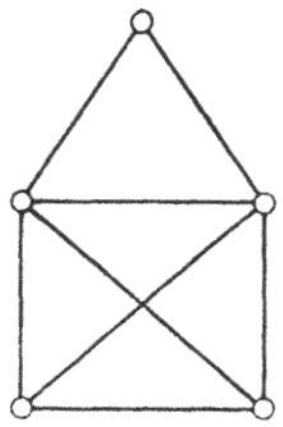

Fig. 0.1

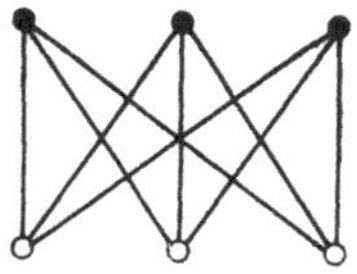

Fig. 0.2

By assigning a suitable graph to the problem, we transform the task into the following: assign a node to each player, connect each node with each of the others by an edge and then try to split up the resulting graph, which is a complete graph with seven nodes, in such a way that in each part there are exactly three edges and six nodes. In the language of graph theory: try to decompose the graph into seven matchings.

To the credit of the chess-players, it should be said that in fact they solved this problem long ago, for any number of players. In the case of an odd number of players, each of them plays exactly as many times as black, and in the case of an even number of players, half of them have white once more often than they have black and the other half have black once more often than they have white.

Let us now define some terms which we will use more frequently:

By a graph $G = G(\mathfrak{X}, \mathfrak{U})$, we mean a set $\mathfrak{U}$ of *edges* (or *arcs*), a set $\mathfrak{X}$ of *vertices* and an *incidence function* f assigning to each of the edges $u \in \mathfrak{U}$ an ordered or an unordered pair (X, Y) of vertices X and Y from $\mathfrak{X}$. X and Y are called the *end-points* of the edge u; if $X = Y$, u is called a *loop*. If only ordered pairs of vertices are assigned to them, the graph is called *directed* or *oriented*, otherwise *undirected*. In the case of a directed graph G and $f(u) = (X, Y)$, X is called the *initial vertex* and Y the *terminal vertex* of the arc u.

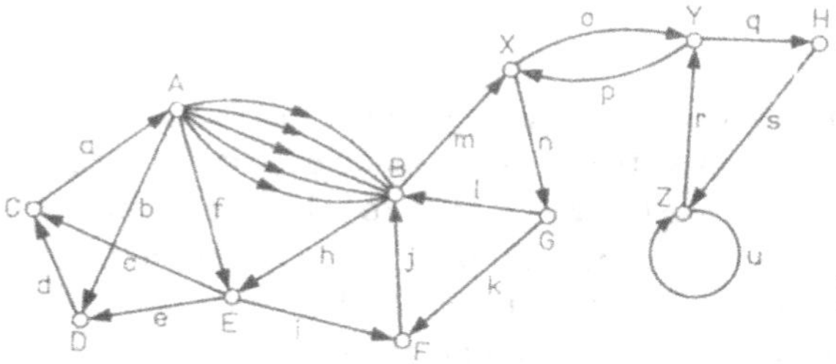

Fig. 0.3

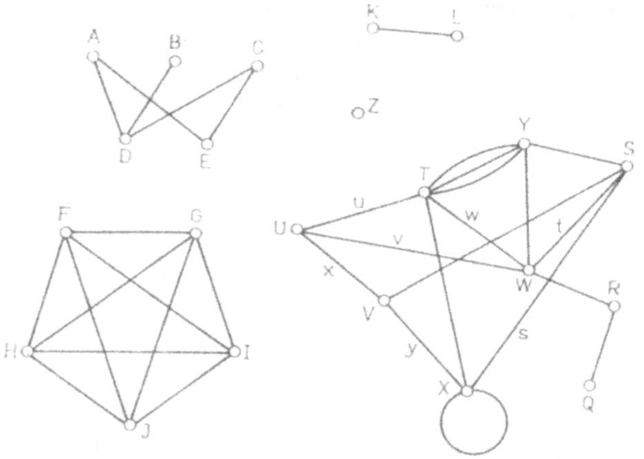

Fig. 0.4

Directed graphs will be represented as in Fig. 0.3, the undirected ones as in Fig. 0.4. Graphs which are partially directed and partially undirected will not be considered here. The elements of $\mathfrak{X}$ and $\mathfrak{U}$ will often be chosen from the set of natural numbers as shown in Fig. 0.5.

If several edges of an undirected graph have the same end-points, for example, the three edges of Fig. 0:4 having the end-points T and Y, the set of those edges is called a *multiple edge*. (In the case of the example cited, a *triple edge*.) Correspondingly, arcs which have the same initial and terminal vertices are called *multiple arcs*. Thus, in Fig. 0.3, the arcs with the initial point A and the terminal point B

form a *quintuple arc*, the two arcs with the terminal points X and Y, respectively, do not form a double arc, because the arcs are distinctly oriented.

Let $u_1, u_2, \ldots, u_r$ be edges of an undirected graph G, with for each subscript i ($i = 2, 3, \ldots, r - 1$) the edge u_{i-1} having one of its end-points in common with u_{i-1} and the other in common with u_{i+1}. Then we call $W = (u_1, u_2, \ldots, u_r)$ a ***chain*** of G. Thus, u, v, w, u, x, y in Fig. 0.4 form a chain. A chain is called a *simple chain* when it does not use the same edge twice, for example (w, u, v, t). A simple chain that does not encounter the same vertex twice is called *elementary*, for example (t, w, u, x). If a simple chain is closed, without the same vertex being used twice, we call it a *circuit*, as for example in Fig. 0.4 (t, w, u, x, y, s).

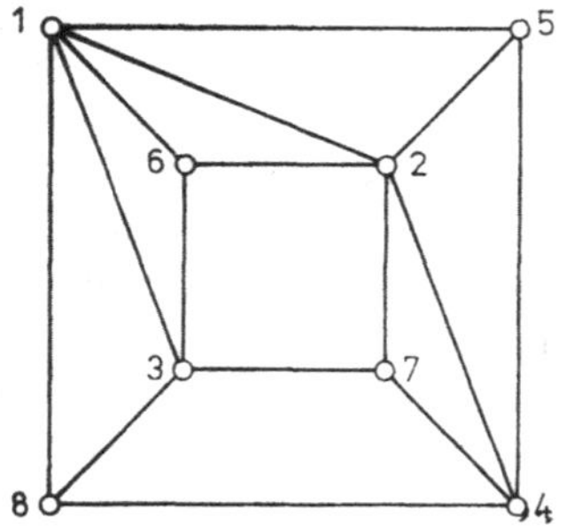

Fig. 0.5

A graph G is called *connected* if there exists an elementary chain for each pair of distinct vertices of G. The graphs in Figs. 0.1, 0.2 and 0.5 are connected; that in Fig. 0.4 is not connected, it possesses five components. A *component* is a maximal connected subgraph of a graph.

Two vertices X, Y are called *neighbouring* or *adjacent* if there is an edge u with $f(u) = (X, Y)$; X and Y are said to be *incident with u*. A vertex that is not incident with any edge is called *isolated*, as it is, for example, vertex Z in Fig. 0.4. The number of edges incident with a vertex X is called *degree* $v(X)$ of X. In Fig. 0.4, $v(S) = 4$ holds. Furthermore, we put $v(X) = 5$ (the incidence with a loop shall yield an increase of the degree of 2).

Let us consider some further concepts relating to directed graphs (details will be given in the first chapter):

A sequence $u_1, u_2, \ldots, u_r$ of arcs forms an *elementary chain* (or simply, ***chain***) if it goes over into a path of the resulting undirected graph, with the orientations of the arcs being ignored. If all arcs are oriented in the direction of the traversing of a chain, then the chain is a *simple path* (or an elementary chain). In Fig. 0.3, a, d, e, h, m form a chain, but not a simple path, connecting the vertices A and X, and p, n, l, h, c form a simple path connecting the vertices Y and C.

A directed graph is called *connected* if there exists, for each pair of vertices, a chain joining these vertices, if the undirected graph resulting from ignoring all orientations is connected. We call it *strongly connected* if for each pair of vertices there exists a simple path joining these two points.

The graph represented in Fig. 0.3 is strongly connected, that of Fig. 0.6 only connected.

Explaining the vertices as states of a Markov's chain and joining X and Y by an arc — if it should be possible to go from state X over to state Y with positive probability — would mean that the resulting graph would be strongly connected so that the states of the Markov's chain form a class of essential states.

We shall deal exclusively with finite graphs. These are graphs with a finite number of vertices, edges or arcs.

Finally, let us explain two more concepts: We shall frequently use the terms *maximal* and *maximum* (correspondingly *minimal* and *minimum*). A clear distinction has to be made between them. We always use maximal in the sense of relatively maximal, whereas maximum is used in the sense of absolutely maximal. We explain this by means of an example which is typical for the following expositions: We consider the graph G in Fig. 0.6. The vertices X, Y, Z span a maximal strongly connected subgraph G' since there is no "larger" strongly connected subgraph of G containing G'. But because of the strong connectivity, the graph G' is not a maximum subgraph, since the subgraph spanned by the vertices A, B, C, D, E is also strongly connected and contains five vertices.

We take another example (cf. Fig. 0.5): We search for those sets of vertices which *represent all circuits*, i.e., for a set $\mathfrak{Y}$ of vertices containing at least one vertex from each circuit.

Evidently, {5, 6, 7, 8} represents all circuits. This set is even minimal, because there is no proper subset representing *all* circuits. On the other hand, {2, 3, 5} or {2, 3, 4} forms a minimum set representing all circuits.

If we think, however, of a *vertex valuation* $w(X)$ (*weighting*) to be given, say in the form $w(X) = v(X)$, i.e. if its degree is assigned as valuation to each vertex, then for the above minimal sets such valuations result that {5, 6, 7, 8} with a total weight of 12 becomes a minimum set, just like {2, 3, 5}.

In the course of our expositions, we shall constantly come across such *weightings* or *valuations* of vertices (say by *potentials*) or of edges or arcs (by *flows*, *tensions*, *lengths*, *capacities*, *costs*, etc.), and it is through the valuation of the elements of a graph that we leave the *hard graph theory* as denoted by G. A. Dirac and come to the *network theory* or to the *application of graph theory*.

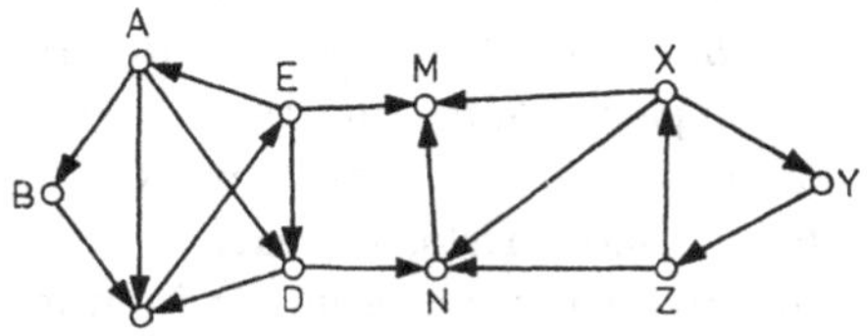

Fig. 0.6

Chapter 1

Flows and tensions on networks

1.1. BASIC CONCEPTS

For setting up a theory of flows and tensions the concepts of cycle and cocycle are of fundamental significance. Therefore, we want to put them at the top of our considerations.

Let $G(\mathfrak{X}, \mathfrak{U})$ be a directed graph with m arcs $u_1, \ldots, u_m$. A *cycle* μ is a cyclically ordered set of arcs $u_{i_1}, \ldots, u_{i_k}$ of G that are pairwise different, with the property that the arc u_{i_j} has one of its end-points in common with one of the end-points of $u_{i_{j-1}}$, and the other of its end-points in common with one of the end-points of $u_{i_{j+1}}$. Here, j is to be reduced modulo k (Fig. 1.1).

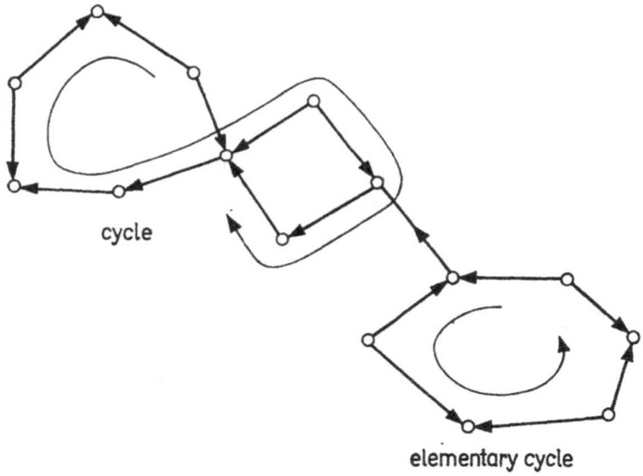

Fig. 1.1

We denote by μ^* the set of all arcs in μ, without taking into consideration the order:

$$\mu^* = \{u_{i_1}, \ldots, u_{i_k}\}.$$

By the cyclic order of the arcs the cycle μ is given a direction of traversing decomposing μ^* into two classes μ^+ and μ^-. The set μ^+ contains all arcs in μ^* that are in the direction in which the cycle μ is traversed, and the set μ^- contains all arcs in μ^* that are in the opposite direction in which μ is traversed. Thus, we get the relation

$$\mu^* = \mu^+ \cup \mu^- .$$

Let $\boldsymbol{\mu}^+$ and $\boldsymbol{\mu}^-$ be the characteristic vectors of μ^+ and μ^-, respectively. We assign to the cycle μ the vector

$$\boldsymbol{\mu} = \boldsymbol{\mu}^+ - \boldsymbol{\mu}^-$$

such that if we put

$$\boldsymbol{\mu} = (\mu_1, \mu_2, \ldots, \mu_m),$$

it holds:

$$\mu_i = \begin{cases} 1 & \text{if } u_i \text{ is contained in cycle } \mu \text{ and is in the direction in which this cycle is traversed,} \\ -1 & \text{if } u_i \text{ is contained in cycle } \mu \text{ and is in the opposite direction in which this cycle is traversed,} \\ 0 & \text{if } u_i \text{ is not contained in cycle } \mu. \end{cases}$$

If the direction of traversing is inverted, a cycle μ will be transformed into a cycle $\tilde{\mu}$. Evidently, it holds for the related vectors:

$$\tilde{\boldsymbol{\mu}}^+ = \boldsymbol{\mu}^-, \qquad \tilde{\boldsymbol{\mu}}^- = \boldsymbol{\mu}^+, \qquad \tilde{\boldsymbol{\mu}} = -\boldsymbol{\mu}. \tag{1}$$

We can also write

$$\tilde{\boldsymbol{\mu}} = -\boldsymbol{\mu}.$$

The set of the terminal vertices of the arcs of a cycle μ is called the *set of vertices of* μ.

A cycle μ is called an *elementary cycle* if, when traversing it, no vertex is encountered more than once. A cycle μ is called *minimal* if the set μ^* of its arcs contains no proper subset the arcs of which may be arranged themselves to form a cycle The following theorem is evident (task!):

THEOREM 1.1. *A cycle is minimal if and only if it is an elementary cycle.*

If for a given cycle μ pairwise arc-disjoint cycles $\mu^1, \ldots, \mu^q$ with $q \geqq 1$ can be found such that the appertaining vectors satisfy the equation

$$\boldsymbol{\mu} = \boldsymbol{\mu}^1 + \boldsymbol{\mu}^2 + \cdots + \boldsymbol{\mu}^q,$$

it is said that μ can be partitioned into the cycles $\mu^1, \mu^2, \ldots, \mu^q$.

Furthermore, it holds:

THEOREM 1.2. *Each cycle μ can be partitioned into elementary cycles.*

Proof. Follow a traversing of μ and split off an elementary cycle μ^1 when reaching for the first time a vertex already traversed. If μ is not identical to μ^1, then pursue the traversing and split off an elementary cycle μ^2 correspondingly, etc., until all arcs of μ are traversed (Fig. 1.2).

REMARK. Fig. 1.2 shows that, in general, the partition into elementary cycles is not unique.

A cycle is called a *conformally directed cycle* μ (*c-cycle* for short) if all of its arcs are directed in the sense of the orientation of μ (cf. Fig. 1.3). Thus, it holds $\mu_i \neq -1$ for all i's. We call a conformally directed elementary cycle an *elementary circuit* (shortly also, *circuit* (cf. Fig. 1.3)).

In order to get the concept of a cocycle, we proceed as follows. Let the set $\mathfrak{X}$ of the vertices of $G(\mathfrak{X}, \mathfrak{U})$ be partitioned into two non-empty classes $\mathfrak{A}$, $\mathfrak{B}$ such that the following relations hold:

$$\mathfrak{A} \cup \mathfrak{B} = \mathfrak{X}, \qquad \mathfrak{A} \cap \mathfrak{B} = \emptyset, \qquad \mathfrak{A} \neq \emptyset, \qquad \mathfrak{B} \neq \emptyset.$$

Furthermore, let ω^* be the set of those arcs that have one of their end-points in $\mathfrak{A}$ and the other end-point in $\mathfrak{B}$ (we assume that $\omega^* \neq \emptyset$). Considering the classes

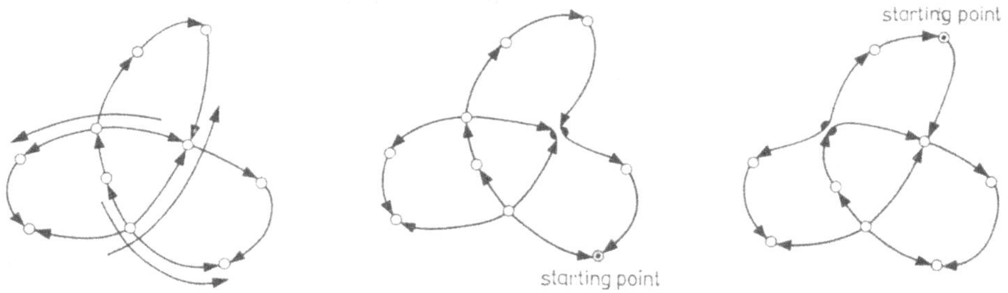

Fig. 1.2

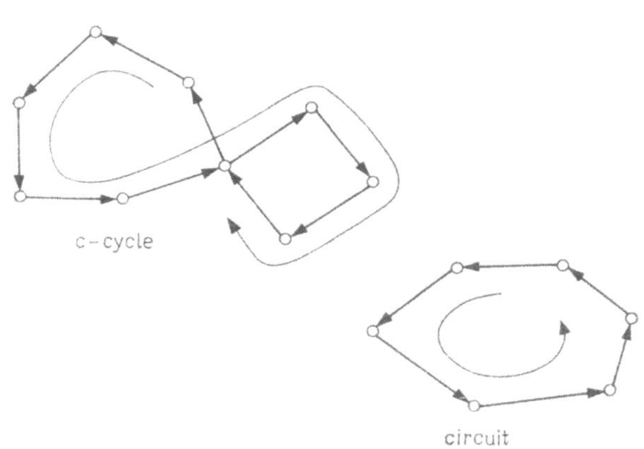

Fig. 1.3

$\mathfrak{A}$, $\mathfrak{B} = \mathfrak{X} - \mathfrak{A}$ to be an ordered pair we fix an orientation σ ("from $\mathfrak{A}$ to $\mathfrak{B}$"). Together with this orientation σ we denote the set ω^* to be the *cocycle* $\omega = \omega(\mathfrak{A})$ generated by $\mathfrak{A}$ (cf. Fig. 1.4).

By giving an orientation σ to the cocycle ω, the set ω^* of arcs is divided into two classes ω^+ and ω^-, with ω^+ and ω^- containing exactly those arcs of ω^* which are directed in the sense and in the countersense, respectively, of the orientation σ. According to the relations for cycles, it follows

$$\omega^* = \omega^+ \cup \omega^- .$$

Let $\boldsymbol{\omega}^+$ and $\boldsymbol{\omega}^-$ be the characteristic vectors of ω^+ and ω^-, respectively. Assign to the cocycle ω the vector

$$\boldsymbol{\omega} = \boldsymbol{\omega}^+ - \boldsymbol{\omega}^-$$

such that, when putting $\boldsymbol{\omega} = (\omega_1, \ldots, \omega_m)$, it holds for the components:

$$\omega_i = \begin{cases} 1 & \text{if } u_i \text{ is contained in the cocycle } \omega \text{ and directed in the sense of its orientation,} \\ -1 & \text{if } u_i \text{ is contained in the cocycle } \omega \text{ and directed in the countersense of its orientation,} \\ 0 & \text{if } u_i \text{ is not contained in the cocycle } \omega. \end{cases}$$

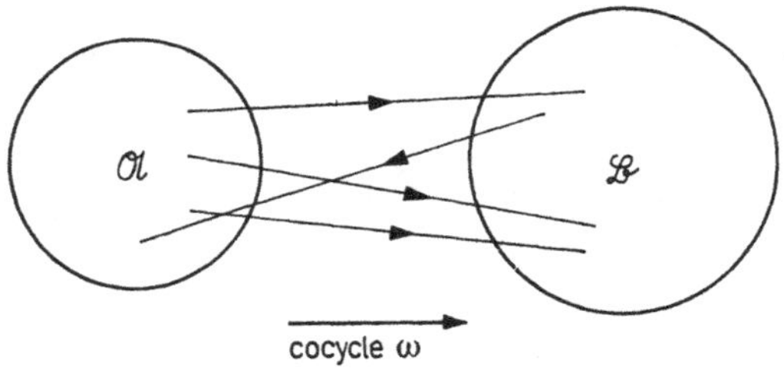

Fig. 1.4

Evidently, by reversing its orientation a cocycle $\omega = \omega(\mathfrak{A})$ is transformed into the cocycle $\tilde{\omega}$ generated by $\mathfrak{B} = \mathfrak{X} - \mathfrak{A}$, and analogously to the equations (1) for cycles, it holds:

$$\tilde{\boldsymbol{\omega}}^+ = \boldsymbol{\omega}^-, \qquad \tilde{\boldsymbol{\omega}}^- = \boldsymbol{\omega}^+, \qquad \tilde{\boldsymbol{\omega}} = -\boldsymbol{\omega}. \tag{2}$$

We can also write

$$\tilde{\omega} = -\omega$$

or

$$\omega(\mathfrak{B}) = \omega(\mathfrak{X} - \mathfrak{A}) = -\omega(\mathfrak{A}).$$

If only a single vertex X is contained in set $\mathfrak{A}$, then instead of $\omega(\{X\})$ write simply $\omega(X)$.

A clear idea of the structure of the cocycles may be got in the following way (cf. Fig. 1.5):

Let $\omega = \omega(\mathfrak{A})$ be a cocycle of the graph $\boldsymbol{G}$, and let $\hat{\boldsymbol{G}}$ be that graph resulting from $\boldsymbol{G}$ by deleting the arcs of ω. Evidently, $\hat{\boldsymbol{G}}$ represents the union of the two (disjoint) graphs $\boldsymbol{G}_{\mathfrak{A}}$ and $\boldsymbol{G}_{\mathfrak{B}}$ generated (spanned) in $\boldsymbol{G}$ by $\mathfrak{A}$ and by $\mathfrak{B} = \mathfrak{X} - \mathfrak{A}$. Let $\boldsymbol{G}_{\mathfrak{A}}$ have the components $\boldsymbol{A}_1, \boldsymbol{A}_2, \ldots, \boldsymbol{A}_s$ and $\boldsymbol{G}_{\mathfrak{B}}$ the components $\boldsymbol{B}_1$, $\boldsymbol{B}_2, \ldots, \boldsymbol{B}_t$. Let

$$\boldsymbol{A}_i = (\mathfrak{A}_i, \mathfrak{U}_i) \quad \text{and} \quad \boldsymbol{B}_j = (\mathfrak{B}_j, \mathfrak{V}_j) \qquad (i = 1, \ldots, s;\ j = 1, \ldots, t).$$

Thus, it holds:

$$\mathfrak{A} = \bigcup_{i=1}^{s} \mathfrak{A}_i, \qquad \mathfrak{B} = \bigcup_{j=1}^{t} \mathfrak{B}_j, \qquad \mathfrak{U} = \bigcup_{i=1}^{s} \mathfrak{U}_i \cup \bigcup_{j=1}^{t} \mathfrak{V}_j \cup \omega^*.$$

A cocycle ω is called *elementary* if all of its arcs connect one and the same component $\boldsymbol{A}_i$ of $\boldsymbol{G}_{\mathfrak{A}}$ with one and the same component $\boldsymbol{B}_j$ of $\boldsymbol{G}_{\mathfrak{B}}$ (cf. Fig. 1.6). This can be expressed also in the following way: The cocycle ω is *elementary* if and only if the number $p(\boldsymbol{G})$ of the components of $\boldsymbol{G}$ is increased by exactly 1 by deleting the arcs of ω, i.e.:

$$p(\boldsymbol{G} - \omega) = p(\boldsymbol{G}) + 1.$$

A cocycle ω is called *mininal* if the set ω^* of its arcs does not contain a proper subset, the arcs of which also form a cocycle.

Let us prove a lemma for cocycles that corresponds to lemma 1.1.

THEOREM 1.3. *A cocycle is minimal if and only if it is an elementary cocycle.*

Proof. Let ω be an elementary cocycle. We shall show that ω is minimal.

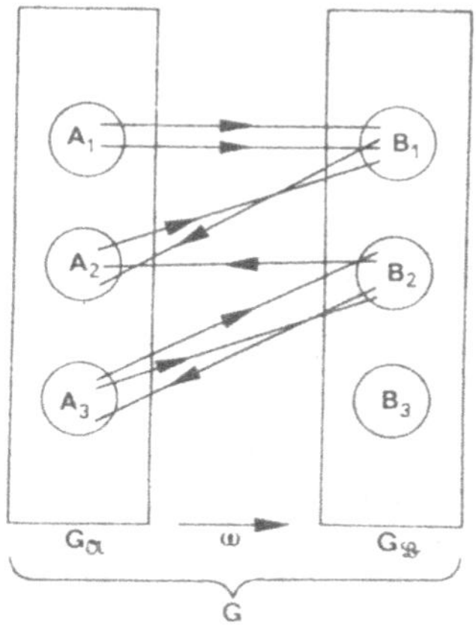

Fig. 1.5

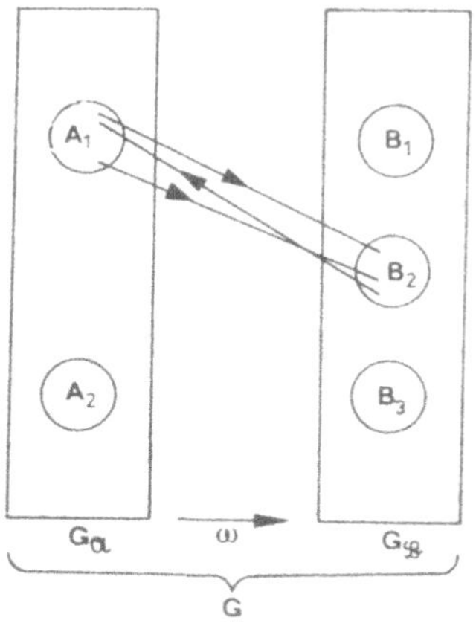

Fig. 1.6

Without restriction of generality we assume that each arc of ω connects one vertex of $\boldsymbol{A}_1$ with one vertex of $\boldsymbol{B}_1$. Evidently, it is sufficient to consider the subgraph $\boldsymbol{G}_1$ generated by $\mathfrak{X}_1 = \mathfrak{A}_1 \cup \mathfrak{B}_1$ in $\boldsymbol{G}$ and to take ω as cocycle ω_1 of $\boldsymbol{G}_1$. Cocycle ω_1 is generated in $\boldsymbol{G}_1$ by the set $\mathfrak{A}_1$ of vertices. If then $\omega_1' = \omega_1'(\mathfrak{A}_1')$ is any other cocycle of $\boldsymbol{G}_1$, it results either from ω_1 by reversing the orientation, in which case $\omega_1'^*$ is not properly contained in ω_1^*, or there exists a pair $X \in \mathfrak{A}_1'$ and $Y \in \mathfrak{B}_1' = \mathfrak{X}_1 - \mathfrak{A}_1'$ of vertices which belong both to $\mathfrak{A}_1$ or both to $\mathfrak{B}_1$. We may assume that X and Y are both contained in $\mathfrak{A}_1$. Since $\boldsymbol{A}_1$ is connected, there is in $\boldsymbol{A}_1$ a chain of arcs connecting X and Y. No arc of this chain belongs to ω_1, but at least one of its arcs belongs to ω_1', because $X \in \mathfrak{A}_1'$ and $Y \in \mathfrak{B}_1'$. From this, it follows that $\omega_1'^*$ is not contained in ω_1^*, where the assertion results from.

Conversely, let ω be a non-elementary cocycle. We shall show that ω is not minimal in this case.

Since ω is not elementary, there are either in $\boldsymbol{G}_{\mathfrak{A}}$ two components $\boldsymbol{A}_{i_1}$, $\boldsymbol{A}_{i_2}$ or in $\boldsymbol{G}_{\mathfrak{B}}$ two of them, say $\boldsymbol{B}_{j_1}$, $\boldsymbol{B}_{j_2}$ containing end-points of arcs which belong to ω. We may assume that this applies say to $\boldsymbol{B}_1$ and $\boldsymbol{B}_2$ (cf. Fig. 1.5). Obviously, the arcs of ω^*, one of the end-points of which lies in $\boldsymbol{B}_2$, but not the other, are then forming a cocycle, because, when these arcs are deleted, $\boldsymbol{B}_2$ becomes a new component which was not contained in $\boldsymbol{G}$. Then, however, ω was not minimal. This proves lemma 1.3.

Next we show:

THEOREM 1.4. *Each cocycle ω may be decomposed into elementary cocycles.*

Analogously to the decomposition of a cycle, we mean by the *decomposition of a cocycle ω into cocycles $\omega^1, \ldots, \omega^q$* a decomposition of the set ω^* of arcs, with each one, ω^{i*}, being a cocycle.

Proof. Show that there are pairwise arc-disjoint elementary cocycles $\omega^1, \ldots, \omega^q$ such that for the vectors assigned to them the equation

$$\boldsymbol{\omega} = \boldsymbol{\omega}^1 + \boldsymbol{\omega}^2 + \cdots + \boldsymbol{\omega}^q$$

holds.

Evidently,

$$\boldsymbol{\omega} = \boldsymbol{\omega}(\mathfrak{A}) = \boldsymbol{\omega}(\mathfrak{A}_1) + \boldsymbol{\omega}(\mathfrak{A}_2) + \cdots + \boldsymbol{\omega}(\mathfrak{A}_s)$$

with the cocycles $\omega(\mathfrak{A}_i)$ being pairwise arc-disjoint (Fig. 1.5). Therefore, it is sufficient to show that each of the single cocycles $\omega(\mathfrak{A}_i)$ can be decomposed into pairwise arc-disjoint elementary cocycles. We prove this for $\omega(\mathfrak{A}_1)$. Let $\mathfrak{A}' = \mathfrak{A}_1$ and $\mathfrak{B}' = \mathfrak{X} - \mathfrak{A}_1$. Then the graph $\boldsymbol{G} - \omega(\mathfrak{A}_1)$ splits into the components $\boldsymbol{B}_k'$ $= (\mathfrak{B}_k', \mathfrak{B}_k')$ with $k = 1, \ldots, r$ $(r \geqq 1)$. Here, we may assume that the numbering has been made in such a way that exactly the components $\boldsymbol{B}_1', \boldsymbol{B}_2', \ldots, \boldsymbol{B}_l'$ $(1 \leqq l \leqq r)$ in $\boldsymbol{G}$ are connected with $\boldsymbol{A}_1$ by at least one arc each (Fig. 1.7). Then, all of the cocycles generated by the $\mathfrak{B}_\nu'$'s $(\nu = 1, \ldots, l)$ are elementary and pair-

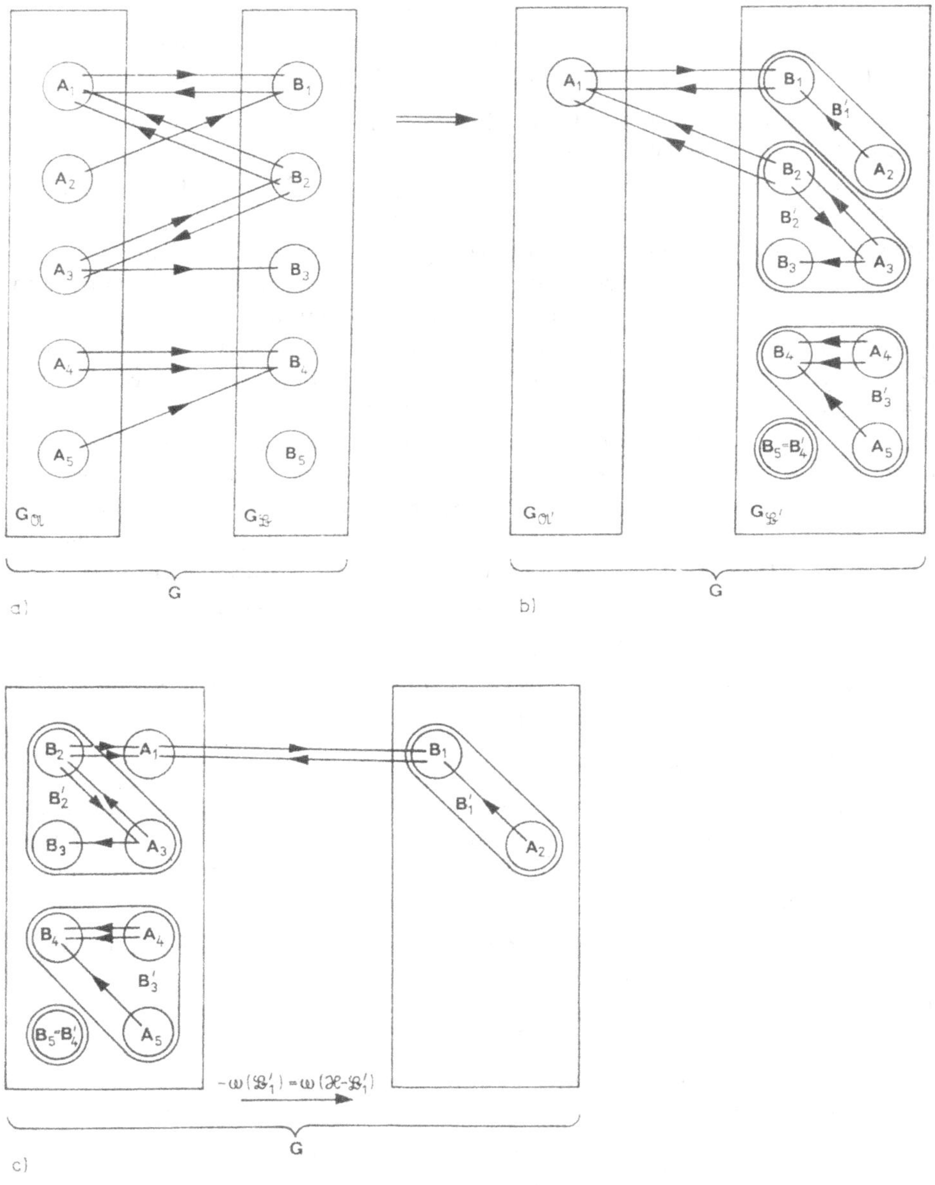

Fig. 1.7

wise arc-disjoint. Furthermore, it holds:

$$\omega(\mathfrak{A}_1) = -\omega(\mathfrak{B}_1') - \omega(\mathfrak{B}_2') - \cdots - \omega(\mathfrak{B}_l').$$

That means that $\omega(\mathfrak{A}_1)$ is decomposed into the pairwise arc-disjoint elementary cocycles $-\omega(\mathfrak{B}_\nu') = \omega(\mathfrak{X} - \mathfrak{B}_\nu')$ with $\nu = 1, \ldots, l$. This proves theorem 1.4.

The reader will have noticed the "duality" existing between the concepts of *cycle* and *cocycle* (to which also the designation refers). This is, however, an essential phenomenon, and as we shall see, the analogies may be carried much further: consider, for example, the field of electrical engineering where in particular the "duality" known between current and voltage can be based on it. This is, nevertheless, not duality in the true sense, i.e., in the sense that it would be possible to base the whole theory on a system of axioms being dual in itself. If this were the case, then it would also be possible to dualize the proofs, and it would be hard to see why proving theorems 1.3 and 1.4 for cocycles should give more trouble than the proofs of the analogous theorems 1.1 and 1.2 for cycles. But if we restrict ourselves only to planar graphs (cf. p. 29), duality can actually be encountered in the true sense. In a general case we can only speak of a "partial duality". Dealing with these questions in more detail leads to the *Theory of Matroids* established by H. Whitney [19] and developed to an essential degree in particular by W. T. Tutte [15—17]. The interested reader's attention is drawn to the book of W. T. Tutte [18] and to the fundamental work of G. J. Minty [11].

For our further studies let us explain some other concepts.

DEFINITION. A *conformally directed cocycle* ω (simply called *c-cocycle*) is a cocycle, all edges of which are directed in the sense of the orientation of ω.

Thus, for the components ω_i of a *c*-cocycle, it holds: $\omega_i \neq -1$.

DEFINITIONS. An elementary *c*-cocycle is called a *cocircuit*. A *spanning tree* (*cotree*) of a connected graph is a partial graph without cycles (or cocycles) having the property of forming a cycle (or cocycle) after any other arc has been added.

Let us first formulate and prove an important lemma.

LEMMA of Minty (*Arc Colouring Lemma*). *Consider a graph G with any m arcs* $1, \ldots, m$. *Colour arc* 1 *black, and arbitrarily colour the remaining arcs* $2, 3, \ldots, m$ *black, green or red.*

One of the following conditions holds:

a) *There is an elementary cycle containing arc* 1 *and only red and black arcs with the property that all black arcs are directed in the sense of the cycle*[1]).

b) *There is an elementary cocycle containing arc* 1 *and only green and black arcs with the property that all black arcs are directed in the sense of the cocycle*[1]).

---

[1]) Containing black *and* red arcs and black *and* green arcs, respectively, also means that red or green need not necessarily occur.

Proof. Let arc $1 = (X, Y)$ be directed from X to Y. Successively, label the vertices using the following procedure:

(i) Label Y.

(ii) Label vertex Z if there is either
- a black arc (U, Z) with labelled U or
- a red arc (U, Z) with labelled U or
- a red arc (Z, U) with labelled U.

Because of the finiteness of the graph, the labelling procedure stops and the following two cases may occur:

Case 1: X has been labelled. According to the labelling procedure, there will be a cycle containing arc 1 and only black and red arcs, with all black arcs being directed in the sense of the cycle. Following theorem 1.2, this cycle splits into elementary cycles among which there is obviously one elementary cycle μ satisfying the condition a). Suppose arc 1 to be contained in an elementary cocycle ω according to b), then ω separates X from Y, i.e., at least one arc of each path between X and Y is contained in ω. Thus, there must be still another arc of μ lying in ω. But this is either a red one or a black arc directed in the sense of ω and opposite to arc 1. This, however, contradicts our assumption. If there exists an elementary cycle according to a) through 1, there will be no elementary cocycle according to b) through 1.

Case 2: X is not labelled. Let $\mathfrak{A}$ be the set of the labelled and $\mathfrak{B}$ be the set of the unlabelled vertices of the graph. The arcs between $\mathfrak{A}$ and $\mathfrak{B}$ form a cocycle ω containing arc 1. According to the labelling procedure, ω contains no red arc and a black one if and only if this arc is directed from $\mathfrak{B}$ to $\mathfrak{A}$. Thus, we have found a cocycle according to b). Then, using theorem 1.4, there will also be an elementary cocycle according to b). Suppose there exists an elementary cycle μ according to a). Following our considerations in case 1, this assumption can be proved to be a contradiction.

The lemma is thus proved.

COROLLARY. *Let u be any arc of a graph* $\mathbf{G}$. *Then, u belongs either to a circuit or to a cocircuit, but never to both.*

To prove this, colour all arcs black and apply the lemma of Minty.

We now turn to the "Partial duality" mentioned above. The following theorem holds:

THEOREM 1.5. *Let* $\mathbf{G}$ *be a connected graph,* $\mathbf{V}$ *and* $\mathbf{W}$ *any spanning tree and spanning cotree, respectively, of* $\mathbf{G}$. *Then, the arcs of* $\mathbf{G}$ *that do not belong to* $\mathbf{V}$ *and* $\mathbf{W}$, *respectively, form a spanning cotree and a spanning tree, respectively, of* $\mathbf{G}$.

Proof. Let $\boldsymbol{V} = (\mathfrak{X}, \mathfrak{B})$ be a spanning tree of $\boldsymbol{G} = (\mathfrak{X}, \mathfrak{U})$. Suppose the graph $\boldsymbol{G}' = (\mathfrak{X}, \mathfrak{U} - \mathfrak{B})$ has a cocycle $\omega(\mathfrak{A})$ of $\boldsymbol{G}$. Because there exists a chain of arcs in $\boldsymbol{V}$ directed from $\mathfrak{A}$ to $\mathfrak{X} - \mathfrak{A}$, at least one of the arcs from $\omega(\mathfrak{A})$ lies in $\mathfrak{B}$. This yields, however, a contradiction to the definition of a cocycle in $\boldsymbol{G}'$. Adding any arc of $\mathfrak{B}$ to $\boldsymbol{G}'$ leads to the decomposition of $\boldsymbol{V}$. The resulting graph thus contains a cocycle of $\boldsymbol{G}$.

Conversely, let $\boldsymbol{W}$ be a spanning cotree of $\boldsymbol{G}$, say $\boldsymbol{W} = (\mathfrak{X}, \mathfrak{W})$. $\boldsymbol{W}$ does not contain a cocycle, but such a cocycle arises after adding any other arc. Consider the set of arcs $\mathfrak{B}^* = \mathfrak{U} - \mathfrak{W}$, choose any arc in $\mathfrak{B}^*$ and denote it by 1. Now, colour the arcs of $\boldsymbol{G}$. The arc 1 is coloured black, the remaining arcs of $\mathfrak{B}^*$ are coloured red. The arcs of $\mathfrak{W}$ are coloured green.

Since $\boldsymbol{W}$ is a cotree, $\mathfrak{W} \cup \{1\}$ contains a cocycle going through 1 which is coloured black and green in conformity with the colouring procedure. According to the lemma of Minty, no cycle coloured black and red will then go through arc 1, thus $\mathfrak{B}^*$ contains no cycle through arc 1. Since arc 1 has been arbitrarily chosen, there will be no cycle in $\mathfrak{B}^*$ at all.

It remains to show that a cycle arises when adding any arc to $\mathfrak{B}^*$.

Let 1 be any arc not contained in $\mathfrak{B}^*$, i.e., let $1 \in \mathfrak{W}$. Colour the graph. The arc 1 is coloured black, the arcs from $\mathfrak{B}^*$ are coloured red, those from $\mathfrak{W} - \{1\}$ are coloured green.

Since $\boldsymbol{W}$ is a cotree, no subset of $\mathfrak{W}$ is a cocycle, thus, $\boldsymbol{G}$ contains no cocycle coloured green and black going through arc 1. Then, according to the lemma of Minty, there exists in $\boldsymbol{G}$ a cycle coloured black and red which is formed by arcs of $\mathfrak{B}^* \cup \{1\}$ following the colouring procedure. Since here again arc 1 has been arbitrarily chosen, all is proved.

Q.E.D.

Let us now come to a concept which is of essential importance for the theory of flows and tensions.

DEFINITION. The cycles $\mu^1, \mu^2, \ldots, \mu^p$ (and the cocycles $\omega^1, \omega^2, \ldots, \omega^p$, respectively) are called *independent* if the systems of the associated vectors $\boldsymbol{\mu}^1, \boldsymbol{\mu}^2, \ldots, \boldsymbol{\mu}^p$ (and $\boldsymbol{\omega}^1, \boldsymbol{\omega}^2, \ldots, \boldsymbol{\omega}^p$, respectively) are linearly independent.

We may continue the theorems confirming the "partial duality".

THEOREM 1.6. *Let $\boldsymbol{G}$ be a connected graph with m arcs and n vertices. Then, it holds:*

a) *There exist $m - n + 1$, but not more independent elementary cycles.*

b) *There exist $n - 1$, but not more independent elementary cocycles.*

Proof. Obviously, any spanning tree $\boldsymbol{V}$ of $\boldsymbol{G}$ contains $n - 1$ arcs. If we select arbitrarily one of the remaining $m - (n - 1)$ arcs and add it to $\boldsymbol{V}$, one elementary cycle is generated the orientation of which is thought to be in the sense of the arc added.

In this way it is possible to generate exactly $m - n + 1$ elementary cycles which are obviously independent, since each of the vectors assigned to these cycles has a unity in a component in which all other vectors have a zero.

Now, we show that there exist $n - 1$ independent elementary cocycles. For this, consider any spanning cotree $\boldsymbol{W}$. According to the definition, $\boldsymbol{W}$ contains no cocycle. If one adds, however, any arc, a cocycle will arise that is obviously elementary. Using theorem 1.5 there are exactly $n - 1$ arcs not contained in $\boldsymbol{W}$. It is easy to see that these $n - 1$ cocycles are independent, since the vectors being assigned to the elementary cocycles also have the property that each of them has a unity in a component in which all others have a zero.

Finally, we still show that for any cycle μ and for any cocycle ω the relation

$$\langle \mu, \omega \rangle = 0$$

holds, i.e., the scalar product of the vectors assigned to them equals zero. That means, however, (since the sum of the dimensions of the two subspaces spanned by the cycles and the cocycles, respectively, cannot be greater than m) that there are neither more than $m - n + 1$ independent elementary cycles (and thus cycles) nor more than $n - 1$ independent elementary cocycles.

We consider the scalar product

$$\langle \mu, \omega \rangle = \mu_1\omega_1 + \mu_2\omega_2 + \cdots + \mu_m\omega_m$$

with only those components being of interest in case of which the arcs corresponding to them are contained in both μ and ω because in the other cases, at least one of the two factors μ_j, ω_j is equal to zero. Furthermore, consider two successive arcs k, l in μ also lying in ω. The following four cases may occur:

1. $\omega_k = 1 \Rightarrow \mu_k = 1$ or $\mu_k = -1$
   $\omega_l = 1 \quad\ \mu_l = -1 \quad\ \mu_l = 1$
2. $\omega_k = 1 \Rightarrow \mu_k = 1$ or $\mu_k = -1$
   $\omega_l = -1 \quad\ \mu_l = 1 \quad\ \mu_l = -1$
3. $\omega_k = -1 \Rightarrow \mu_k = 1$ or $\mu_k = -1$
   $\omega_l = 1 \quad\ \mu_l = 1 \quad\ \mu_l = -1$
4. $\omega_k = -1 \Rightarrow \mu_k = 1$ or $\mu_k = -1$
   $\omega_l = -1 \quad\ \mu_l = -1 \quad\ \mu_l = 1$

In each of the four cases it holds:

$$\omega_k\mu_k + \omega_l\mu_l = 0.$$

Since the number of the arcs contained in μ and ω is certainly even, we can divide the totality of all these arcs into pairs of successive arcs in μ. For each pair the share in the scalar product equals zero, thus also the whole scalar product, as stated.

The above remarks prove theorem 1.6.

Hence we see that the elementary cycles and cocycles constructed for this proof form a basis of the corresponding vector space.

The following theorem can also be proved using some adequate reflections.

THEOREM 1.6′. *If the graph* $\boldsymbol{G}$ *has exactly* p *components, then it holds:*

a) *There exist* $m - n + p$ *independent elementary cycles, but not more.*

b) *There exist* $n - p$ *independent elementary cocycles, but not more.*

As for maps, for example, a cycle basis may be given in a simple form.

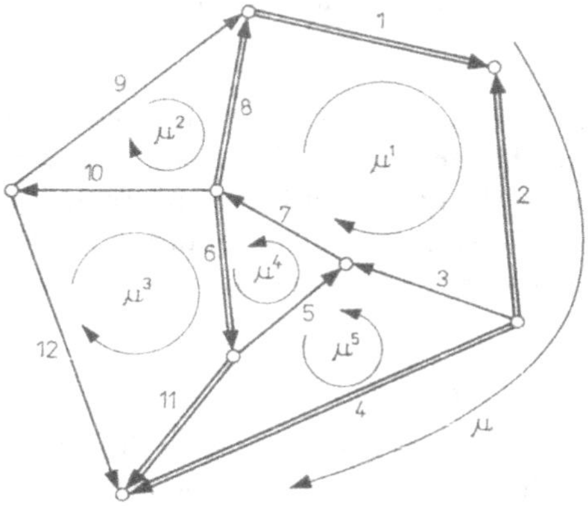

Fig. 1.8

Assume a graph $\boldsymbol{G}$ drawn in the plane, with no arcs crossing one another, and assume that this graph remains connected even after removing any vertex. An exact introduction of the concept of a map may be found by the reader in Sachs [12]. Then, the following theorem, which we give without proof, holds:

THEOREM 1.7. *The outlines of the finite areas of a map form a system of independent cycles, sometimes even a cycle basis.*

Let us explain this theorem by means of an example (cf. Fig. 1.8). Naturally, the orientation of the arcs is of no consequence to the planarity itself. Writing the vectors μ^i line for line, the result for our example is:

Arc	1	2	3	4	5	6	7	8	9	10	11	12
μ^1	1	−1	1	0	0	0	1	1	0	0	0	0
μ^2	0	0	0	0	0	0	0	−1	1	1	0	0
μ^3	0	0	0	0	0	1	0	0	0	−1	1	−1
μ^4	0	0	0	0	1	1	1	0	0	0	0	0
μ^5	0	0	1	−1	−1	0	0	0	0	0	1	0

For example, in the case of the doubly drawn cycle μ of Fig. 1.8, it follows:

$$\mu = \mu^1 - \mu^5 - \mu^4.$$

Let us now turn to the important concept of a strongly connected graph.

DEFINITION. A path

$$(X_{i_1}, X_{i_2}), (X_{i_2}, X_{i_3}), \ldots, (X_{i_k}, X_{i_{k+1}})$$

with $X_{i_r} \neq X_{i_s}$ for $i_r \neq i_s$ (with pairwise distinct i_j) is called a *simple path* from X_{i_1} to $X_{i_{k+1}}$. A graph G is called *strongly connected* if there exists between any two different vertices X and Y a simple path going from X to Y.

Fig. 1.9 shows a strongly connected graph and a path (doubly drawn) from X to Y.

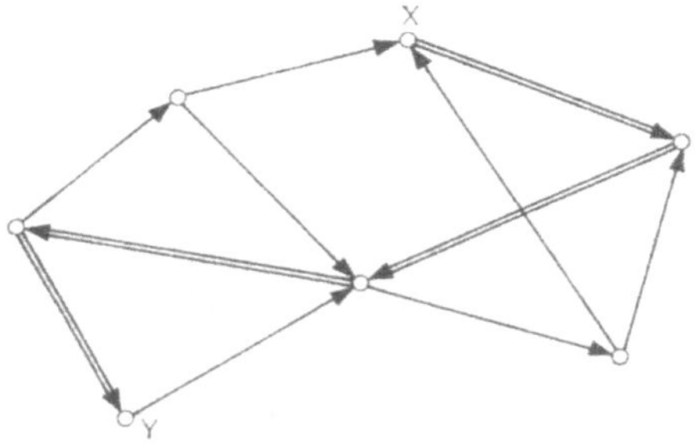

Fig. 1.9

THEOREM 1.8. *Let* G *be a connected graph with at least one arc.*

a) *The following statements are equivalent:*

(1) *Every arc of* G *lies on a circuit.*

(2) *No arc of* G *lies on a cocircuit.*

(3) G *is strongly connected.*

b) *The following statements are equivalent:*

(1′) *Each arc of* G *is contained in a cocircuit.*

(2′) *No arc of* G *lies on a circuit.*

Proof. a) We show: (2) follows from (1). Assume that the arc (X, Y) is contained in a cocircuit. By colouring all arcs black and using the lemma of Minty we see that there is no circuit containing (X, Y). This contradicts (1).

Analogously, it can be proved that (1) follows from (2) (task!).

Now, we show that (1) follows from (3). Assume that (X, Y) is not contained in any circuit. Then, there will be no simple path going from Y to X. This contradicts (3).

From (1) there follows (3): Assume that there exists no simple path going from X to Y. Consider the set $\mathfrak{A}$ of the vertices that can be reached from X via simple paths. Obviously, all arcs of $\omega(\mathfrak{A})$ are directed to $\mathfrak{A}$. Since G is connected, it holds: $\omega(\mathfrak{A}) \neq \emptyset$. Then, however, no arc of $\omega(\mathfrak{A})$ can be contained in a circuit. This contradicts (1).

b) From (1′) there follows (2′). Assume that the arc (X, Y) is contained in a circuit. By colouring all arcs black and applying the lemma of Minty, we recognize that the arc (X, Y) is not contained in any cocircuit, which contradicts (1′).

Analogously, (1′) follows from (2′). Thus, theorem 1.8 is proved.

We should point out that there exists for the concept of "strongly connected" no (reasonable) dual concept used in the sense of the "partial duality" described above.

One problem arising fairly often in graph theory applications is to decide whether or not a given graph is strongly connected. The following simple algorithm makes such a decision possible.

Algorithm 1.1 (for deciding whether or not a graph is strongly connected).

Choose from G any vertex Z and mark it with "+" and "−". Now, mark a vertex X with "+" if there exists an arc (Y, X) with Y marked with "+", and similarly, mark a vertex U with "−" if there exists an arc (U, Y) with Y marked with "−".

It can easily be seen that G is strongly connected if and only if at the end of the labelling procedure each vertex is marked both with "+" and with "−". If not all of the vertices are labelled with both marks during this labelling procedure, then those vertices labelled with both marks form the strongly connected component of G containing Z.

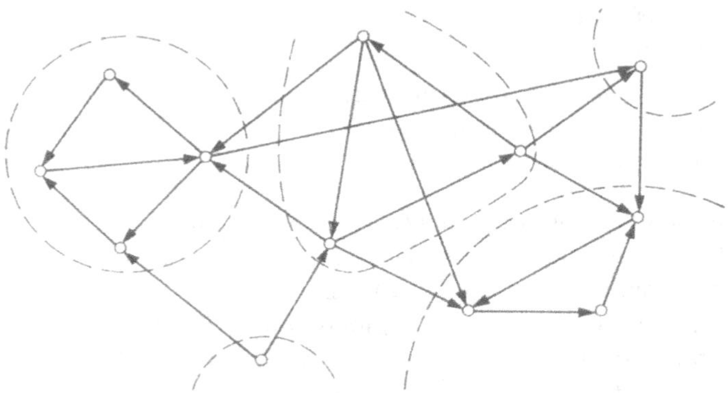

Fig. 1.10

If one chooses successively each vertex of G as the beginning of the labelling procedure Z, then the graph splits into strongly connected components. Of course, such an equivalence class may possibly contain only one single vertex, too.

To understand this equivalence relation, look at the example shown in Fig. 1.10.

Before becoming acquainted with the next algorithm let us set the following task for the reader.

TASK 1. Let G be a graph (containing at least one arc) with no circuit contained in it. Then, G possesses at least one source (one vertex into which no arc enters) and at least one sink (a vertex which no arc leaves).

Algorithm 1.2 (for deciding whether or not a graph is circuit-free).

If there exists a source in G, we remove this and all arcs being incident with it (leaving it). If there exists a source in the remaining graph, we remove this and all incident arcs, etc.

If the algorithm breaks off when there is no more vertex, the graph is circuit-free. If, nevertheless, a circuit was contained, one would have removed a vertex of this circuit in the course of the algorithm which was not a source, since some arcs (one of the circuit) entered it and at least one arc (of the circuit) also left it.

If the algorithm ends before all of the vertices are removed, we start out from any vertex X_1 in the remaining graph G', follow any arc entering X_1 in the reverse direction and arrive at a vertex X_2 which at least one arc enters, too. Again we follow this arc in the reverse direction, say to X_3, etc. Because of the finiteness of graph G (and thus G') we finally arrive at a vertex that we have already reached on our reverse way (not necessarily X_1). In such a case, however, we have found a circuit.

Let us now make a statement concerning the bases of the spaces of the cycles and cocycles.

THEOREM 1.9. *Let G be a connected graph with m arcs and n vertices.*

a) *If G is a strongly connected graph, there exist $m - n + 1$ independent circuits in G, i.e., the vector space of the cycles possesses a circuit basis.*

b) *If G possesses no circuit, then there exist $n - 1$ independent cocircuits in G, i.e., the space of the cocycles possesses a cocircuit basis.*

Proof. First, we prove a) in showing that any cycle may be represented as a sum of circuits. The following lemma is true:

LEMMA. *Each closed sequence of arcs directed in the same sense* (including a repetition of arcs) *may be split into circuits.*

The lemma may easily be proved by means of mathematical induction according to the number of arcs by traversing the closed sequence of arcs until a vertex is traversed twice. Then, one will have traversed a circuit that can be split off; the remaining part splits into components, with each of them satisfying the assumption of the lemma.

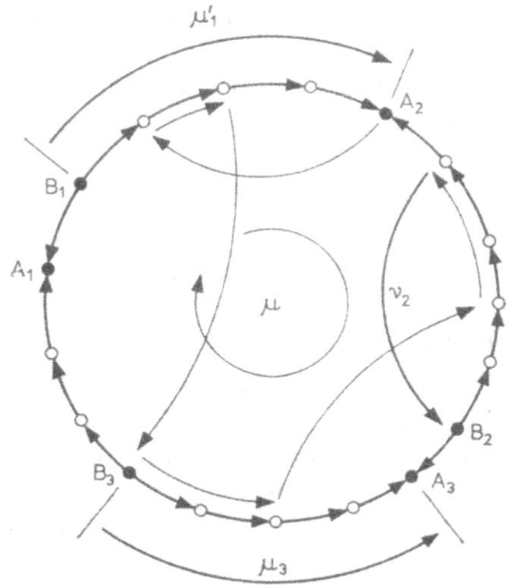

Fig. 1.11

The idea of the proof of a) is the following: We partition a cycle μ into two sets of arcs. One of these sets contains all those arcs oriented in the direction of μ, the other one will be completed to form circuits, which is possible because of the strong connectivity. All sets of arcs which have been added, together with the arcs of μ oriented in the direction of μ, also form circuits (satisfying the assumptions of the lemma). According to theorem 1.2, we may confine ourselves to the partition of elementary cycles. We partition μ into maximal paths (cf. Fig. 1.11). Denote the vertices such that the simple paths run along between B_i and A_{i+1} in the direction of μ (if s maximal paths run along in the direction of μ, we have to calculate modulo s; then, also s maximal paths run along in the opposite direction of μ) and in the opposite direction of μ from B_i to A_i. The simple path on μ connecting B_i with A_{i+1} is denoted by μ_i', and thus it has the direction of μ. The simple path on μ connecting B_i with A_i is denoted by μ_i. The simple path (which exists because of the strong connectivity of the graph) connecting A_i with B_i is denoted by ν_i. Let us call a closed sequence of arcs directed in the same sense a *c-cycloid*, only when carrying out the proof. This designation is made on the model of "conformally directed cycle" where the arcs, however, may be contained several times.

Let us first consider the c-cycloid λ that is generated in the following way (cf. Fig. 1.11):

$$\lambda = \mu_1' \cup \nu_2 \cup \mu_2' \cup \nu_3 \cup \ldots \cup \mu_n' \cup \nu_1.$$

We follow μ provided that the arcs have the same orientation as μ, leave out those parts of the elementary cycle μ on which the arcs are oriented in the opposite direction of μ and pass along on the simple paths ν_i from A_i to B_i, which exist because of the strong connectivity. Now we have traversed also the ν_i, which is too much, but we have also left out all μ_i, which is too little.

But c-cycloids are just formed by the

$$\lambda_i = \mu_i \cup \nu_i$$

and since the arcs of μ_i are traversed in the countersense of μ, we get the following expression using the vectorial method of writing for sets of arcs (characteristic function):

$$\mu = \lambda - \sum_{i=1}^{n} \lambda_i.$$

Since c-cycloids may be decomposed into circuits, as we have seen in the lemma, we have proved assertion a).

Now, let us prove b). Since G has no circuit, a complete order of the vertices may be given as follows: We choose any source in G and denote it by A_1. Now, remove A_1 and all arcs incident with A_1 (i.e. leaving arcs) and denote by A_2 a source of the remaining graph, etc. Obviously, the set of those arcs separating $\mathfrak{A} = \{A_1, \ldots, A_i\}$ from $\mathfrak{B} = \{A_{i+1}, \ldots, A_n\}$ forms a cocircuit because all arcs are going from $\mathfrak{A}$ to $\mathfrak{B}$. The independence of these $n - 1$ cocircuits is guaranteed, since in each of them there will be contained an arc not lying on any other, provided that there are no isolated vertices occurring during decomposition. In case an isolated vertex X appears during decomposition after the removal of A_i (i.e., after the removal of A_{i-1} only one arc (A_i, X) enters X), we choose this one to be the cocircuit. In this way, the connectivity of G assures that $n - 1$ independent cocircuits can be found (task).

This proves theorem 1.9.

COROLLARY from a). *A graph G without cocircuits having p components possesses $m - n + p$ independent circuits.*

COROLLARY from b). *A graph G without circuits having p components possesses a cocyclic basis consisting of $n - p$ cocircuits.*

Now, we can still deduce an interesting corollary from the lemma of Minty.

THEOREM 1.10 (Camion).

a) *Let G be a graph in which each cycle possesses as many components $+1$ as -1. Then, the set of arcs of G splits into pairwise disjoint cocircuits.*

b) *Let G be a graph in which each cocycle possesses as many components $+1$ as -1. Then, the set of arcs splits into pairwise disjoint circuits.*

Proof. a) We imagine the arcs of disjoint cocircuits $\omega_1, \ldots \omega_r$, to be successively removed from G. The resulting graph G' will no longer contain a cocircuit. If G' no longer contains an arc, then the proof is obvious. We thus may assume that there still exists at least one arc in G'. Colour all arcs black that belong to G' and assign to any one of them the number 1. Colour the arcs of $\omega_1^* \cup \omega_2^* \cup \cdots \cup \omega_r^*$ red. Since there exist no green arcs and G' does not contain a cocircuit, then, according to the lemma of Minty, there is in G a cycle μ coloured black and red which contains the arc 1, with all black arcs being directed conformally. Consider any red arc u of μ. Since u lies in one of the cocircuits $\omega_1, \omega_2, \ldots, \omega_r$, say in ω_1, there is a red arc v lying in μ which is equally directed with u relative to ω_1, but oppositely directed relative to μ. That means that μ has as many red arcs of the direction of μ as red arcs of the opposite direction of μ. Thus, in the components corresponding to the red arcs of μ, $+1$ occurs as often as -1. The black arcs of μ, however, have all either $+1$ or -1. Because the set of the black arcs of μ is not empty, there exists a cycle μ in G whose number of components $+1$ does not equal the number of components -1, which contradicts the assumption of the theorem 1.10a). Statement b) may be proved by the reader himself.

The following theorem follows:

THEOREM 1.11. *A c-cycle containing all arcs of a graph* G *exists if and only if* G *is connected and each cocycle contains as many components* $+1$ *as* -1.

Proof. Given a c-cycle containing all arcs (such a chain is also called a ***Euler line***). Obviously, G is connected. Choose any cocycle ω. The number of the arcs of ω is obviously even, and they are traversed half in the direction of the c-cycle and half in the opposite direction. Conversely, let each cocycle of G have the same number of positive as negative components. According to the theorem of Camion, the set of arcs splits into disjoint circuits. Since G is connected, these circuits may be composed to form a c-cycle such that all arcs are traversed in the same direction.

This proves the theorem.

In conclusion of this section, the following few remarks are made.

1. If we imagine the graph to be undirected, then we get as an equivalent condition for the existence of a Euler line (which is now undirected): G is connected, and each *cut* (corresponding to the cocycle when directed) possesses an even number of edges. It is sufficient to claim that the cuts generated by the vertices have an even number of edges. That, however, means that the degree of each vertex must be even.

2. The "partially dual" statement is obtained in the following form. A cut containing all edges of G exists if and only if each circuit (more precisely, each closed

chain) has an even length, i.e., possesses an even number of edges. This is equivalent to the already known statement that a cut containing all edges of G exists if and only if G is *bipartite.* (G is said to be bipartite if the vertices of G may be partitioned into two classes such that the vertices contained in the same class are not adjacent.)

3. The reader may construct an example showing that there exist graphs in which each cycle contains as many positive as negative components, but in which there exists no cocircuit containing all arcs.

1.2. PROPERTIES OF FLOWS AND TENSIONS

DEFINITIONS. Imagine the arcs of a graph to be numbered from 1 to m. We mean by an arc function $\psi(u)$ a vector of the G-dimensional vector space R^m with, in general, real components. For many problems it will be seen that one may assume the components to be integers. In principle, the components may also be complex, but for most applications, however, real components $\psi_i = \psi(u_i)$ of the vector $\boldsymbol{\psi}$ can be presupposed. For an arc $u = (X, Y)$ we also write, instead of $\psi\big((u)\big)$, simply $\psi(u) = \psi(X, Y)$. An arc function φ is called *flow* if for each vertex X the following relation holds:

$$\sum_{u\in\omega^+(X)} \varphi(u) - \sum_{u\in\omega^-(X)} \varphi(u) = 0.$$

The ith component of the vector $\boldsymbol{\varphi}$ is called *flux* through the arc i or flux through the arc u_i, respectively, if the natural numbers are used for other purposes and confusions might occur.

The definition of a flow obviously includes the well-known *Kirchhoff's node rule* for electrical networks. Using another way of writing we get:

An arc function φ is a flow if and only if for any vertex $X \in \mathfrak{X}$ of a graph $G = (\mathfrak{X}, \mathfrak{U})$ the scalar product equals zero:

$$\langle \boldsymbol{\omega}(x), \boldsymbol{\varphi} \rangle = 0.$$

In other words:

$$\sum_{i=1}^{m} \omega_i \varphi_i = 0 \quad \text{with} \quad \boldsymbol{\omega}(X) = (\omega_1, \ldots, \omega_m).$$

The reader may prove the following theorem:

THEOREM 1.12. *For any subset $\mathfrak{A}$ of the set $\mathfrak{X}$ of vertices of a graph G and for any flow φ defined on G, it holds:* $\langle \omega(\mathfrak{A}), \varphi \rangle = 0$, i.e.,

$$\sum_{u\in\omega^+(\mathfrak{A})} \varphi(u) - \sum_{u\in\omega^-(\mathfrak{A})} \varphi(u) = 0 \quad \text{for all} \quad \mathfrak{A} \subseteqq \mathfrak{X}.$$

DEFINITION. If for an arc function ϑ and for any elementary cycle μ the relation

$$\sum_{u \in \mu^+} \vartheta(u) - \sum_{u \in \mu^-} \vartheta(u) = 0$$

holds, then ϑ is called a *tension* on the graph $\boldsymbol{G}$. The value $\vartheta_i = \vartheta(u_i)$ of the ith component of $\boldsymbol{\vartheta}$ is called *partial tension* on u_i.

The definition of the tension obviously contains the *Kirchhoff's mesh rule* for electrical networks. From this it can be seen that an arc function ϑ is a tension if and only if it holds for each cycle:

$$\langle \mu, \boldsymbol{\vartheta} \rangle = 0 \quad \text{or} \quad \sum_{i=1}^{m} \mu_i \vartheta_i = 0.$$

DEFINITION. A function that assigns to each vertex of a graph a (generally real) number is said to be a *node function* or a *potential.*

It is really not characteristic of a mathematician to introduce for a term already given (node function) yet another one (potential). As will be shown by the following theorem, we may not, however, ignore the close connection between tension and potential which is already known to us from the field of electrical engineering.

THEOREM 1.13.

a) *Let t be a potential defined on any graph $\boldsymbol{G}$. Then, the arc function d assigning to each arc $u = (P, Q)$ the potential difference $d(u) = d(P, Q) = t(Q) - t(P)$ is a tension.*

b) *Let $\boldsymbol{G}$ be a connected graph and ϑ any tension defined on $\boldsymbol{G}$. Then, there exists a potential t, which is uniquely defined, except for an additive constant, such that the potential difference obtained from t is in conformity with ϑ, i.e., for each arc $u = (P, Q)$ the relation $\vartheta(u) = t(Q) - t(P)$ holds.*

Proof.

a) We only have to show that d with $d(P, Q) = t(Q) - t(P)$ is a tension. Consider any elementary cycle μ, with $P_1, \ldots, P_k, P_{k+1} = P_1$ being the vertices of μ as they are met successively when traversing μ. Choose two successive vertices P_i, P_{i+1}. If there exists in μ the arc (P_i, P_{i+1}), it holds:

$$\mu(P_i, P_{i+1}) = 1 \quad \text{and} \quad d(P_i, P_{i+1}) = t(P_{i+1}) - t(P_i),$$

i.e.,

$$\mu(P_i, P_{i+1}) \cdot d(P_i, P_{i+1}) = t(P_{i+1}) - t(P_i).$$

If, however, the arc (P_{i+1}, P_i) lies in μ, then

$$\mu(P_{i+1}, P_i) = -1 \quad \text{and} \quad d(P_{i+1}, P_i) = -t(P_{i+1}) + t(P_i)$$

In this case, too, we get:

$$\mu(P_{i+1}, P_i) \cdot d(P_{i+1}, P_i) = t(P_{i+1}) - t(P_i).$$

Forming the scalar product leads to:

$$\langle \mu, d \rangle = \sum_{i=1}^{k} \mu_i \cdot d_i$$

$$= t(P_2) - t(P_1) + t(P_3) - t(P_2) + - \cdots - + t(P_1) - t(P_k) = 0.$$

This proves that d is a tension.

b) A node function t is defined as follows. Let $P \in \mathfrak{X}$ be any fixed vertex. Let $t(P) = t_0$. Let $t(U)$ be already determined, and let (U, V) exist with $t(V)$ not yet determined. Then, put $t(V) = t(U) + \vartheta(U, V)$. But if $t(U)$ is already known, if the arc (V, U) exists and if $t(V)$ is not yet determined, we put $t(V) = t(U) - \vartheta(V, U)$. We show that this explains a node function. In other words: If $t(P) = t_0$ is already given, then all $t(X)$ are uniquely determined independently of the order in which the X are being chosen.

Assume that it would be possible to assign to a vertex Q (if the vertices are labelled in different order) two different values (say $t(Q)$ and $t'(Q)$). Then, there will be two different labelling paths, namely

$$\mathfrak{W}_1 = (P = P_0, P_1, \ldots, P_{q-1}, P_q = Q)$$

and

$$\mathfrak{W}_2 = (P = Q_0, Q_1, \ldots, Q_{s-1}, Q_s = Q).$$

Let $P_{i_1}, P_{i_2}, \ldots, P_{i_r}$ be the vertices lying on $\mathfrak{W}_1$ and $\mathfrak{W}_2$ in the order in which they are encountered on $\mathfrak{W}_1$. Certainly, $t(P_{i_1}) = t'(P_{i_1})$ and $t(P_{i_r}) \neq t'(P_{i_r})$. Choose k such that $t(P_{i_j}) = t'(P_{i_j})$ for $j \leq k$ and that $t(P_{i_{k+1}}) \neq t'(P_{i_{k+1}})$. The arcs lying on $\mathfrak{W}_1$ between P_{i_k} and $P_{i_{k+1}}$ form, together with certain arcs from $\mathfrak{W}_2$, an elementary cycle along which the tension drop does not equal zero, and this contradicts the assumption. The proof of theorem 1.13 is evident.

From this there results immediately the

COROLLARY. *The set of all tensions on a graph* $\mathbf{G}$ *with* n *vertices and* p *components depends exactly on* $n - p$ *paramters.*

Now, we come to some essential properties of flows and tensions.

THEOREM 1.14.

a) *Each linear combination of flows is a flow, too. The flows thus form a subspace* $\boldsymbol{\Phi}$ *of* R^m.

b) *Each linear combination of tensions is a tension, too. The tensions thus form a subspace* $\boldsymbol{\Theta}$ *of* R^m.

Proof. a) It is to be shown: If for any vertex X the relation $\langle\omega(X), \varphi^i\rangle = 0$ holds, then for any numbers c_i the relation $\langle\omega(X), \sum \varphi^i c_i\rangle = 0$ is true. This, however, is clear because of the linearity of the scalar product, since it holds:

$$\langle\omega(X), \sum \varphi^i c_i\rangle = \sum c_i\langle\omega(X), \varphi^i\rangle = \sum c_i \cdot 0 = 0.$$

The proof of b) is to be established analogously to that of a).

Now, we turn to a theorem the form of which may be, however, somewhat surprising.

THEOREM 1.15.

a) *Each cycle is a flow.*

b) *Each cocycle is a tension.*

This theorem is to be understood in the following way. An arc function (its characteristic function) had been assigned to a cycle (cocycle). This arc function satisfies the conditions of a flow (or a tension, respectively) function.

Proof.

a) First, we show that an elementary cycle is a flow, too. In the proof of theorem 1.6 we saw that for any cycle μ and for any cocycle ω of a graph the relation $\langle\mu, \omega\rangle = 0$ holds. It holds in particular also for the special cocycles $\omega(X)$ with $X \in \mathfrak{X}$. According to the definition of a flow, μ is thus a flow.

If there is any cycle given, we decompose it into elementary cycles; each of these cycles is a flow, and even the original cycle itself is a flow because of theorem 1.14.

The reader may prove b) in an analogous way.

This proves theorem 1.15.

We have seen that the space of the cycles is contained in the space of the flows (is thus subspace) and that the space of the cocycles is subspace of the space of the tensions. All these spaces are subspaces of R^m, if we consider the set of the graphs with m arcs. In the first paragraph, we also saw that the spaces of the cycles and the cocycles are orthogonal to each other.

This enables us to formulate the principal theorem of flows and tensions.

THEOREM 1.16. *The space $\boldsymbol{\Phi}$ of the flows is identical to the space $\boldsymbol{M}$ generated by the elementary cycles. The space $\boldsymbol{\Theta}$ of the tensions is identical to the space $\boldsymbol{\Omega}$ formed by the elementary cocycles.*

Proof. We have already shown by our last remarks that the space of the flows contains that of the elementary cycles. We shall see in the following that this is true also in the reverse case. Let φ be any flow, then, for each elementary cocycle $\omega(X)$, it holds $\langle\omega(X), \varphi\rangle = 0$. But this meas that ϑ lies in the orthogonal complement of $\boldsymbol{\Omega}$. Thus, φ lies in $\boldsymbol{M}$, as was to be shown, since $\boldsymbol{\Omega}$ and $\boldsymbol{M}$ are orthogonal complements in R^m.

The second part of the theorem may be proved in quite an analogous way.

REMARK. *Φ and Θ are thus orthogonal complements in R^m.*

Consider now any spanning tree (maximal subgraph without cycles) $\boldsymbol{H}$ of a connected graph $\boldsymbol{G}$ with the arcs $v_1, \ldots, v_{n-1}$ of the spanning tree and with the remaining arcs $u_1, \ldots, u_{m-n+1}$, which naturally form a cotree (according to theorem 1.5). Together with $\boldsymbol{H}$, exactly one elementary cycle μ_i corresponds to each arc u_i of the cotree. The elementary cycles μ_i are independent (cf. proof of theorem 1.6). The following proposition is true:

THEOREM 1.17. *With the assumptions mentioned it holds: Any flow φ may be linearly combined from the μ_i in exactly one way, and it holds:*

$$\boldsymbol{\varphi} = \varphi(u_1)\, \boldsymbol{\mu}_1 + \cdots + \varphi(u_z)\, \boldsymbol{\mu}_z$$

with $z = m - n + 1$.

Proof. Let us first recall that the orientation of the elementary cycle μ_i had been chosen such that the orientation of u_i corresponds to that of μ_i.

If there actually exists such a representation (given by the theorem) of a flow, then it will certainly be unique, since it is only μ_i that has a non-zero value on the component belonging to the arc μ_i. Thus, φ is certainly represented on the arcs of the cotree. It remains to show that also on the arcs v_j of the spanning tree $\boldsymbol{H}$ the flow φ is represented in accordance with the theorem. For this, we form a new flow

$$\boldsymbol{\varphi}' = \boldsymbol{\varphi} - \varphi(u_1)\, \boldsymbol{\mu}_1 - \cdots - \varphi(u_z)\, \boldsymbol{\mu}_z.$$

As already said, the relation $\boldsymbol{\varphi}' = 0$ is valid on the arcs of the cotree. Choose now any vertex X that is incident with only one arc within the spanning tree (there exist at least two of this kind! Cf. Tasks). Since on all other arcs incident with X (these are the arcs of cotrees), the relation $\boldsymbol{\varphi}' = 0$ holds, then, also $\varphi'(v) = 0$ for the arc v incident with X, because of the flow condition. Carrying on these reflections (mathematical induction!) will lead us to the conclusion that $\boldsymbol{\varphi}'$ has disappeared also for the arcs of the spanning tree.

This proves theorem 1.17.

From this theorem there result some corollaries. The reader may be recommended to solve them.

TASK 2. A flow on a graph is already determined if the fluxes on the arcs of a cotree are known.

TASK 3. The elementary cycles $\mu_1, \ldots, \mu_z$ $(z = m - n + 1)$ form a basis of the vector space $\boldsymbol{\Phi}$ of all flows, consequently it holds: $\boldsymbol{\Phi} = \boldsymbol{M}$ and thus $\dim \boldsymbol{\Phi} = z$.
We call $z = m - n + 1$ the *cyclomatic number* of the connected graph $\boldsymbol{G}$.

We also may formulate a theorem that is "dual" to the last one, but let us first give some reminders.

Let $\boldsymbol{G}$ be a connected graph, $\boldsymbol{H}$ any spanning tree of $\boldsymbol{G}$ and $v_1, \ldots, v_{n-1}$ the arcs of $\boldsymbol{H}$. As is well-known, each arc v_j of $\boldsymbol{H}$ uniquely defines an elementary cocycle ω_j. We thus get $n - 1$ independent elementary cocycles $\omega_1, \ldots, \omega_{n-1}$.

THEOREM 1.18. *Let ϑ be a tension on a connected graph $\boldsymbol{G}$. Then, ϑ may be linearly combined from the $c = n - 1$ independent elementary cocycles $\omega_1, \ldots, \omega_c$ in exactly one way, and it holds:*

$$\boldsymbol{\vartheta} = \vartheta(v_1)\,\omega_1 + \cdots + \vartheta(v_c)\,\omega_c.$$

Proof. Let us first recall that the orientation of the cocycle ω_j corresponds to that of the arc v_j.

If it is actually possible to represent ϑ in the form shown in the theorem, then this representation will certainly be unique, since the component corresponding to the arc v_j is equal to 1 only in $\boldsymbol{\omega}_j$ and equals zero in all other $\boldsymbol{\omega}_i$. The partial tensions $\vartheta(v_j)$ on the arcs v_j of the spanning tree $\boldsymbol{H}$ are represented in the form given by the theorem. It remains to show that the representation is also true for the remaining arcs.

Consider the tension ϑ' with

$$\boldsymbol{\vartheta}' = \boldsymbol{\vartheta} - \vartheta(v_1)\,\omega_1 - \cdots - \vartheta(v_c)\,\omega_c.$$

Certainly, for all arcs v_j of the spanning tree, it holds: $\vartheta'(v_j) = 0$.

Now we define a potential t' on the vertices of $\boldsymbol{G}$ as follows. To any fixed vertex P_1 there is assigned the potential $t'(P_1) = 0$ (or a, with a being any real number). To the other vertices there are assigned the potentials according to the rule in theorem 1.13b), with the valuation being made along the spanning tree $\boldsymbol{H}$. Evidently, all vertices get the zero potential (or the value a), because on all arcs of the spanning tree, the tension is equal to zero. Thus, the value of the tension on any arc becomes zero, representing the difference of the potential values of those vertices that are incident with this arc. That means that, in fact, $\boldsymbol{\vartheta}' = 0$ holds, which was to be shown.

The proof of theorem 1.18 is evident.

As it was the case already in theorem 1.17, some conclusions may also be drawn which we recommend to the reader.

TASK 4. A tension ϑ on a connected graph is uniquely defined if the partial tensions on the arcs of a spanning tree are known.

TASK 5. The elementary cocycles $\boldsymbol{\omega}_1, \ldots, \boldsymbol{\omega}_c$ form a basis in the space $\boldsymbol{\Theta}$ of all tensions, and it consequently holds: $\boldsymbol{\Theta} = \boldsymbol{\Omega}$ and thus, $\dim \boldsymbol{\Theta} = c = n - 1$.

Another conclusion is the already well-known fact that each flow is othogonal to each tension, and that the space $\boldsymbol{\Phi}$ of all flows and the space $\boldsymbol{\Theta}$ of all tensions are orthogonal to each other and complementary with respect to R^m. The number $c = n - 1 = \dim \Theta$ is called *cocyclomatic number* of $\boldsymbol{G}$.

TASK 6. How large are the values for the cyclomatic and for the cocyclomatic number in case of a graph having p components?

We sum up the results that we have obtained so far.

— *The vector spaces Φ and Θ are orthogonal to each other and complementary subspaces of the space R^m of all arc functions.*

— *The scalar product of any flow and any tension is equal to zero.*

— *An arc function is a flow (a tension) if and only if it is orthogonal to all tensions (flows).*

— *Any arc function is the unique sum of a certain tension and a certain flow (the components result as orthogonal projections into the spaces Φ and Θ).*

To conclude this paragraph, let us make a few more statements about *planar graphs*, that is, about graphs that can be embedded in a plane without edge crossing.

Let us first recall theorem 1.7, which said that the boundary of the finite faces of a planar graph form a cycle basis and thus, according to what we have learned in this paragraph, a basis for the space Φ of all flows.

We now get the well-known *Euler's polyhedron formula* for connected planar graphs:

Let f be the number of all faces of a planar graph $\boldsymbol{G}$. Then, the number of all finite faces will be equal to $f - 1$, and this number is equal to the dimension of the space of all flows, i.e., equal to the cyclomatic number $z = m - n + 1$, that is, $n - m + f = 2$.

Of course, this may also be proved in a direct way by means of mathematical induction (cf. Sachs [12]).

TASK 7. In a planar graph without loops and without multiple edges there exist vertices of a degree $\leqq 5$.

Let $\boldsymbol{G}$ be a plane graph (i.e., a planar graph embedded in a plane). Assign to the graph $\boldsymbol{G}$ another plane graph $\boldsymbol{G}' = \boldsymbol{DG}$ (*dual graph*) as follows: A vertex M of $\boldsymbol{DG}$ is assigned to each (elementary) boundary cycle μ of $\boldsymbol{G}$ and this is considered to be mathematically positively oriented for each finite face. Two vertices M_1, M_2 are connected in $\boldsymbol{DG}$ if and only if the elementary cycles μ_1, μ_2 corresponding to them in $\boldsymbol{G}$ possess a common boundary arc (P_1, P_2). In other words, we draw in $\boldsymbol{DG}$ the edge (M_1, M_2) if (P_1, P_2) lies in the direction of cycle μ_1. Otherwise we draw the edge (M_2, M_1) (cf. Fig. 1.12).

The reader will easily ascertain that $\boldsymbol{DDG} = (\boldsymbol{G}')'$ is a planar graph which is obtained from $\boldsymbol{G}$ by reversing all arcs in $\boldsymbol{G}$.

If a graph $\boldsymbol{G}$ has a loop which forms the boundary of a face, then $\boldsymbol{DG}$ has a vertex of degree 1 corresponding to this loop. If a domain borders on itself in $\boldsymbol{G}$, then the vertex corresponding to this domain is incident with a loop in $\boldsymbol{DG}$ (i.e., is adjacent to itself), etc.

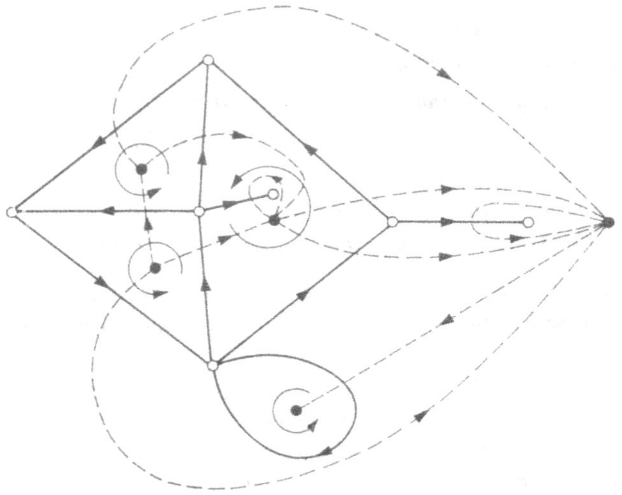

Fig. 1.12

THEOREM 1.19. *An elementary cocycle (cocircuit or elementary cycle or circuit, respectively) in* $\boldsymbol{DG}$ *corresponds to an elementary cycle in* $\boldsymbol{G}$ *(circuit, elementary cocycle, cocircuit).*

Proof. We only prove one of the assertions, the other proofs are to be made analogously. Let $\mu = \{u_1, u_2, \ldots, u_r\}$ be any elementary cycle of $\boldsymbol{G}$ and $\mathfrak{H}'$ the set of the vertices of $\boldsymbol{DG}$ lying in the interior of μ (i.e., to put it more exactly: we consider the set of the elementary cycles of $\boldsymbol{G}$ forming the boundaries of a domain and lying within μ).

Consider the cocycle $\omega(\mathfrak{H}')$. The subgraph of $\boldsymbol{DG}$, which is spanned by $\mathfrak{H}'$, is connected, since one can get from any domain within μ to any other domain within it without crossing μ. The same applies, however, to the vertices of $\boldsymbol{DG}$ that do not belong to $\mathfrak{H}'$. Thus, $\omega(\mathfrak{H}')$ is an elementary cocycle.

TASK 8. Prove the other three assertions of theorem 1.19.

COROLLARY. *The cyclomatic number of* $\boldsymbol{G}$ *equals the cocyclomatic number of* $\boldsymbol{DG}$*, and vice versa, which is due to the fact that the cycle in* $\boldsymbol{G}$ *and the cocycle in* $\boldsymbol{DG}$

uniquely correspond to each other (Task!), *in particular also maximal sets of independent cycles and cocycles, respectively.*

Exactly one cotree in $\boldsymbol{DG}$ *corresponds to one spanning tree of* $\boldsymbol{G}$, *and vice versa.*

The proof of the last statement may be established by the reader.

1.3. THE MAXIMUM FLOW PROBLEM

1.3.1. Introduction

In this chapter, we shall deal with the following problem. Given a directed graph $\boldsymbol{G}(\mathfrak{X}, \mathfrak{U})$ with valuated arcs. The arc valuations $c(u) \geqq 0$ with $u \in \mathfrak{U}$ are called *capacities* of the arcs. Let two vertices Q and S (*source* and *sink*, respectively) be marked in $\boldsymbol{G}$, in assuming that an arc (S, Q) does not exist in $\mathfrak{U}$. Find an arc function $\varphi(u)$ which

1. is compatible with the capacities, i.e., the relation $0 \leqq \varphi(u) \leqq c(u)$ holds for all $u \in \mathfrak{U}$,

2. satisfies in each vertex P, perhaps with the exception of Q and S, the *Kirchhoff's node rule*:

$$\sum_{u \in \omega^+(P)} \varphi(u) = \sum_{u \in \omega^-(P)} \varphi(u) \quad \text{for all} \quad P \in \mathfrak{X} \setminus \{Q, S\},$$

3. maximizes the *value* of Q among all arc functions satisfying the conditions 1 and 2, i.e.:

$$\sum_{u \in \omega^+(Q)} \varphi(u) - \sum_{u \in \omega^-(Q)} \varphi(u) \to \text{Max!}$$

The condition $c(u) \geqq 0$ for all $u \in \mathfrak{U}$ guarantees the existence of a flow satisfying the conditions 1 and 2, that is, $\varphi(u) = 0$ for all $u \in \mathfrak{U}$. We call a flow satisfying the conditions 1 and 2 a *feasible* flow, which is done on the model of the designations occurring in optimization theory.

Since with the possible exception of Q and S the Kirchhoff's node condition is satisfied in each vertex, we can find among all flows $\{\varphi(u)\}$ maximizing the value of Q such a flow — say $\varphi_0(u)$ — that $\varphi_0(u) = 0$ holds for each arc u entering Q and also for each arc u leaving S. Without restriction of generality we thus may assume that no arc enters Q (in other words, we may omit such arcs without causing a change in the maximum flow distribution) and no arc leaves S.

In order to make the formulation of our optimization problem more simple and to ensure that the node condition is satisfied in each vertex, we introduce an auxiliary arc $u_0 = (S, Q)$ with $c(u_0) = \infty$ and turn to the following task:

Given a directed graph $\boldsymbol{G}(\mathfrak{X}, \mathfrak{U})$ with arc valuations $c(u) \geqq 0$ for all $u \in \mathfrak{U}$. Let two vertices Q, S be marked in $\boldsymbol{G}$ as well as one arc $u_0 = (S, Q)$ with $c(u_0) = \infty$ (u_0 is called a *return arc*). Let also $\omega^-(Q) = \omega^+(S) = \{u_0\}$ be true. (Such graphs will be called *networks*, too.) Find an arc function (a flow) $\varphi(u)$ for which the following relations are true:

1. $0 \leqq \varphi(u) \leqq c(u)$ for all $u \in \mathfrak{U}$,
2. $\sum_{u \in \omega^+(P)} \varphi(u) = \sum_{u \in \omega^-(P)} \varphi(u)$ for all $P \in \mathfrak{X}$,
3. $\varphi(u_0)$ is maximum.

If several source points Q_i $(i = 1, \ldots, r)$ or several sink points S_j $(j = 1, \ldots, s)$ are given by the problem, then this task can be transformed into the above-mentioned one by adding two auxiliary points Q' and S' and by adding the arcs $u_0 = (S', Q')$, (Q', Q_i) $(i = 1, \ldots, r)$ and (S_j, S') $(j = 1, \ldots, s)$ with a capacity of ∞ each. (Note, however, that the sum of the sets of flows going from the Q_i to the S_j is to be maximized, with the capacity restrictions being taken into consideration.)

Before trying to solve the problem, let us explain it by means of a simple example (cf. Fig. 1.13).

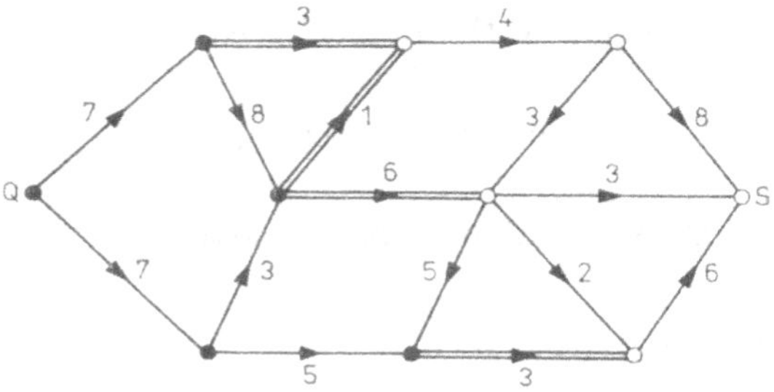

Fig. 1.13

Let the numbers assigned to the arcs be the capacities of them. (The meaning of the doubly drawn arcs will be explained in the next chapter.) One can see that a maximum flow may assume the value 14 at the most since more cannot leave the source. Furthermore, one can easily ascertain that the sum of the capacities of any *cut* (of a set of arcs after the removal of which there will no longer exist a directed path — a simple path — from Q to S) represents an upper bound for the value of the maximum flow.

The theorem of Ford and Fulkerson to be proved in the next chapter will then show that the maximal flow value is equal to the cut capacity (i.e., equal to the sum of the capacities of the arcs of a cut) of a minimal cut.

The following problem is investigated in 1.3.3: Given a directed graph $\boldsymbol{G}(\mathfrak{X}, \mathfrak{U})$, with two numbers $b(u_i) = b_i$ and $c(u_i) = c_i$ with $-\infty \leqq b_i \leqq c_i \leqq \infty$ being assigned to each arc u_i. Furthermore, let two vertices Q and S and one arc $u_0 = (S, Q)$ be marked in $\boldsymbol{G}$, with $b(u_0) = b_0 = -\infty$ and $c(u_0) = c_0 = \infty$ as well as $\omega^-(Q) = \omega^+(S) = \{u_0\}$ being valid.

Find an arc function (a flow) $\varphi(u_i) = \varphi_i$ for which the following conditions are fulfilled:

1. $b_i \leqq \varphi(u_i) \leqq c_i$ for all $u_i \in \mathfrak{U}$,
2. $\sum_{u \in \omega^+(P)} \varphi(u) = \sum_{u \in \omega^-(P)} \varphi(u)$ for all $P \in \mathfrak{X}$,
3. $\varphi(u_0) = \varphi_0$ is maximum.

For the case $b_i = 0$ $(i = 0, \ldots, m)$ the problem obviously becomes the "classic" problem of Ford and Fulkerson.

In paragraph 1.3.4, we will treat the question of how it is possible, in an undirected graph with the edges valuated (by capacities), to calculate at the same time for all pairs of vertices P_i and P_j the maximal flow quantities which can be transported from P_i to P_j. There will also be given the algorithm found by Gomory and Hu for the solution of this problem.

1.3.2. The theorem of Ford and Fulkerson

Let $G(\mathfrak{X}, \mathfrak{U})$ be a network, i.e., a directed graph in which to each arc u_i there is assigned one arc valuation $c(u_i) \geqq 0$, which is called the *capacity* of u_i. Furthermore, let two marked vertices Q, S exist in $\mathfrak{X}$ as well as one marked arc $u_0 = (S, Q) \in \mathfrak{U}$ with $c(u_0) = \infty$. Here, u_0 is the only arc entering Q and also the only one leaving S, i.e., thus $\omega^+(S) = \omega^-(Q) = \{u_0\}$.

DEFINITION. A set $\mathfrak{S} \subseteqq \mathfrak{U}$ of arcs is called a *cut of G* if the vertices of G may be subdivided into two disjoint classes $\mathfrak{V}$ and $\mathfrak{W}$ such that $Q \in \mathfrak{V}$, $S \in \mathfrak{W}$ and $\mathfrak{S}$ is composed of all arcs $u = (A, B)$ with $A \in \mathfrak{V}$ and $B \in \mathfrak{W}$.

In other words: A *cut* $\mathfrak{T}$ is such a set of arcs that no path (no directed simple path) going from Q to S exists any longer in the graph after the removal of the arcs of $\mathfrak{T}$, or: for some set $\mathfrak{A}$ of vertices containing Q and not containing S, it holds $\mathfrak{T} = \omega^+(\mathfrak{A})$. In network theory, especially in the case of transportation and flow problems, the concept of a cut is usually used.

DEFINITION. The *capacity $c(\mathfrak{S})$ of a cut* $\mathfrak{S} = (\mathfrak{V}/\mathfrak{W})$ is the sum of the capacities of all arcs of $\mathfrak{S}$, i.e.,

$$c(\mathfrak{S}) = c(\mathfrak{V}/\mathfrak{W}) = \sum_{\substack{A \in \mathfrak{V} \\ B \in \mathfrak{W}}} c(A, B)$$

if $c(A, B)$ denotes the capacity of the arc (A, B). (If $\mathfrak{S} = \emptyset$, then $c(\mathfrak{S}) = 0$.)

The doubly drawn arcs in Fig. 1.13 form a cut $\mathfrak{S} = (\mathfrak{V}/\mathfrak{W})$, with $\mathfrak{V}$ containing all filled vertices and $\mathfrak{W}$ all those that are not filled. Obviously, the capacity of the cut equals 13.

In a network, some special cuts form the sets of arcs $\omega^+(Q)$ and $\omega^-(S)$.

Before formulating the theorem of Ford and Fulkerson, we shall give a few lemmata.

LEMMA 1. *Let* φ *be a flow on a network* $\boldsymbol{G}$ $(\mathfrak{X}, \mathfrak{U})$ *with* $\varphi(u_0) = \varphi(S, Q) = \varphi_0$. *If* $\mathfrak{S} = (\mathfrak{B}/\mathfrak{W})$ *is any cut of* $\boldsymbol{G}$, *then it holds:*

$$\sum_{u \in \omega^+(\mathfrak{B})} \varphi(u) = \sum_{\substack{u \in \omega^-(\mathfrak{B}) \\ u \neq u_0}} \varphi(u) + \varphi_0 .$$

The proof is direct. Because the flow condition is satisfied in each vertex, the sum of the flows going from vertices of $\mathfrak{B}$ to vertices of $\mathfrak{W}$ equals the sum of the flows going from vertices of $\mathfrak{W}$ to vertices of $\mathfrak{B}$.

LEMMA 2. *Let* φ *be a feasible flow in a network* $\boldsymbol{G}(\mathfrak{X}, \mathfrak{U})$ *of the value* φ_0 (i.e., $\varphi(u_0) = \varphi_0$). *The capacity* c $(\mathfrak{B}/\mathfrak{W})$ *of a cut* $\mathfrak{S} = (\mathfrak{B}/\mathfrak{W})$ *will then be given by the relation* $\varphi_0 \leqq c(\mathfrak{B}/\mathfrak{W})$.

This statement is evident since the quantity of flow which can be transported from vertices of $\mathfrak{B}$ with $Q \in \mathfrak{B}$ to vertices of $\mathfrak{W}$ with $S \in \mathfrak{W}$ cannot exceed the capacity of any cut.

LEMMA 3. *If for a special cut* $\mathfrak{S}_0 = (\mathfrak{B}_0/\mathfrak{W}_0)$ *and a flow* φ *the equality* $\varphi_0 = c(\mathfrak{B}_0/\mathfrak{W}_0)$ *holds, then* φ_0 *is a maximum flow.*

This lemma is clear because of lemma 2.

DEFINITION. A cut $\mathfrak{S}_0 = (\mathfrak{B}_0/\mathfrak{W}_0)$ for which $c(\mathfrak{S}_0) \leqq c(\mathfrak{S})$ for any cut $\mathfrak{S}$ is true is called *minimal cut*.

Interpreting the statements of lemmata 2 and 3 from a physical point of view will lead us to the clear result that the quantity which may be transported via a pipe line system cannot be greater than that passing through its tightest section.

LEMMA 4. *Let* $\boldsymbol{G}$ $(\mathfrak{X}, \mathfrak{U})$ *be a network and* $\mathfrak{S}$ *a cut of finite capacity, i.e.,* $c(\mathfrak{S}) < \infty$. *Then, there exists on* $\boldsymbol{G}$ *a maximum flow* φ_0 (*that is, for any feasible flow* φ *it holds* $\varphi(u_0) \leqq \varphi_0(u_0)$).

Proof. Since $\boldsymbol{G}$ is finite, there is only a finite number of cuts. Since the existence of a finite cut (that is, a cut of finite capacity) is being assumed, there is also a minimal cut $\mathfrak{S}_0 = (\mathfrak{B}_0/\mathfrak{W}_0)$. The fact that there exists on $\boldsymbol{G}$ also a flow φ_0 satisfying the condition

$$\varphi_0(u_0) = c(\mathfrak{B}_0/\mathfrak{W}_0)$$

will later be proved in connection with the algorithm of Ford and Fulkerson.

The existence of a maximum flow may also directly be proved by means of the theory of linear optimization, which is due to the fact that the flow problem may be represented in the form of a linear program, too, and the flows are also finite because of the existence of a finite cut.

To solve our problem by means of the algorithms set up in optimization theory (for example the *simplex method*) would, however, not be useful, since it is not possible to take the specific structure of the task into account as is the case in the following algorithm.

Let us assume, from now on, all capacities to be integers or ∞. This is not so essential a restriction for practical applications, especially with a view to the use of digital computers.

Algorithm of Ford and Fulkerson

Given a feasible flow φ_1 on $\boldsymbol{G}$ (e.g. $\varphi_1 = 0$ is feasible), that is, for all $u \in \mathfrak{U}$ the relation $0 \leqq \varphi_1(u) \leqq c(u)$ holds. Since the capacities are integers, we also may assume the arc flows to be integers.

1. *Labelling procedure*

(i) Label source Q.

(ii) Let the vertex P_k be labelled and let an arc (P_k, P_j) exist with $\varphi_1(P_k, P_j) < c(P_k, P_j)$. Then label P_j.

(iii) Let P_k be labelled and let an arc (P_j, P_k) exist with $\varphi_1(P_j, P_k) > 0$. Then label P_j.

2. *Augmenting step*

If sink S is labelled according to the labelling procedure, the flow can be augmented (i.e., it can be changed in such a way that the flow quantity of the return arc u_0 is larger than $\varphi_1(u_0)$). This augmentation is made along a chain from Q to S by at least 1. Let $\boldsymbol{K} = (Q = P_0, P_1, \ldots, P_r = S)$ be a labelling chain (cf. the doubly drawn arcs in Fig. 1.14).

(i') If (P_i, P_{i+1}) in $\boldsymbol{K}$ is traversed in the sense of its orientation, that is, if P_{i+1} is labelled according to (ii), then we put

$$\varphi_2(P_i, P_{i+1}) = \varphi_1(P_i, P_{i+1}) + 1.$$

(ii') If (P_i, P_{i+1}) in $\boldsymbol{K}$ is traversed in its opposite direction, i.e., if it is labelled according to (iii), then we put

$$\varphi_2(P_i, P_{i+1}) = \varphi_1(P_i, P_{i+1}) - 1.$$

(iii') Put $\varphi_2(u_0) = \varphi_1(u_0) + 1$.

(iv') For all those arcs u not lying on $\boldsymbol{K}$ put $\varphi_2(u) = \varphi_1(u)$.

One can easily make sure that in each vertex the flow condition (node rule) remains satisfied.

Let now, in Fig. 1.14, the capacity of an arc be indicated by the first number on it and let the number in parentheses indicate a feasible flow φ_1. Small crosses at the vertices serve as labels. If labelling took place according to (ii), then the cross would be found above, if it was done according to (iii), then the cross would be made below.

As can be seen, the flow on u_0 may be augmented according to the labelling chain described (doubly drawn arcs!). The example shows that it can even be augmented by the value 2.

TASK 9. Find out the least value by which the flow along a labelling chain may be augmented.

In Fig. 1.15 the last possible augmentation is carried out and also the new labelling is given. The doubly drawn arcs in this figure possess the property that

— the initial point of such an arc is labelled,

— the end-point of such an arc is not labelled,

— $\varphi(u) = c(u)$ holds for such an arc.

One can also recognize that these arcs form a cut (which is still to be proved for the general case) and that this cut forms a minimum cut because of the relation

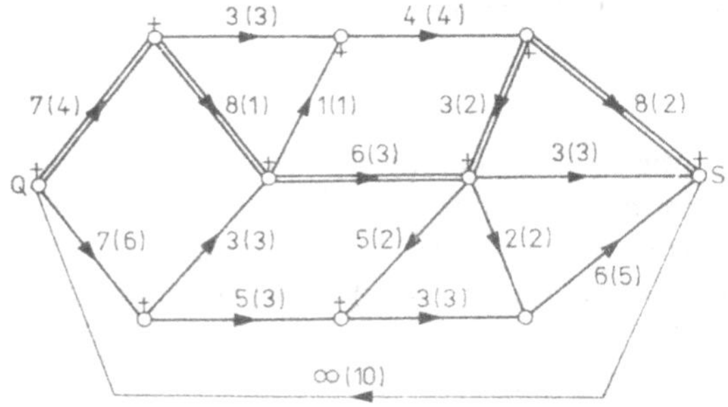

Fig. 1.14

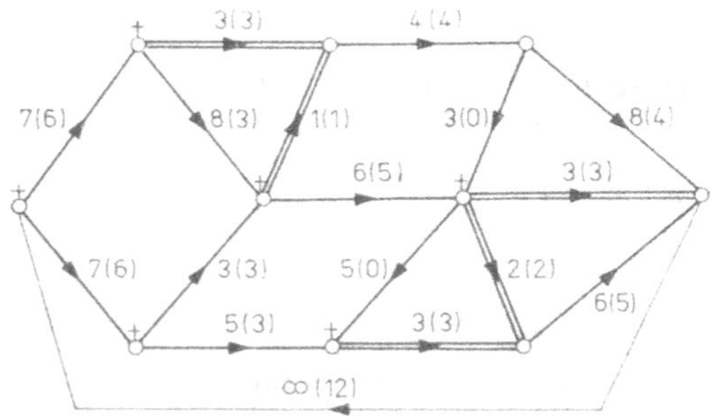

Fig. 1.15

$\varphi(u) = c(u)$ holding for each arc of the cut. That means, however, that we have found a maximum flow.

Consider the limiting arcs, i.e., all those arcs which have one labelled end-point and one unlabelled. Let $\mathfrak{V}$ be the set of the labelled and $\mathfrak{W}$ the set of the unlabelled vertices. Evidently, the relations $Q \in \mathfrak{V}$ and $S \in \mathfrak{W}$ are true.

a) For an arc $u = (A, B)$ with $A \in \mathfrak{V}$ and $B \in \mathfrak{W}$, it holds $\varphi(u) = c(u)$; otherwise, B would have been labelled according to (ii), i.e., $B \in \mathfrak{V}$ would be valid;

b) For an arc $u = (A, B)$ with $A \in \mathfrak{W}$ and $B \in \mathfrak{V}$, it holds $\varphi(u) = 0$; otherwise, A would have been labelled according to (iii).

Now, we show that the set $\mathfrak{S}$ of all arcs $u = (A, B)$ with labelled A and unlabelled B forms a cut.

It is clear that the subdivision of the vertices into two classes (with $\mathfrak{V}$ containing all labelled and $\mathfrak{W}$ all unlabelled vertices) represents in fact a decomposition (the source Q lies in $\mathfrak{V}$ and the sink S in $\mathfrak{W}$). The other property of a cut, however, may directly be seen from the labelling rule.

Because of the validity of the Kirchhoff's node theorem, the flow $\varphi(u_0)$ equals the cut capacity of $\mathfrak{S}$. The flow found in this way is maximal by lemma 3.

The results achieved may now be summed up; in the case of non-integer capacities, we refer to the textbook of Sachs [12]:

FUNDAMENTAL THEOREM 1.20. *If the network $G(\mathfrak{X}, \mathfrak{U})$ possesses a cut of finite capacity, then the maximum flow $\varphi_0(u_0)$ of all flows compatible with the capacities of the network equals the minimum cut capacity* (minimal under all cuts); *if the network does not possess a cut of finite capacity, then there exist flows of any value being compatible with the capacities of the network.*

Using the algorithm of Ford and Fulkerson it is possible, in the case of the existence of a finite cut, to find both a maximum flow and a cut of minimal capacity.

To conclude, let us give another example from the field of simple mathematics concerning the application of the fundamental theorem mentioned above.

In a group of n boys and n girls let each boy be acquainted with exactly m girls and each girl be acquainted with exactly m boys. Is it possible that in the course of m dancing sets each boy dances exactly once with each of his acquaintances, and each girl dances exactly once with each of her acquaintances, too?

Construct a graph $G(\mathfrak{X}, \mathfrak{U})$ in the following way (cf. Fig. 1.16): $\mathfrak{X}$ consists of two classes

$$\mathfrak{M} = \{M_1, \ldots, M_n\}, \qquad \mathfrak{J} = \{J_1, \ldots, J_n\}.$$

An arc u is of the form $u = (J_i, M_j)$. Such an arc u will be drawn if the boy J_i is acquainted with the girl M_j. Furthermore, there will be inserted two auxiliary points Q and S and all arcs of the form (Q, J_i) and (M_j, S) $(i, j = 1, \ldots, n)$ as well as the return arc (S, Q). The arcs (Q, J_i) and (M_j, S) are provided with the capacity 1, all other arcs with ∞. Evidently, a minimal cut will then have the capacity n (in our example, this is 6) and contains either all arcs leaving Q or all arcs entering S. Because the capacities are integers, one can find such a maximum flow of the value $\varphi(u_0) = n$, by using the above algorithm, that there are only flows of the value zero or 1, except for the return arc u_0.

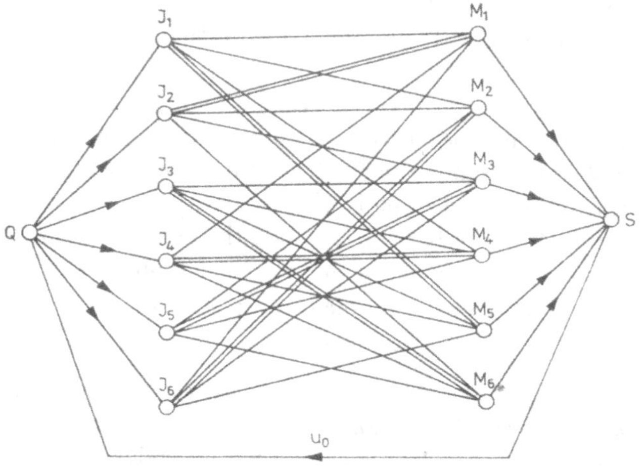

Fig. 1.16

The arcs (J_i, M_j) with $\varphi(J_i, M_j) = 1$ represent just one dancing set. (According to the graph theory terminology, these arcs are forming a *linear factor*.) The other dancing sets can be found out by eliminating the arcs forming the first dancing set, applying once again the algorithm, etc.

The reader who wants to familiarize himself in more detail with the existence of factors in graphs should consult the textbook of Sachs [12].

1.3.3. Generalized theorem of Ford and Fulkerson

This chapter deals with the following problem.

Given a simple[1]) graph $\boldsymbol{G}(\mathfrak{X}, \mathfrak{U})$. Let two numbers $b(u_i) = b_i$ and $c(u_i) = c_i$ with $-\infty \leqq b_i \leqq c_i \leqq \infty$ be assigned to each arc u_i $(i = 1, \ldots, m)$. Furthermore, let two vertices Q, S and an arc $(S, Q) = u_0$ with $b_0 = -\infty$, $c_0 = \infty$ be marked.

---

[1]) A graph is called *simple* if it contains neither loops nor multiple arcs.

Let the relation $\omega^-(Q) = \omega^+(S) = \{u_0\}$ be true. Find an arc function φ (a flow) for which

1. $b_i \leqq \varphi_i \leqq c_i \quad (i = 1, \ldots, m)$,
2. $\sum_{u \in \omega^+(P)} \varphi(u) = \sum_{u \in \omega^-(P)} \varphi(u)$ for all $P \in \mathfrak{X}$,
3. $\varphi(u_0) = \varphi_0$ is maximal.

A few remarks are still to be made.

a) The case $b_i = 0$ for each i and $c_i \geqq 0$ remains the most important application. This case then represents the problem dealt with in point 1.3.2, where there will always be the possibility of indicating a feasible initial flow $\varphi \equiv 0$.

b) As for the problem just formulated, the question of the existence of a feasible initial flow is of great importance. Let us give here a necessary and sufficient condition for the existence of a flow that is compatible with the capacity restrictions.

THEOREM 1.21. *Given numbers b_i, c_i with $-\infty \leqq b_i \leqq c_i \leqq \infty$ $(i = 1, 2, \ldots, m)$. Then, for each i there will exist a flow $\varphi = \{\varphi_1, \ldots, \varphi_m\}$ with $b_i \leqq \varphi_i \leqq c_i$ if and only if for each elementary cocycle ω the relations*

$$\sum_{u_i \in \omega^-} c_i \geqq \sum_{u_i \in \omega^+} b_i \quad \textit{and} \quad \sum_{u_i \in \omega^+} c_i \geqq \sum_{u_i \in \omega^-} b_i$$

are true.

The reader will recognize the necessity of these conditions if he takes into account that the b_i are demanding a minimum quantity of the flows to pass through the arc u_i, whereas the c_i require maximum quantities to flow.

Now, we show the sufficiency of these conditions and make the proof by means of mathematical induction via the number of vertices.

For a graph containing $n = 1$ vertices, the condition may easily be satisfied (in the absence of cocycles), since if any arc exists, it will always be a loop. Nevertheless, it is possible for any flow to pass through these (eventual) loops as far as it is compatible with the bounds b_i and c_i.

Let the theorem be proved for all graphs with n $(n = 1, 2, \ldots, N - 1;\ N \geqq 2)$ vertices.

Let $\boldsymbol{G}$ be any graph containing N vertices. If $\boldsymbol{G}$ contains no cocycle (that is, only loops at the most), then it will be easy to give a feasible flow (see case $n = 1$). Let now ω be any cocycle of $\boldsymbol{G}$. Consider any arc $u_j = (P, R) \in \omega$. By removing u_j from $\boldsymbol{G}$ and identifying P and R, the resulting graph $\boldsymbol{G}^+$ contains $N - 1$ vertices. According to the induction hypothesis, there exists in $\boldsymbol{G}^+$ a feasible flow φ^+ with $b_i \leqq \varphi_i^+ \leqq c_i$ for each arc $u_i \neq u_j$ of $\boldsymbol{G}$.

In the chapter on cocycles, we had assigned to a cocycle $\omega(P)$ a vector having m components $\omega_i(P)$ according to

$$\omega_i(P) = \begin{cases} 0 & \text{if } u_i \text{ is not incident with } P, \\ 1 & \text{if } u_i \text{ is going away from } P, \\ -1 & \text{if } u_i \text{ is going to } P. \end{cases}$$

Putting

$$\varphi_i'' = \varphi_i^+ \quad \text{for} \quad i \neq j,$$

$$\varphi_j'' = -\sum_{i \neq j} \omega_i(P) \cdot \varphi_i^+ \left(= \sum_{k \neq j} \omega_k(R)\, \varphi_k^+\right)$$

we get a flow φ'' for which, however, the relation $b_j \leqq \varphi_j'' \leqq c_j$ is not necessarily true.

Let us define some certain intervals $\boldsymbol{J}$. We assign the interval $\boldsymbol{J}^i := [b_i, c_i]$ to each arc u_i. Assign to each cocycle ω an interval $\boldsymbol{J}_\omega$ with

$$\boldsymbol{J}_\omega := \Big[\sum_{u_i \in \omega^+} b_i - \sum_{u_i \in \omega^-} c_i, \sum_{u_i \in \omega^+} c_i - \sum_{u_i \in \omega^-} b_i\Big].$$

To each cocycle ω containing the arc u_j there will be assigned an interval $\boldsymbol{J}_\omega^j$ with

$$\boldsymbol{J}_\omega^j := \begin{cases} \Big[\sum_{u_i \in \omega^-} b_i - \sum_{\substack{u_i \in \omega^+ \\ i \neq j}} c_i, \sum_{u_i \in \omega^-} c_i - \sum_{\substack{u_i \in \omega^+ \\ i \neq j}} b_i\Big] & \text{if} \quad u_j \in \omega^+, \\ \Big[\sum_{u_i \in \omega^+} b_i - \sum_{\substack{u_i \in \omega^- \\ i \neq j}} c_i, \sum_{u_i \in \omega^+} c_i - \sum_{\substack{u_i \in \omega^- \\ i \neq j}} b_i\Big] & \text{if} \quad u_j \in \omega^-. \end{cases}$$

We show first that these are really intervals, and furthermore, that for each cocycle ω containing u_j the relation $\varphi_j'' \in \boldsymbol{J}_\omega^j$ is true.

As for $\boldsymbol{J}^i$ and $\boldsymbol{J}_\omega$ there is nothing to show since this is directly included in the hypothesis.

Thus, we show that $\boldsymbol{J}_\omega{}^j$ is an interval for which $\varphi_j'' \in \boldsymbol{J}_\omega{}^j$ holds. By taking the definitions of φ^+ and φ'' into account, we distinguish between two cases.

1. $u_j \in \omega^+$, then

$$\sum_{u_i \in \omega^-} b_i - \sum_{\substack{u_i \in \omega^+ \\ i \neq j}} c_i \leqq \sum_{u_i \in \omega^-} \varphi_i'' - \sum_{u_i \in \omega^+} \varphi_i'' + \varphi_j'' = \varphi_j'' \leqq \sum_{u_i \in \omega^-} c_i - \sum_{\substack{u_i \in \omega^+ \\ i \neq j}} b_i.$$

2. $u_j \in \omega^-$, then

$$\sum_{u_i \in \omega^+} b_i - \sum_{\substack{u_i \in \omega^- \\ i \neq j}} c_i \leqq \sum_{u_i \in \omega^+} \varphi_i'' - \sum_{u_i \in \omega^-} \varphi_i'' + \varphi_j'' = \varphi_j'' \leqq \sum_{u_i \in \omega^+} c_i - \sum_{u_i \in \omega^-} b_i.$$

Since for any cocycle ω containing u_j it holds that φ_j'' lies in $\boldsymbol{J}_\omega{}^j$, we have proved at the same time that

$$\bigcap_{\substack{\omega \\ u_j \in \omega}} \boldsymbol{J}_\omega{}^j \neq \emptyset$$

is true.

Now, let us show that the relation $\boldsymbol{J}_\omega{}^j \cap \boldsymbol{J}^j \neq \emptyset$ holds. If we denote the initial point of the interval $\boldsymbol{J}_\omega{}^j$ by $\gamma_\omega{}^j$ and the terminal point by $\delta_\omega{}^j$, that is, $\boldsymbol{J}_\omega{}^j = [\gamma_\omega{}^j, \delta_\omega{}^j]$, then we get, with two cases being distinguished again:

1. $u_j \in \omega^+$: The conditions of our theorem result in

$$\sum_{u_i \in \omega^-} b_i - \sum_{u_i \in \omega^+} c_i \leqq 0,$$

that is,

$$\gamma_\omega{}^j = \sum_{u_i \in \omega^-} b_i - \sum_{\substack{u_i \in \omega^+ \\ i \neq j}} c_i \leqq c_j$$

and because of $\sum_{u_i \in \omega^+} b_i - \sum_{u_i \in \omega^-} c_i \leqq 0$ the relation

$$\delta_\omega{}^j = \sum_{u_i \in \omega^-} c_i - \sum_{\substack{u_i \in \omega^+ \\ i \neq j}} b_i \geqq b_j$$

holds. From the respective position of $\gamma_\omega{}^j, \delta_\omega{}^j, b_j, c_j$, however, there follows directly the assertion

$$\boldsymbol{J}_\omega{}^j \cap \boldsymbol{J}^j \neq \emptyset.$$

The case

2. $u_j \in \omega^-$

may be treated in quite an analogous way, which proves the assertion.

Thus, we have shown that all of the intervals $\boldsymbol{J}^j$ and $\boldsymbol{J}_\omega{}^j$, with ω passing through all cocycles containing u_j, intersect. That is, there is a number y_j that lies in each of the intervals $\boldsymbol{J}^j, \boldsymbol{J}_\omega{}^j$ $(u_j \in \omega)$.

Now, we define some other intervals:

$$\hat{\boldsymbol{J}}^i := [\hat{b}_i, \hat{c}_i] = [b_i, c_i] = \boldsymbol{J}^i \quad \text{for} \quad i \neq j,$$

$$\hat{\boldsymbol{J}}^j := [\hat{b}_j, \hat{c}_j] = [y_j, y_j].$$

We will show that with the new interval end-points $\hat{b}_i, \hat{c}_i$ the conditions of the theorem are satisfied, too.

1. $u_j \in \omega^+$: Since $\hat{b}_j = \hat{c}_j = y_j \in \boldsymbol{J}_\omega{}^j$, it follows

$$\sum_{u_i \in \omega^-} \hat{b}_i - \sum_{\substack{u_i \in \omega^+ \\ i \neq j}} \hat{c}_i = \sum_{u_i \in \omega^-} b_i - \sum_{\substack{u_i \in \omega^+ \\ i \neq j}} c_i \leqq \hat{b}_j = \hat{c}_j$$

$$\leqq \sum_{u_i \in \omega^-} c_i - \sum_{\substack{u_i \in \omega^+ \\ i \neq j}} b_i = \sum_{u_i \in \omega^-} \hat{c}_i - \sum_{\substack{u_i \in \omega^+ \\ i \neq j}} \hat{b}_i.$$

Consequently, the relation

$$\sum_{u_i \in \omega^-} \hat{b}_i - \sum_{u_i \in \omega^+} \hat{c}_i \leqq 0 \leqq \sum_{u_i \in \omega^-} \hat{c}_i - \sum_{u_i \in \omega^+} \hat{b}_i$$

is true.

2. $u_j \in \omega^-$: This case is to be dealt with analogously.

From this there results the relation

$$\sum_{u_i \in \omega^+} \hat{b}_i - \sum_{u_i \in \omega^-} \hat{c}_i \leqq 0 \leqq \sum_{u_i \in \omega^+} \hat{c}_i - \sum_{u_i \in \omega^-} \hat{b}_i .$$

Thus, we can state that the conditions of the theorem are fulfilled by the $\hat{b}_i$, $\hat{c}_i$ in both cases.

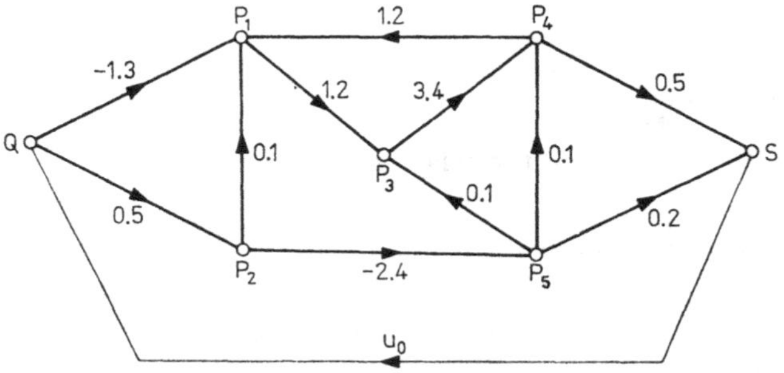

Fig. 1.17

Thus, we have managed to contract one of the given intervals $[b_j, c_j]$ to a point without violating the premises of the theorem. We make this contraction for each arc in succession and in doing so, the premises of the theorem are never violated. Having achieved this, then there will be just the flow condition (the components of this flow are exactly the y_j-values calculated) in place of the condition assumed. That means that we have found a flow also for a graph with N vertices.

Q.E.D.

By this, the theorem of the existence of a flow is entirely proved.[1])

The reader may solve the following tasks relating to the existence theorem proved just now:

TASK 10. For each arc u_i, there exists a flow φ with $\varphi_i \geqq b_i$ if and only if the condition $\sum_{u_i \in \omega} b_i \leqq 0$ is fulfilled for each cocircuit ω.

TASK 11. A flow φ with $\varphi_i \leqq c_i$ exists if and only if the condition $\sum_{\omega_i \in \omega} c_i = 0$ is fulfilled for each cocircuit ω.

---

[1]) A generalization of theorem 1.21 and a simpler proof of it are given by L. Hempel and D. Würbach in their article published in the periodical "Mathematische Operationsforschung und Statistik", series "Optimization".

Here again we consider only the most important case, i.e., if the b_i and the c_i and thus also the fluxes φ_i are integers. Let us assume as well as the simplicity of the network, although the proof may be made even without this restriction by making some slight modifications only.

We will explain the problem to be described and the following algorithm by means of a small example (cf. Fig. 1.17). Let the first number of the number pair at the arcs u_i denote the value of b_i and the second one that of c_i. A feasible flow is easy to find (task!), that is, the necessary and sufficient conditions indicated above are satisfied, although it is in general very difficult to verify them.

The algorithm which has been mentioned before and which is necessary for the solution of the given task is decomposed into two parts as follows.

1. **Algorithm** for finding a feasible flow

From the graph $G(\mathfrak{X}, \mathfrak{U})$ with the capacity restrictions b_i and c_i we go over to a graph $G'(\mathfrak{X}', \mathfrak{U}')$. For the capacity restrictions in G', the conditions $b_i' = 0$ and $c_i' \geqq 0$ will be true.

All vertices and all arcs lying in G also lie in G', and for these arcs $u_i' = u_i$ we put

$$b_i' = b(u_i') = 0 \quad \text{and} \quad c_i' = c(u_i') = c_i - b_i.$$

Add now in G' two more vertices being marked, namely Q' and S', and some more arcs marked as well, according to the following rule (cf. Figs. 1.17 and 1.18):

a) If $u_i = (P_k, P_l)$ is an arc of G with $b_i \geqq 0$, then two arcs (Q', P_l) and (P_k, S') with the capacity restrictions $c'(Q', P_l) = c'(P_k, S') = b_i$ will be added. (For $b_i = 0$ these arcs need not necessarily be added, since no flow at all can pass through these arcs.)

b) Let $u_i = (P_k, P_l)$ be an arc of G with $b_i < 0$. Here again, two new arcs (Q', P_k) and (P_l, S') will be added, with the capacity restrictions $c'(Q', P_k) = c'(P_l, S') = -b_i$ being valid. For all arcs u_i of G, except for the return arc $u_0 = (S, Q)$, new arcs are then added.

The new graph may contain double arcs. There will, for example, result two arcs going from P_1 to S', since one arc of the form (P_1, S') with $c'(P_1, S') = 1$ is provided by the arc (P_1, P_3) — cf. case a) and another arc (P_1, S') with $c'(P_1, S') = 1$, too, is being provided by the arc (Q, P_1). Such multiple arcs may be avoided by adding only one arc and assigning to it as capacity the sum of all capacities of the arcs virtually resulting from the construction of G'. Figure 1.18 shows the result of the generation of G'; there we have also inserted the return arc marked (S', Q') with a capacity of ∞. The number at an arc indicates the newly established capacity.

In the new graph G', we have to search for such a flow φ' that the arcs leaving Q' (and thus, also the arcs entering S') are saturated. We shall prove a theorem

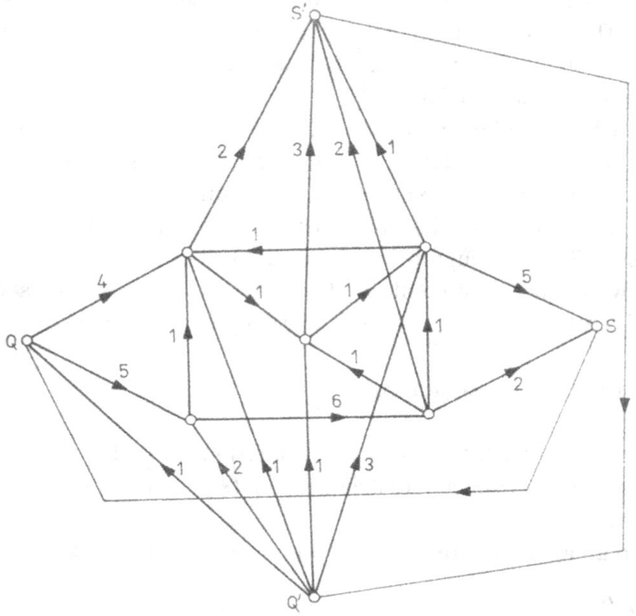

Fig. 1.18

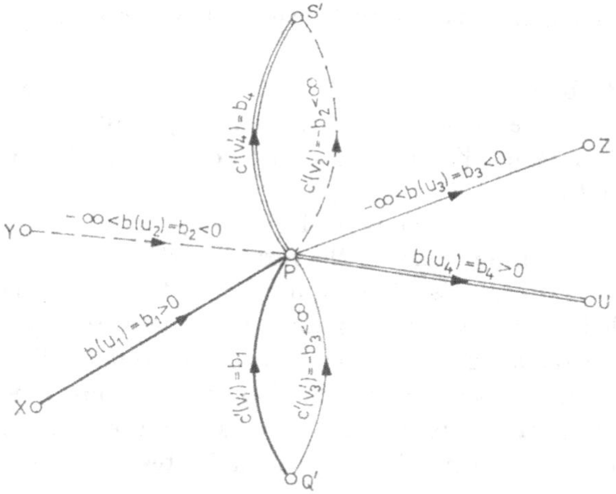

Fig. 1.19

which assures, on the basis of the knowledge of a maximum flow in G', the existence of a feasible flow in G.

THEOREM 1.22. *To each flow φ in G compatible with the capacity intervals $[b(u), c(u)]$ there corresponds in G' a flow φ' compatible with the capacity intervals $[0, c'(u')]$ that saturates the outgoing arcs (P, S'), and vice versa. It holds*

$$\varphi_i = \varphi_i' + b_i \quad (i = 1, \ldots, m), \qquad \varphi(u_0) = \varphi'(u_0).$$

Proof. Let φ' be a flow in G' saturating all arcs leaving Q' (and thus, all arcs entering S'). Show that a feasible flow φ in G may be obtained from φ'. Let P be any vertex of G (cf. Fig. 1.19). Four different types of arcs may be incident with P:

1. $u_1 = (X, P)$ with $b_1 > 0$,
2. $u_2 = (Y, P)$ with $b_2 < 0$,
3. $u_3 = (P, Z)$ with $b_3 < 0$,
4. $u_4 = (P, U)$ with $b_4 > 0$.

According to the formation rule for G', there exist arcs v_i' as follows (analogously to the above order).

1. $v_1' = (Q', P)$ with $c'(v_1') = b_1$,
2. $v_2' = (P, S')$ with $c'(v_2') = -b_2$,
3. $v_3' = (Q', P)$ with $c'(v_3') = -b_3$,
4. $v_4' = (P, S')$ with $c'(v_4') = b_4$.

Now, one can easily see that a feasible flow φ is induced on G by a flow φ' saturating all arcs leaving Q' and also all arcs entering S'.

1. The flux $\varphi_1 = \varphi(u_1)$ on u_1 must satisfy in G the condition $b_1 \leqq \varphi_1 \leqq c_1$ with the flux going to P. If there is a flow φ' from Q' to P which saturates the arcs leaving Q', then the flow quantity in G' will be $\varphi'(v_1') = b_1$. If on v_1' the flux is reduced by b_1, then it may be increased on u_1 by b_1 (thus, the flux is feasible in this arc for the initial graph G). Furthermore, the arc v_1' is without flux and may be omitted.

Correspondingly, consider the remaining three cases.

These reflections being made for each vertex, they result in a feasible flow φ on G, with $\varphi(u_i) = \varphi'(u_i) + b_i$ for each arc u_i in G.

Now, we show that the relation $\varphi(u_0) = \varphi'(u_0)$ holds for the return arc u_0.

Consider vertex S (cf. Fig. 1.20). To an arc u_2 with $b(u_2) = b_2 < 0$ there was added in G' an arc v_2' with $c'(v_2') = -b_2$ for which $\varphi'(v_2') = -b_2$ holds according to the premise. If the flux on v_2' is reduced by the amount $-b_2$ and if it is also reduced on arc u_2 by this amount, then the flow condition remains fulfilled in S, there is no flux any longer on the auxiliary arc v_2' (it may thus be omitted), and the flux $\varphi(u_2)$, which has resulted in this way, satisfies the required conditions $b_2 \leqq \varphi(u_2) \leqq c_2$ on arc u_2.

Correspondingly, we increase the flux by the amount b_1 on an arc u_1 leading to S with $b(u_1) = b_1 > 0$, and we reduce it by the amount b_1 on the arc v_1' leading in G' from Q' to S. Thus, this arc becomes free of fluxes, whereas the flux on u_1 satisfies just the capacity restrictions. On the returning arc $u_0 = (S, Q)$, the flux does not change since we have explicitly failed to introduce auxiliary arcs for it. This proves, however, that $\varphi(u_0) = \varphi'(u_0)$ is valid.

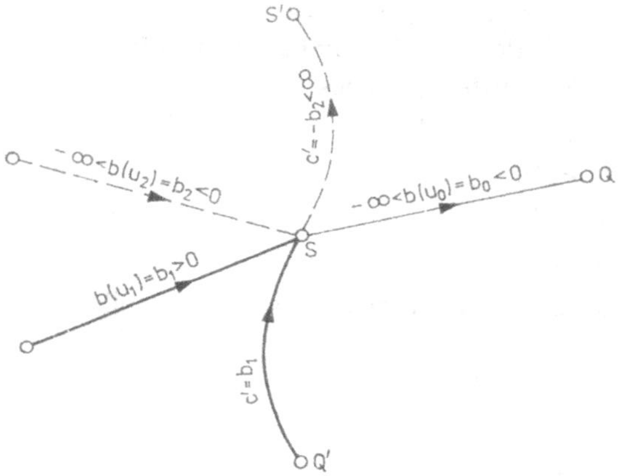

Fig. 1.20

Now, we turn to the second part of the proof of theorem 1.22.

Let φ be a feasible flow on G. Show that the graph G' which has been generated according to our construction rule has such a structure that a flow φ' may be found on it saturating the arcs leaving Q' (and entering S'). Let $u_1 = (X, Y) \neq u_0$ be any arc of G with $b_1 > 0$. In addition to this arc, two arcs (cf. Fig. 1.21) $w_1' = (Q', Y)$ and $v_1' = (X, S')$ are introduced in G', with the capacity $c'(w_1') = c'(v_1') = b_1$ being assigned to them. If on u_1 the flux $\varphi(u_1)$ is reduced by the amount b_1 (the capacity limits b_1 and c_1 are reduced at the same time by the amount b_1 each), then the arc u_1 in G' will have the capacity limits 0 and $c_1 - b_1$, and the flux $\varphi'(u_1)$ lies within these bounds. In order to ensure that the flow condition remains satisfied in the vertices X and Y, we let a flux of the value b_1 pass from X via arc v_1' to S' and a flux of the value b_1 from Q' via w_1' to Y. This however, makes, these two arcs saturated. The reader may carry on by himself an analogous reflection for an arc $u_2 = (X, Y)$ with $b_2 < 0$.

Since we have not introduced any new arcs for the return arc u_0, we obtain the relation $\varphi'(u_0) = \varphi(u_0)$ by making some similar reflections as in the first part of this proof.

Thus, the proof of theorem 1.22 is complete.

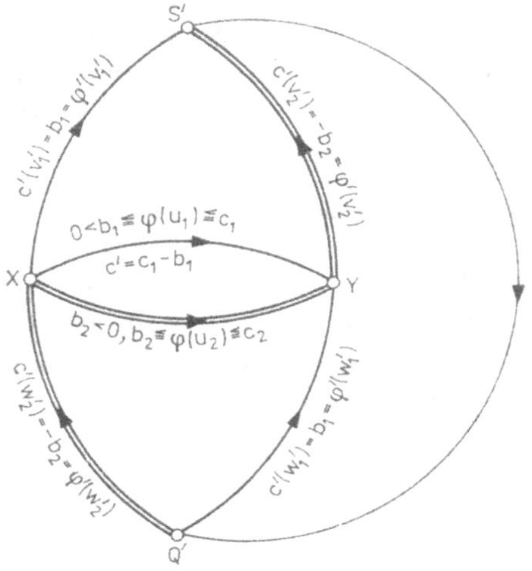

Fig. 1.21

As for our problem, the following task results for finding a feasible flow in $\boldsymbol{G}$:

(i) Construct $\boldsymbol{G}'$ on the basis of $\boldsymbol{G}$ and determine the capacity restrictions c_i'.

(ii) Find a maximum flow in $\boldsymbol{G}'$, i.e., a flow for which the value on the arc (S', Q') becomes maximal.

(iii) If all arcs leaving Q' are saturated, we will find a feasible flow for each arc $u \neq u_0 = (S, Q)$ and $\varphi(u_0) = \varphi'(u_0)$ of the graph $\boldsymbol{G}$ according to the relation

$$\varphi(u) = \varphi'(u) + b(u).$$

(iv) If a maximum flow in $\boldsymbol{G}'$ does not saturate all arcs leaving Q', then there is no feasible flow in $\boldsymbol{G}$.

Now, we come to the construction of a maximum flow in $\boldsymbol{G}$ (as far as we have found a feasible flow in $\boldsymbol{G}'$).

2. **Algorithm** for determining a maximum flow in $\boldsymbol{G}$

(i) Label the source Q in $\boldsymbol{G}$.

(ii) If P_j is labelled and P_k is not, the latter will also be labelled, in so far as there exists an arc (P_j, P_k) with $\varphi(P_j, P_k) < c(P_j, P_k)$.

(iii) If P_j is labelled and P_k is not, then we label also P_k, in so far as there exists an arc (P_k, P_j) with $\varphi(P_k, P_j) > b(P_k, P_j)$.

If we succeed in this procedure in labelling the sink S, then we can augment the flux on the returning arc (according to the algorithm of Ford and Fulkerson dealt with in section 1.3.2.) by at least 1 (with the capacity restrictions being assumed as integers). In this case, repeat the algorithm. If it is not possible to label S, then the flux on the return arc (S, Q) will be maximum.

Let $\mathfrak{S}$ be the set of all those vertices being labelled after having broken off the labelling rule (i.e., $Q \in \mathfrak{S}$, $S \notin \mathfrak{S}$). Then, the relation $\varphi(u) = c(u)$ holds for any arc $u \in \omega^+(\mathfrak{S})$, and $\varphi(u) = b(v)$ holds for any arc $v \in \omega^-(\mathfrak{S})$. Thus, it results that the flow quantity from Q to S (or on the return arc (S, Q), too) cannot be further augmented since no greater quantity of flow may pass through the arcs of $\omega^+(\mathfrak{S})$ because the upper capacity limits have been reached.

For the maximum flow $\varphi_0 = \varphi(u_0)$ obtained in this way, we get:

$$\varphi_0 = \sum_{u_i \in \omega^+(\mathfrak{S})} \varphi(u_i) - \sum_{u_i \in \omega^-(\mathfrak{S})} \varphi(u_i) = \sum_{u_i \in \omega^+(\mathfrak{S})} c_i - \sum_{u_i \in \omega^-(\mathfrak{S})} b_i .$$

These considerations result in the following generalized theorem of Ford and Fulkerson:

THEOREM 1.23. *If in a graph* $\mathbf{G}$ *with capacity limits* b_i, c_i *for the arcs* u_i *there exists a flow* φ, *i.e.*

$$b_i \leqq \varphi(u_i) \leqq c_i \quad (i = 1, \ldots, m),$$

then the maximum flow φ_0 *on the return arc* $u_0 = (S, Q)$ *has the value*

$$\varphi_0 = \min \left\{ \sum_{u_i \in \omega^+(\mathfrak{S})} c_i - \sum_{u_i \in \omega^-(\mathfrak{S})} b_i \right\}$$

with the minimum being formed over all sets $\mathfrak{S}$ *of vertices for which* $Q \in \mathfrak{S}$, $S \notin \mathfrak{S}$ *hold.*

Thus, for the problem indicated at the beginning of this chapter the solution assumes the value five for the maximum flow (cf. Fig. 1.22).

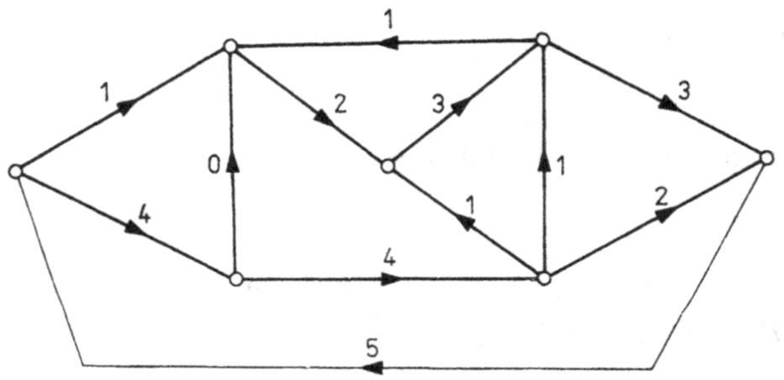

Fig. 1.22

1.3.4. *The multi-terminal problem*

In this section, we restrict ourselves to undirected graphs or to *symmetric* graphs (there is no substantial difference between them). Symmetric graphs are graphs in which there exists an arc (X, Y) if and only if (Y, X) exists. Loops and also parallel edges will not be allowed. Let a capacity $c(P_i, P_j) = c_{ij} = c_{ji}$ be assigned to each arc (P_i, P_j). Here again, we only consider integer values of the c_{ij}. As required we will use either the directed symmetric graph $\boldsymbol{G}$ introduced just now or the undirected graph $\boldsymbol{G}'$ thus resulting from $\boldsymbol{G}$ so that we substitute the (undirected) edge (P_i, P_j) having a capacity c_{ij} for the two arcs (P_i, P_j) and (P_j, P_i).

Let us now examine which flow quantity can be sent through the graph, and that from any chosen vertex P_k to any other vertex P_j. We denote this maximum flow quantity by f_{kj}, and because of the symmetry of the graph, it is clear that the relation $f_{ij} = f_{ji}$ holds for any vertices P_i, P_j. In addition, suppose that $f_{ii} = \infty$ for each vertex P_i.

In this paragraph, we try to find these f_{ij} all at once by a suitable algorithm (which has been found by R. E. Gomory and T. C. Hu) for each couple of vertices. Before beginning, note that the quantities f_{ij} and f_{kl} may, in general, not be transported at the same time.

The following theorem reveals a simple connection between the f_{ij}.

THEOREM 1.24. *A set* $\mathfrak{F} = \{f_{ij} : f_{ij} = f_{ji}, i, j, = 1, \ldots, n\}$ *may be considered to be the set of the maximum flow quantities of a certain network* $\boldsymbol{G}$ *if and only if the relation*

$$f_{ik} \geqq \min (f_{ij}, f_{jk}) \tag{1}$$

holds for all i, j, k.

Proof.

1. *Necessity.* According to the theorem of Ford and Fulkerson, there exists in $\boldsymbol{G}$ a cut $\mathfrak{S} = (\mathfrak{V}/\mathfrak{W})$ with $P_i \in \mathfrak{V}$, $P_k \in \mathfrak{W}$ and $f_{ik} = c(\mathfrak{V}/\mathfrak{W})$, i.e., a cut of minimum capacity separating P_i from P_k. Let P_j be any third vertex of $\boldsymbol{G}$. Then there will be two possibilities:

a) $P_j \in \mathfrak{V}$: Since $P_k \in \mathfrak{W}$, the set $\mathfrak{S} = (\mathfrak{V}/\mathfrak{W})$ also turns out to be a cut separating P_j from P_k, i.e., it holds:

$$f_{jk} \leqq c(\mathfrak{V}/\mathfrak{W}) = f_{ik}.$$

b) $P_j \in \mathfrak{W}$: Since $P_i \in \mathfrak{V}$, the set $\mathfrak{S}$ is also separating the vertices P_i from P_j, i.e., it holds:

$$f_{ij} = c(\mathfrak{V}/\mathfrak{W}) = f_{ik}.$$

One of the two cases a) or b) occurs, i.e., the inequality (1) holds.

2. *Sufficiency*. Consider a set of $\binom{n}{2}$ numbers f_{ij} (we had assumed $f_{ii} = \infty$ and $f_{ij} = f_{ji}$ for all i, j) satisfying the inequality (1). Let G be a symmetric, complete graph with n vertices $P_1, \ldots, P_n$ (complete in the sense that for any two different vertices P_i, P_j there exists not only the arc (P_i, P_j), but also the arc (P_j, P_i)). We assign to each of the arcs (P_i, P_j) a valuation f_{ij} and call f_{ij} the *length* of the arc assigned to it. We search for a spanning tree M in G in which the sum of the lengths of the arcs (or of the edges when undirected) of M is maximum.

Let $u_{is} = (P_i, P_s)$ be an edge of G not contained in M. Then, there is in M a unique path $W = (u_{ij}, u_{jk}, \ldots, u_{rs})$ going from P_i to P_s. We will show that

$$f_{is} \leqq \min (f_{ij}, f_{jk}, \ldots, f_{rs}).$$

Suppose that there exists an edge u_{pq} on W the length of which is shorter than that of u_{is}. If we remove from M the edge u_{pq} and add u_{is}, a spanning tree M' of G will develop which is longer than M which contradicts the maximality of M.

On the other hand, it is easy to see (mathematical induction!) that theorem 1.24 may be extended, namely that the condition (1) is equivalent to the condition

$$f_{is} \geqq \min (f_{ij}, f_{jk}, \ldots, f_{rs}). \tag{2}$$

From the last two inequalities, however, it results

$$f_{is} = \min (f_{ij}, f_{jk}, \ldots, f_{rs}) \tag{3}$$

with $(u_{ij}, u_{jk}, \ldots, u_{rs})$ being that path between P_i and P_s which has been uniquely fixed by the maximum spanning tree.

Because of (3), the maximum spanning tree M of G with the capacities $c_{ij} = f_{ij}$ for all $(i, j) \in M$ is a graph possessing property (1).

REMARKS.

1. Let P_i, P_j, P_k be any three vertices of G. Then, at least two of the three maximum flow values f_{ij}, f_{jk}, f_{ki} are equal, which follows immediately from (3).

2. Of the $\binom{n}{2}$ maximum flow values f_{ij} there differ from each other $n - 1$ values at the most, and this also follows directly from (3). There exist $n - 1$ values of the f_{ij} differing from each other only if the lengths of the edges of a maximal spanning tree M are pairwise different.

Now, we come to Gomory's and Hu's algorithm for finding all f_{ij} at once. Here, we take into consideration that it is sufficient to construct a spanning tree with maximum length, since the formula (3) offers the possibility of calculating the remaining maximum flow values.

Algorithm of Gomory and Hu

Let us first specify the concept of *condensation of a set of vertices*. Let $\mathfrak{A} \subseteq \mathfrak{X}$ be a subset of the set $\mathfrak{X}$ of vertices of an undirected graph $\boldsymbol{G}$.

We substitute a vertex A for $\mathfrak{A}$, and the edges to be newly introduced have to be drawn according to Fig. 1.23. If in $\boldsymbol{G}$, for example, the vertices $A_1, \ldots, A_r$ of $\mathfrak{A}$ are connected with a vertex $B \notin \mathfrak{A}$, then an edge (A, B) with the capacity $\sum_{i=1}^{r} c(A_i, B)$ will be substituted for the edges (A_i, B), $i = 1, \ldots, r$ in the condensation graph $\boldsymbol{G}^+$. The capacities of those edges having no end-point in $\mathfrak{A}$ remain unchanged.

(i) Choose any two vertices P_i, P_j in $\boldsymbol{G}$ and determine a maximum flow of the value f_{ij} flowing between them, by means of the algorithm of Ford and Fulkerson. The algorithm supplies a minimum cut $\mathfrak{S}_1 = (\mathfrak{B}_1/\mathfrak{W}_1)$ with $P_i \in \mathfrak{B}_1$ and $P_j \in \mathfrak{W}_1$, and $f_{ij} = c(\mathfrak{B}_1/\mathfrak{W}_1)$. For the sake of brevity we put $c_1 = c(\mathfrak{S}_1)$ and represent the result in Fig. 1.24a.

(ii) If one of the two classes $\mathfrak{B}_1$, $\mathfrak{W}_1$ contains more than one vertex (this is say $\mathfrak{W}_1$), then a maximum flow will again be determined between any two vertices P_k and P_l of $\mathfrak{W}_1$ by means of the algorithm of Ford and Fulkerson. This maximum flow in turn induces a minimum cut $\mathfrak{S}_2 = (\mathfrak{B}_2/\mathfrak{W}_2)$ ($\mathfrak{B}_1$ may remain condensed

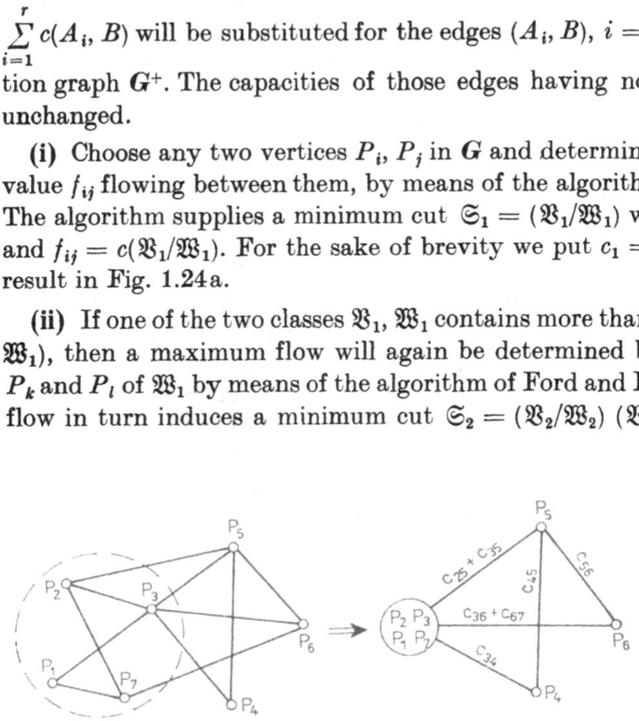

Fig. 1.23

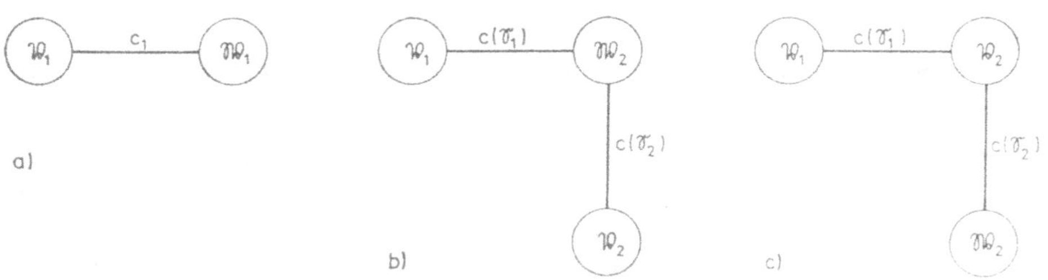

Fig. 1.24

in this case). This process is represented in Fig. 1.24b. Here, we decompose $\mathfrak{W}_1$ into two classes and join them to each other by an edge with the valuation $c_2 = c(\mathfrak{S}_2)$. If in $\boldsymbol{G}$ the cut $\mathfrak{S}_2$ disconnects the sets $\mathfrak{V}_2$ and $\mathfrak{V}_1$, then the "node" $\mathfrak{W}_2$ will be connected with the "node" $\mathfrak{V}_1$. But if $\mathfrak{S}_2$ disconnects the sets $\mathfrak{W}_2$ and $\mathfrak{V}_1$ from each other, we connect $\mathfrak{V}_2$ and $\mathfrak{V}_1$.

(iii) Now, find among the three classes of nodes $\mathfrak{V}_1$, $\mathfrak{W}_1$, $\mathfrak{V}_2$ such one containing more than one vertex (if there is none, we have finished). This class, however, will be decomposed according to (ii). The procedure will be continued as long as there can still be found classes of nodes comprising more than one element.

Let us look at an example (cf. Fig. 1.25). The numbers at the edges stand for the capacities assigned to them. We try to find the maximum flow values occurring between each couple of vertices of this network. Six vertices can be found in the network $\boldsymbol{G}$; that means, we have to determine five edges of the spanning tree with "maximum lengths", i.e., to solve five maximum flow problems.

1. Choose arbitrarily two vertices, say P_2 and P_3, and determine a maximum flow going between them, and also a cut of minimum capacity that separates these two vertices from each other. We get for the maximum flow a value of $f_{23} = 14$. In such a minor case, this may be found by trying out, but in general the algorithm of Ford and Fulkerson will have to be applied. Here, each edge u is to be replaced by two oppositely oriented arcs having the same capacity like u, two auxiliary points Q and S are to be introduced, with one arc going from Q to P_2, the second from P_3 to S and the third from S to Q; to these three arcs there must be assigned a capacity of ∞ and the flow going on the arc (S, Q) is to maximize. A cut of minimum capacity disconnects the vertices P_1, P_2, P_6 from P_3, P_4, P_5 (cf. Fig. 1.25b).

2. We go on decomposing for example $\mathfrak{W}_1 = \{P_1, P_2, P_6\}$ and condense $\mathfrak{V}_1 = \{P_3, P_4, P_5\}$ to form a vertex (Fig. 1.25c). If we now look for a maximum flow between P_1 and P_2, we can do this in the network of Fig. 1.25c. We get for the maximum flow a value of $f_{12} = 15$ and as minimum cut $(\{P_2\}/\{P_1, P_3, P_4, P_5, P_6\})$. An edge with the valuation of 15 is drawn from $\{P_2\}$ to $\{P_1, P_6\}$, since P_2 is disconnected from the remaining vertices when this minimum cut is removed (Fig. 1.25d).

3. Now, we search for a maximum flow going from P_1 to P_6, etc. (cf. Fig. 1.25e to j).

Figure 1.25j shows furthermore the single minimum cuts (interrupted line), whereas the maximum flow tree is represented in Fig. 1.25i. If we want to find, for example, the maximum flow from P_2 to P_4, we have to seek the path joining P_2 to P_4 in this tree. In this way, we get for the maximum flow the value $f_{24} = 14$, because the "shortest" edge on this path has the value 14.

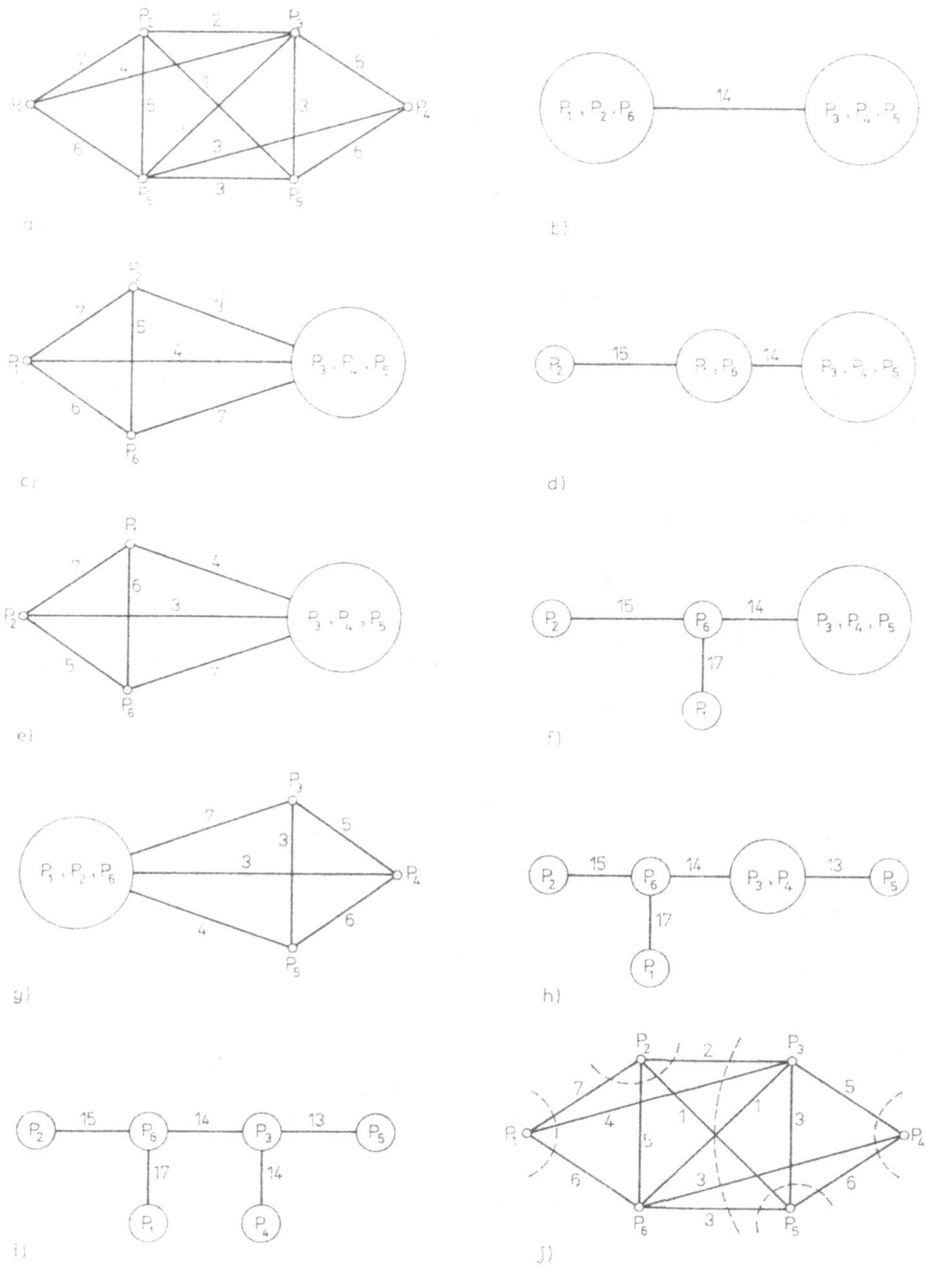

Fig. 1.25

The following maximum flow matrix results for our example:

$$\begin{array}{c} \\ P_1 \\ P_2 \\ P_3 \\ P_4 \\ P_5 \\ P_6 \end{array} \begin{array}{c} \begin{array}{cccccc} P_1 & P_2 & P_3 & P_4 & P_5 & P_6 \end{array} \\ \begin{bmatrix} \infty & 15 & 14 & 14 & 13 & 17 \\ 15 & \infty & 14 & 14 & 13 & 15 \\ 14 & 14 & \infty & 14 & 13 & 14 \\ 14 & 14 & 14 & \infty & 13 & 14 \\ 13 & 13 & 13 & 13 & \infty & 13 \\ 17 & 15 & 14 & 14 & 13 & \infty \end{bmatrix} \end{array}$$

Further maximum flow problems will be treated in chapters 3 and 4. The interested reader, however, may find other flow problems in the books of Hu [9], Ghouila-Houri/Berge [7], Busacker/Saaty [1], Ford/Fulkerson [5], Lawler [20] and Carré [21].

1.4. THE MAXIMUM TENSION PROBLEM

1.4.1. The existence theorem for a tension

In this section, we shall discuss the following problem:

Let $G(\mathfrak{X}, \mathfrak{U})$ be a finite directed graph. Let an interval $[k_i, l_i]$ be assigned to each arc $u_i \in \mathfrak{U}$. Find a tension $\vartheta = (\vartheta_1, \ldots, \vartheta_m)$ on G with the restrictions

$$k_i \leqq \vartheta_i \leqq l_i \quad (i = 1, \ldots, m).$$

We first want to give an existence theorem.

THEOREM 1.25. *Given real numbers k_i, l_i with $-\infty \leqq k_i \leqq l_i \leqq \infty$ $(i = 1, \ldots, m)$. A tension ϑ with $k_i \leqq \vartheta_i \leqq l_i$ for each i exists if and only if for each elementary cycle μ the relations*

$$\sum_{u_i \in \mu^-} l_i \geqq \sum_{u_i \in \mu^+} k_i, \qquad \sum_{u_i \in \mu^+} l_i \geqq \sum_{u_i \in \mu^-} k_i \tag{1}$$

hold.

This theorem may be proved analogously to the method applied for proving theorem 1.21. We recommend this to the interested reader.

Furthermore, the reader may also solve the following tasks resulting as corollaries from the theorem.

TASK 12. A tension ϑ with $\vartheta_i \geqq k_i$ $(i = 1, 2, \ldots, m)$ exists if and only if for each circuit μ the relation $\sum_{u_i \in \mu} k_i \leqq 0$ is true.

TASK 13. A tension ϑ with $\vartheta_i \leqq l_i$ $(i = 1, 2, \ldots, m)$ exists if and only if for each circuit μ the relation $\sum_{u_i \in \mu} l_i \geqq 0$ is true.

1.4.2. The problems of the shortest and the longest paths as potential problems

Given a directed graph $\boldsymbol{G}$ with two vertices marked, namely Q (source) and S (sink), and one arc $u_0 = (S, Q)$ also marked, with an arc function d assigning to each arc u a number $d(u) \geqq 0$. Find a potential t (a node function $t(X)$) such that for each arc $u = (X, Y)$ the tension ϑ induced by the potential t satisfies the condition

$$\vartheta(u) = t(Y) - t(X) \leqq d(u).$$

In addition, find among all potentials also a t such that $t(S)$ is maximum, with $t(Q) = 0$ being given.

Putting $d(u_0) = 0$, we can also formulate this problem as follows.

Among all tensions ϑ which are compatible with the arc valuations find a tension for which

$$-\vartheta(u_0) = t(S) - t(Q)$$

is maximum.

If we interpret the arc valuations $d(u)$ as distance between the end-point of u and the initial point of u, then $t(S) - t(Q)$ is just the shortest distance from Q to S (for the case when t is induced by a tension ϑ maximizing $-\vartheta(u_0)$); on no arc, however, the tension is allowed to exceed the arc valuation (i.e., the length), thus, the sum of all tensions along any path oriented from Q to S must not exceed the total length of this path. That means that the sum of the tensions (i.e., the potential difference $t(S) - t(Q)$) on a shortest path is not greater than the length of this shortest path.

We still have to show that a tension may be found, the sum of which along a shortest path equals the length of the shortest path (i.e., the tension along such a path is equal to the arc valuation). Suppose that for the tension maximizing $-\vartheta(u_0)$ and for each path $\boldsymbol{W}$ connecting Q with S it holds

$$t(S) < l(\boldsymbol{W}),$$

with $t(S)$ being the potential value of S (induced by ϑ) and $l(\boldsymbol{W})$ the length of a path $\boldsymbol{W}$, which is supposed to be of minimum length for our further considerations, i.e.,

$$l(\boldsymbol{W}) = \sum_{u \in \boldsymbol{W}} d(u), \qquad t(S) = \sum_{u \in \boldsymbol{W}} \vartheta(u).$$

Then, there will exist on each path going from Q to S at least one arc the tension of which might be augmented without violating the restriction d, but this, however, would contradict the maximality of ϑ.

Before turning to the solution of the shortest path problem, let us formulate the problem of the determination of a longest path between two vertices as a potential problem. Let us suppose the graphs examined to be circuit-free (otherwise it would be possible, when traversing such a circuit sufficiently often, to indicate sequences of arcs between two vertices, the length of which may exceed any value).

Given a circuit-free directed graph with two vertices Q and S being marked, one arc $u_0 = (S, Q)$ marked and an arc valuation assigning to each arc u a number $d(u)$, which is also called *length* of u. Find a tension ϑ (and thus, the potential t assigned to this tension, with two potentials differing from each other only by an additive constant; we put $t(Q) = 0$) which satisfies the following conditions:

a) $\vartheta(u) = t(Y) - t(X) \geqq d(u)$ for each arc $u = (X, Y)$,

b) $\vartheta(u_0) = t(Q) - t(S)$ is to maximize.

Why is this the problem of the determination of a longest path?

On each path going from Q to S (all arcs being oriented in this direction!) the potential is growing from vertex to vertex (provided that the arc valuations are all positive). For the arc u the growth of the potential is at least $d(u)$, that is, the potential growth $t(S) - t(Q)$ on any path $\boldsymbol{W}$ from Q to S is at least $l(\boldsymbol{W}) = \sum_{u \in \boldsymbol{W}} d(u)$. Since this is true for each path going from Q to S, this relation holds also for a longest path from Q to S.

Suppose that there exists a tension ϑ such that for the potential t assigned to ϑ the relation

$$t(S) > \max_{\boldsymbol{W}} l(\boldsymbol{W})$$

holds, with $\boldsymbol{W}$ traversing all paths from Q to S. Say that there is, among all such tensions, one tension ϑ such that $t(S)$ is minimum (for the sake of uniqueness, we have generally put $t(Q) = 0$ for the potentials assigned to the tensions).[1]) Then, on each path going from Q to S there is at least one arc $u = (X, Y)$ with $\vartheta(u) > d(u)$. Consider the set of arcs $\mathfrak{U}^0 = \{u \in \mathfrak{U}\} : \vartheta(u) > d(u)\}$. As we saw, $\mathfrak{U}^0 \neq \emptyset$. Let $u^* \in \mathfrak{U}^0$ be such that

$$\min_{u \in \mathfrak{U}^0} \big(\vartheta(u) - d(u)\big) = \vartheta(u^*) - d(u^*).$$

Since on each path from Q to S there lies at least one arc of $\mathfrak{U}^0$, the set $\mathfrak{U}^0$ contains a cut $\mathfrak{S}^0$ disconnecting Q from S. As u^* has been chosen in this way, it holds for each of the arcs $u = (X, Y) \in \mathfrak{S}^0$:

$$\vartheta(u) - d(u) \geqq \vartheta(u^*) - d(u^*) > 0.$$

We thus can reduce the value $t(Y)$ for each end-point Y of such an arc $u \in \mathfrak{S}^0$, which will also cause $t(S)$ to decrease, by at least $\vartheta(u^*) - d(u^*)$. This, however, would contradict the choice of ϑ and thus the choice of its potential.

---

[1]) The existence of the minimum is guaranteed by well-known theorems of linear optimization.

The problem of the determination of a longest path will be discussed in paragraph 1.5 in connection with an introduction to the problems related to *network analysis*.

Now, we come to an algorithm for determining a shortest path, or more precisely a shortest simple path, in a graph.

1.4.3. Algorithm for determining a shortest simple path

We consider a directed graph $\boldsymbol{G}(\mathfrak{X}, \mathfrak{U})$ with valuated arcs, with the arc valuation $l(u)$ of the arc u being called the *length of u*. In this graph, we imagine two vertices Q and S to be marked and try to determine a shortest simple path (if there is one) going from Q to S, as well as its length. In chapter 3 we shall be dealing with this problem once again, and we shall then determine by means of an algorithm, for each pair of vertices, the lengths of the shortest simple paths between them.

We assume all arc valuations to be non-negative.

Let the sequence of arcs

$$\boldsymbol{B} = \{(Q = A_1, A_2), (A_2, A_3), \ldots, (A_{k-1}, A_k = S)\}$$

denote a simple path from Q to S. Let $l(\boldsymbol{B})$ be the length of the simple path $\boldsymbol{B}$ connecting A_i with A_k. Then, it holds

$$l(\boldsymbol{B}) = \sum_{i=1}^{k-1} l(A_i, A_{i+1}).$$

If we denote by $l(A_r)$ the length of a shortest simple path connecting the vertex $Q = A_1$ with the vertex A_r, then

$$l(A_r) = \min_{\boldsymbol{B}} l(\boldsymbol{B}),$$

with $\boldsymbol{B}$ being taken over all simple paths going from $Q = A_1$ to A_r.

Algorithm for determining a shortest simple path

(i) Put $t(Q) = t(A_1) = 0$.

(ii) For the set $\mathfrak{X}_m$ consisting of m vertices ($m \geqq 1$) let $t(A_i)$ ($A_i \in \mathfrak{X}_m$) be already determined. Among all arcs $u = (X, Y)$ with $X \in \mathfrak{X}_m$, $Y \notin \mathfrak{X}_m$ (if any exist) we choose such an arc for which $t(X) + l(u)$ is minimum (if several such arcs u exist, choose any one of them). If $u' = (X', Y')$ is such an arc, i.e., for which $t(X') + l(u')$ is minimum, then put

$$t(Y') = t(X') + l(u')$$

and

$$\mathfrak{X}_{m+1} = \mathfrak{X}_m \cup \{Y'\}.$$

(iii) If a value t cannot be assigned to any other vertex, then the algorithm terminates.

If to the vertex A_r a value $t(A_r)$ is ascribed, then $t(A_r)$ will be the length of a shortest simple path going from Q to A_r, i.e., $t(A_r) = l(A_r)$.

Before proving the latter statement, note that in each step of the algorithm a tree is constructed which entirely contains its predecessor. It is a rooted tree with the root Q. Relative to Q we also call it *distance tree.* In Fig. 1.26, we have given a distance tree for an undirected graph. If we replace each edge by two oppositely oriented arcs, we get a directed graph on which the distance tree is generated; in this case, it is the distance of all places of the country of Ilmenau from Ilmenau itself. In addition, the distances have been indicated at the vertices.

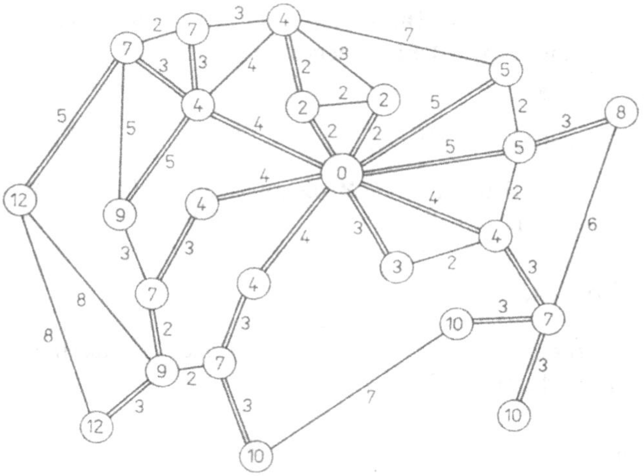

Fig. 1.26

We now show that for a vertex A_k which has been valuated in this procedure with $t(A_k)$ the relation $t(A_k) = l(A_k)$ actually holds.

As the value $t(A_k)$ has been assigned to A_k, there exists a simple path going from $Q = A_1$ to A_k.

The assertion is certainly true for the set $\mathfrak{X}_1$ of vertices, since $\mathfrak{X}_1 = \{A_1\}$, and we have put $t(A_1) = 0$. However, the distance of a point from itself is certainly equal to zero.

Let the assertion be proved for all vertices of $\mathfrak{X}_m$, i.e., $t(A) = l(A)$ for each vertex $A \in \mathfrak{X}_m$. If no other vertex may be found according to (ii), the proof is complete.

Let Y be a vertex found in accordance with (ii), i.e., $\mathfrak{X}_{m+1} = \mathfrak{X}_m \cup \{Y\}$. We suppose that there exists a simple path $\boldsymbol{B}$ going from A_1 to Y the length $l(\boldsymbol{B})$ of which is shorter than $t(Y)$. Let Z be the first vertex found on $\boldsymbol{B}$ and not belonging to $\mathfrak{X}_m$ (it may hold $Z = Y$). Let $u = (X', Z) \in \boldsymbol{B}$ (i.e., $X' \in \mathfrak{X}_m$). Because of the choice of Y according to (ii), it holds that $t(Y) \leqq t(Z)$ (if assignment of the value t is carried out up to the end). This means that, when reaching Z, the simple path $\boldsymbol{B}$ (assumed to be shorter than $t(Y)$) has already a length which is at least as long as $t(Y)$. Since we have supposed all arc valuations (i.e. lengths) to be non-negative, the simple path $\boldsymbol{B}$ will be at least of length $t(Y)$ when reaching Y, which contradicts the assumption that the length of $\boldsymbol{B}$ is shorter than $t(Y)$.

Thus, the proof of the assertion is evident. The algorithm given yields shortest simple paths going from $Q = A_1$ to each vertex that may be reached from Q.

The order of the arcs added according to the algorithm, and thus the order of the vertices added, determines at the same time the order of the vertices with a view to their distance from Q.

Let us give another algorithm making it possible to admit to a limited extent also negative lengths of certain arcs.

Let $\boldsymbol{G}$ be a directed graph without circuits of negative length with valuated arcs. Then, the following algorithm yields shortest simple paths and their lengths from a fixed starting point Q to all vertices which can be reached via simple paths.

We shall not prove it here as the interested reader may find it in the textbook of Sachs [12].

Algorithm

(i') Apply the preceding algorithm which will lead to the arc valuations $t_0(X)$ for all vertices X that can be reached from Q. Let $\boldsymbol{H}_0$ be the tree resulting from this step (in general, this is not the distance tree!).

(ii') Let a tree $\boldsymbol{H}_i$ ($i \geqq 0$) be already constructed and let it have the arc valuations $t_i(X)$. Among all arcs of the graph $\boldsymbol{G}$ which do not belong to $\boldsymbol{H}_i$ choose such an arc $u' = (X', Y')$ (if one exists) for which

$$t_i(Y') > t_i(X') + l(u')$$

is true. Then, remove the arc going to Y' in $\boldsymbol{H}_i$ and add the arc u'. The tree $\boldsymbol{H}_{i+1}$ generated in this way (proof!) is provided with the arc valuations $t_{i+1}(X)$ as follows. Put

$$t_{i+1}(Y') = t_i(X') + l(u')$$

and reduce the valuations of all those vertices of $\boldsymbol{H}_{i+1}$ which can be reached on simple paths from Y' (on simple paths of $\boldsymbol{H}_{i+1}$) by the value

$$t_i(Y') - t_{i+1}(Y').$$

Do not change all the other arc valuations. The valuations which we have got in this way are denoted by $t_{i+1}(X)$.

(iii′) If there exists no arc u' satisfying the conditions in (ii′), the algorithm terminates. When this case occurs after p steps, then we put $\boldsymbol{H} = \boldsymbol{H}_p$ and $t(X) = t_p(X)$.

THEOREM 1.26. *Let $\boldsymbol{G}$ be a directed graph with its arcs being valuated and without circuits of negative length. For any vertex X which can be reached from Q the above-mentioned algorithm yields a shortest simple path $\boldsymbol{B}(X)$ of length $t(X)$ connecting Q with X.*

We subdivide the proof into single tasks which may be recommended to the reader to solve.

TASK 14. Show that the graphs constructed in the course of the algorithm are trees.

TASK 15. For each of the trees $\boldsymbol{H}_i$ it holds:

a) No arc enters Q.

b) Exactly one arc enters any vertex $X \neq Q$.

c) To any vertex $X \neq Q$ there exists exactly one simple path from Q to X of length $t_i(X)$.

TASK 16. Show that the algorithm terminates after a finite number of steps.

TASK 17. Any simple path connecting Q with X in $\boldsymbol{G}$ has a length which is at least as long as the simple path connecting Q with X in $\boldsymbol{H}_p = \boldsymbol{H}$ (that one possessing length $t(X)$).

Figure 1.27 illustrates the second algorithm. Figure 1.27a shows (in double lines) the tree $\boldsymbol{H}_0$ with the valuations $t_0(X)$, whereas in Fig. 1.27b one can see a tree $\boldsymbol{H}_1$ with the valuations $t_1(X)$. Figure 1.27c illustrates the tree $\boldsymbol{H} = \boldsymbol{H}_p$ and the valuations $t(X)$, in fact the distance tree.

1.5. THE CONCEPTION OF NETWORK ANALYSIS

Only a very small insight into the conception of network analysis will be given in this section. We restrict ourselves to a description of the *critical path method* (***CPM***), that is, in graph theory terminology, to the determination of a longest path between two vertices marked in a directed circuit-free graph, with its arcs being valuated. We shall not enlarge on the ***PERT*** method, since the ***CPM*** method suffices to make clear which role is played by graph theory in network analysis. Problems relating to resource planning, etc. will also not be treated. The reader wishing to familiarize himself in more detail with all these problems may study the book of Götzke [8] and [10, 13, 14].

In recent years, network analysis which has stood the test so excellently, for example in the planning of production processes and building projects, has been developing so rapidly, especially by the application of electronic data processing, that it is quite impossible to specify the present status in only one section of a textbook.

Consider the example of a building project. The entire project is decomposed into single ones which are called *activities* or *processes*. The beginning and the end of such activities are called *events*. We assign to each activity an arc (sometimes also called *arrow*), the initial and the end-point of which are called the beginning and the end, respectively, of this activity. An arc thus indicates the direct dependence of events on each other. We mark two vertices Q and S, i.e., the ***beginning*** and the ***end***, respectively, ***of the building project.***

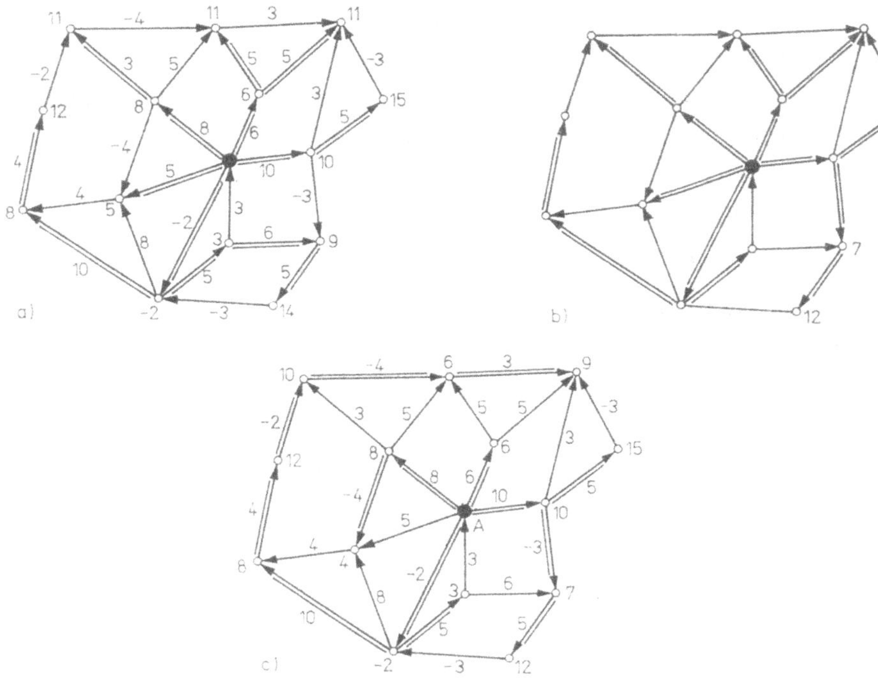

Fig. 1.27

To elaborate a network is, as is known, in general a very difficult job, but we do not wish to treat this problem here. When elaborating a network, the following aspects must be considered:

— What kinds of activities are necessary (listing of these activities)?

— Which activities must be finished before beginning another well-defined activity. For example, in the case of the construction of a house the painting works can be carried out only when the walls have been erected. (Why? — Solve this problem!)

— Which activities may be effected simultaneously (independence)?

— How much time will it take to carry out one activity? (These time durations will later be the arc valuations.)

Because on the whole a circuit-free graph (*network*) must be produced (otherwise it would not be possible either to start or to finish certain activities) in which each event and each activity must be attainable from the beginning Q and also the end S from each event and each activity, it will often be necessary to introduce some so-called *auxiliary activities*. That means that we will have to introduce arcs which only reflect dependences in their chronological order, but do not require themselves any time durations when being "treated". These arcs will thus have the arc valuation zero.

The acyclicity of a network $G(\mathfrak{X}, \mathfrak{U})$ makes it possible to arrange and to number the vertices (events) in a particularly favourable manner:

Algorithm for numbering a circuit-free graph

(i) Q is numbered 0, and we set $\mathfrak{S}_0 = \{Q\}$.

(ii) We remove from G the vertex Q and all arcs which are incident with Q these arcs all have the point Q as initial point because of the acyclicity).

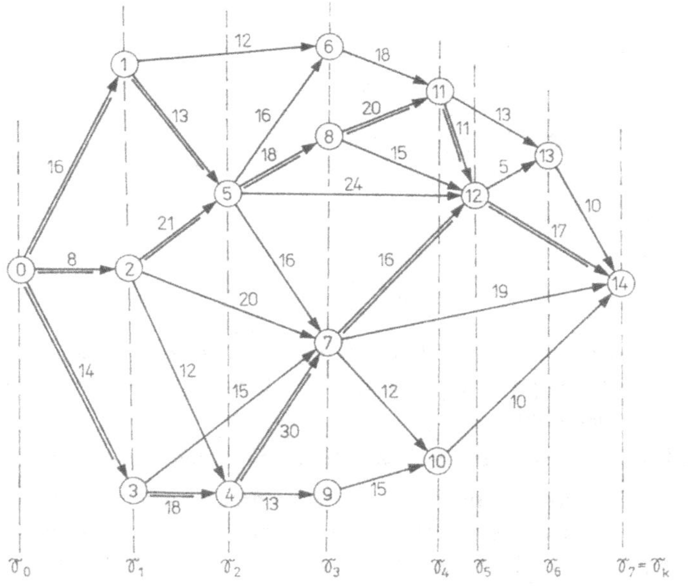

Fig. 1.28

(iii) Let $\mathfrak{S}_1 = \{P_{11}, P_{12}, \ldots, P_{1r_1}\}$ be the set of all source points of $G_1 = G - \mathfrak{S}_0$ (these are the vertices in which no arc enters into the graph G_1). These r_1 vertices are labelled with the numbers 1, 2, ..., r_1 ($r_1 > 0$ is true because of the acyclicity), with the assignment being made in any chosen but fixed order.

(iv) We remove from G_1 all vertices of $\mathfrak{S}_1$ and all arcs which are incident with these vertices.

(v) Let $\mathfrak{S}_2 = \{P_{21}, P_{22}, \ldots, P_{2r_2}\}$ be the set of all source points of $G_2 = G_1 - \mathfrak{S}_1$. The r_2 vertices are labelled, in any chosen but fixed order, with the numbers $r_1 + 1, r_1 + 2, \ldots, r_1 + r_2$, etc.

If G possesses exactly n vertices, there exists a natural number k with $\mathfrak{S}_k = \{S\}$ and $\mathfrak{S}_i = \emptyset$ for $i > k$ and $\mathfrak{S}_i \neq \emptyset$ for $i \leq k$. Furthermore, it holds

$$n = 1 + r_1 + r_2 + \cdots + r_{k-1} + 1 .$$

Each vertex of G lies in exactly one of the classes $\mathfrak{S}_i$, $i \in \{0, 1, \ldots, k\}$. It proves to be practical to arrange the vertices of the same class on a vertical line and the classes with an ascending index from the left to the right (cf. Fig. 1.28).

Let us take the following example, given the following list of activities, with $t(X, Y)$ indicating the duration of the activity (X, Y) which combines the event X with the event Y:

$t(A, B) = 16$	$t(L, M) = 15$
$t(A, C) = 8$	$t(L, E) = 18$
$t(A, L) = 14$	$t(B, I) = 12$
$t(H, M) = 16$	$t(B, H) = 13$
$t(H, I) = 16$	$t(I, F) = 18$
$t(H, K) = 18$	$t(M, D) = 16$
$t(H, D) = 24$	$t(M, P) = 19$
$t(C, G) = 21$	$t(M, G) = 12$
$t(C, M) = 20$	$t(N, G) = 15$
$t(C, E) = 12$	$t(K, F) = 20$
$t(E, M) = 30$	$t(K, D) = 15$
$t(E, N) = 13$	$t(F, O) = 13$
$t(O, P) = 10$	$t(D, O) = 5$
$t(G, P) = 10$	$t(D, P) = 17$
$t(F, D) = 11$	

A possible numbering is illustrated in Fig. 1.28. The numbering has obviously the property that for any activity (i, j) the relation $i < j$ is always true. If we arrange the classes $\mathfrak{S}_i$ of vertices from the left to the right, the end-point of any arc must then lie on the right of the initial point. If this is not the case, the numbering might be wrong or the graph is not circuit-free. In the latter case, the fault will be in the list of activities.

A rectangular scheme may obviously be assigned to a network. Calling it a matrix would not be correct, since certain cells will not be occupied. This may be seen in our example in the table below.

T_f-column		0 A	1 B	2 C	3 L	4 E	5 H	6 I	7 M	8 K	9 N	10 G	11 F	12 D	13 O	14 P
0	$0 = A$		16	8	14											
16	$1 = B$						13	12								
8	$2 = C$					12	21		20							
14	$3 = L$					18			15							
32	$4 = E$								30		13					
29	$5 = H$							16	16	18				24		
	$6 = I$												18			
	$7 = M$											12		16		19
	$8 = K$												20	15		
	$9 = N$											15				
	$10 = G$															10
	$11 = F$													11	13	
	$12 = D$														5	17
	$13 = O$															10
	$14 = P$															
T_s-row												85	67	78	85	95
Buffer								4			25	9			2	

If an activity (i, j) is carried out and if the events are numbered, then the duration $t(i, j)$ of (i, j) will be entered in that cell which is found in the ith row and the jth column. The acyclicity of the network is shown by the fact that no entries are made below the main diagonal. The rectangular scheme representation proves to be advantageous especially when using electronic computers.

Now, the following task is to be solved. Find the longest path (if one takes the time valuations $t(X, Y)$ as the lengths of the arcs (X, Y)) of all paths going from Q to S, for the length of such a longest path indicates the shortest possible duration of the project.

In order to appreciate this, consider two vertices (events) which are joined with each other by an arc (an activity) (i, j) and denoted by the numbers i and j $(i < j)$

Before the event j can take place, the event i must have occurred and the time duration $t(i, j) = t_{ij}$ necessary for realizing the activity (i, j) must have elapsed.

Consider now any directed path $\boldsymbol{W} = (i_1, i_2, \ldots, i_r)$ in $\boldsymbol{G}$. The event i_2 can take place after the event i_1, but not before at least time interval $t_{i_1i_2}$ has gone by, and the event i_3 can take place after the event i_2, but not before at least the time interval $t_{i_2i_3}$ has gone by. That means, the time interval between the events i_3 and i_1 must be at least $t_{i_1i_2} + t_{i_2i_3}$.

This leads us to the result that the event i_m can happen later than the event i_1 after a time period of at least

$$t_{i_1i_2} + t_{i_2i_3} + \cdots + t_{i_{m-1}i_m}.$$

Considering in particular any path linking Q with S, i.e.

$$\boldsymbol{W} = (Q = i_1, i_2, i_3, \ldots, i_{r-1}, i_r = S),$$

the the final event S can take place later than the beginning of the project (e.g. $t_0 = 0$) after a time interval of at least

$$t_{i_1i_2} + t_{i_2i_3} + \cdots + t_{i_{r-1}i_r}.$$

Since this relation holds for each path connecting the source Q with the sink S, we get in particular:

The duration of the project cannot be shorter than a longest path connecting Q with S.

Furthermore, the following theorem which we give below without proof is true:

THEOREM 1.27. *The duration of a project equals the length of a longest path going from Q to S.*

Let us now turn to the problem of finding a longest path connecting Q with S.

Algorithm for determining earliest possible dates

We denote by $T_f(i)$ the *earliest possible date* and by $T_s(i)$ the *latest possible date* for the event i to take place (so that a certain project can be realized within a shortest time).

(i) Put $T_f(0) = 0$, that is, the project is started at $t = 0$.

(ii) Let j be a vertex for all predecessors $j_1, \ldots, j_s$ (these are vertices for which there exists an arc (j_i, j)) of which the earliest possible dates $T_f(j_i)$ have already been determined. Then, we put

$$T_f(j) = \max_{1 \leq i \leq s} [T_f(j_i) + t(j_i, j)].$$

(iii) If the network has exactly n vertices, then $T_f(n - 1)$ will just be the minimum duration of the project.

Before verifying that the algorithm given above supplies in fact the solution for our task, we will apply this algorithm to our example.

According to (i), we get $T_f(0) = 0$.

The earliest possible dates of the predecessor vertices of $\mathfrak{S}_1$ are already known (there is only one predecessor vertex, namely Q, with $T_f(Q) = T_f(0) = 0$). It results:

$$T_f(1) = 16, \qquad T_f(2) = 8, \qquad T_f(3) = 14.$$

We are now able to determine the $T_f(i)$ for the vertices of class $\mathfrak{S}_2$ and get:

$$T_f(4) = \max(14 + 18,\ 8 + 12) = 32,$$

$$T_f(5) = \max(\ 8 + 21, 16 + 13) = 29.$$

It is thus possible to determine successively all $T_f(i)$.

The "matrix" on page 64 contains the values to be found at the vertices assigned to them. We have not filled out some of the cells; this will be left to the reader.

Now, we come to the calculation of the latest possible dates $T_s(i)$.

Algorithm for determining latest possible dates

(i′) For the sink S we put

$$T_s(n-1) = T_f(n-1) \quad \text{(in our example we have 95).}$$

(ii′) Let j be a vertex for all successors $j_1, \ldots, j_r$ of which the $T_s(j_i)$ have already been determined. Then, we put

$$T_s(j) = \min_{1 \leqq i \leqq r} \big(T_s(j_i) - t(j, j_i)\big).$$

As in the case of the determination of the $T_f(i)$, we proceed as follows for finding out the $T_s(i)$. Determine the $T_s(i)$ in the order $n-1, n-2, \ldots, 1, 0$, that is, successively for the classes $\mathfrak{S}_k, \mathfrak{S}_{k-1}, \ldots, \mathfrak{S}_0$. Below the "matrix" you find some of the $T_s(i)$, and the reader himself may find the remaining ones.

It is easy to recognize that the relation

$$T_s(i) - T_f(i) \geqq 0$$

is true for each vertex i.

In network analysis, an essential role is played by those events and activities lying on a so-called *critical path*.

DEFINITION. A path $\boldsymbol{W} = (0 = i_1, i_2, i_3, \ldots, i_r = n-1)$ is called *critical* if for each arc (i_j, i_{j+1}) of $\boldsymbol{W}$ the relation

$$T_f(i_j) = T_s(i_j) \quad \text{for} \quad j = 1, 2, \ldots, r$$

and

$$T_f(i_j) - T_f(i_{j-1}) = t(i_{j-1}, i_j) \quad \text{for} \quad j = 2, \ldots, r$$

is true. Those *activities* and *events* which lie on a critical path are also called *critical.*

The three existing critical paths are marked in our example by doubly drawn arcs (cf. Fig. 1.28).

If an event i does not lie on a critical path, then for this event

$$T_s(i) - T_f(i) = s_i > 0$$

holds. The s_i are called *dummy times* or *buffer times.*

These times may be utilized to a limited extent by making a non-critical activity start not immediately after the beginning of this activity, but without endangering the planned completion of the whole project. In some measure, dummy times reflect the reserves of a project. These dummy times may, in general, not be utilized arbitrarily. A difference is made between them, and in the literature one speaks of *freely available, conditionally available* and *independent* dummy times. Details cannot be treated in this section.

We do not think that it is necessary here to prove explicitly that the algorithm given above really supplies in each case a longest path from Q to S (establishing this proof will be recommended to the reader).

If we want to determine the $T_f(i)$ by means of the network matrix $T = (t_{ij})_{i,j=0,1,\ldots,n-1}$, we proceed as follows.

Insert the value 0 in the first row of the T_f column.

Let the $T_f(0)$, $T_f(1)$, …, $T_f(i-1)$ be determined. These values have been put down in the first, the second, ..., the ith row, respectively, of the T_f column.

$T_f(i)$ is obtained as follows. Search for all cells occupied by a number $t(i_j, i)$ in the $(i+1)$th column (this is the column assigned to the vertex i). Say that these are the cells (i_j, i) with $j = 1, \ldots, r$. Now, find $\max\limits_{i \leq j \leq r} \left(T_f(i_j) + t_{i_j i}\right)$ and insert this value in the $(i+1)$th row of the T_f column as the value $T_f(i)$.

The T_s values can be obtained in a similar way: Insert in the last column of the T_s row the value $T_s(n-1) = T_f(n-1)$ (cf. our example where we have 95). Having determined the $T_s(n-1)$, $T_s(n-2)$, …, $T_s(i+1)$ and having inserted these values in the nth, the $(n-1)$th, …, the $(i+2)$th column, respectively, of the T_s row, we can obtain $T_s(i)$ in the following manner.

Search for all cells occupied by a t_{ii_j} value in the $(i+1)$th row. Say, these are the cells (i, i_1), …, (i, i_r). Now, find $\min\limits_{1 \leq j \leq r} \left(T_s(i_j) - t_{ii_j}\right)$ and put down this value in the $(i+1)$th column of the T_s row as value $T_s(i)$.

At the end, $T_s(0) = 0$ must result as verification.

In conclusion, let us still make some remarks on critical activities. If the activity (i, j) is critical and if it is not possible to observe the planned time period t_{ij} for realizing (i, j), then the project cannot be completed within the shortest possible time, not even if for all other activities the given time delays will be observed. If one aims at reducing the overall duration of the project, then the times along critical paths have to be shortened. In doing this, it is possible for new critical paths to develop.

1.6. BIBLIOGRAPHY

[1] Berge, C., und A. Ghouila-Houri: Programme, Spiele, Transportnetze, 2nd edn., Leipzig 1969. (Translated from French.)

[2] Busacker, R. G., and T. L. Saaty: Finite Graphs and Networks. An Introduction with Applications. New York 1965 (German: Munich/Vienna 1968; Russian: Moscow 1974).

[3] Danzig, G. B.: On the shortest route through a network, Management Sci. **6** (1960), 187–190.

[4] Dück, W., und M. Bliefernich: Operationsforschung, Vol. 3: Mathematische Grundlagen, Methoden und Modelle, Berlin 1973.

[5] Ford, L. R., and D. R. Fulkerson: Maximal flow through a network, Canad. J. Math. **8** (1956), 399–404.

[6] Ford, L. R., and D. R. Fulkerson: Flows in Networks, Princeton, N. J., 1962 (Russian: Moscow 1966).

[7] Ford, L. R., and D. R. Fulkerson: A simple algorithm for finding maximal network flows and an application to the Hitchcock problem, Canad. J. Math. **9** (1957), 210–218.

[8] Götzke, H.: Netzplantechnik — Theorie und Praxis, Berlin 1969.

[9] Hu, T. C.: Integer Programming and Network Flows, Reading, Mass., 1970 (German: Munich/Vienna 1972; Russian: Moscow 1974).

[10] Lewandowski, R.: Zu einer internationalen Bibliographie der Netzplantechnik, Elektron. Datenverarb. **10** (1968), 78–83, 156–162.

[11] Minty, G. J.: On the axiomatic foundations of the theories of directed linear graphs, electrical networks, and network-programming, J. Math. Mech. **15** (1966), 485–520.

[12] Sachs, H.: Einführung in die Theorie der endlichen Graphen, Teil I, II, Leipzig 1970, 1972.

[13] Schreiter, D., D. Stempell und F. Frotscher: Kritischer Weg und PERT, Berlin 1965.

[14] Suchowizki, S. I., und I. A. Radtschik: Mathematische Methoden der Netzplantechnik, 2nd edn., Leipzig 1969. (Translated from Russian.)

[15] Tutte, W. T.: Lectures on matroids, J. Res. Nat. Bur. Stand. **69** (1965), 1–47.

[16] Tutte, W. T.: A homotopy theorem for matroids I, II, Trans. Amer. Math. Soc. **88** (1958), 144–174.

[17] Tutte, W. T.: Matroids and graphs, Trans. Amer. Math. Soc. **90** (1959), 527–552.

[18] Tutte, W. T.: Introduction to the Theory of Matroids, New York 1971.

[19] Whitney, H.: On the abstract properties of linear independence, Amer. J. Math. **57** (1935), 507–533.

[20] Lawler, E.: Combinatorial Optimization: Networks and Matroids, Holt, Rinehart and Winston, New York 1976.

[21] Carré, B.: Graphs and Networks, Clarendon Press, Oxford 1979.

[22] Even, S.: Graph Algorithms, Pitman, London 1979.

[23] Ebert, J.: Effiziente Graphenalgorithmen, Wiesbaden 1981.

Chapter 2

The linear transportation problem

2.1. FORMULATION OF THE PROBLEM

The classical transportation problem as formulated by F. L. Hitchcock in 1941 reads as follows:

Given m producers $X_1, \ldots, X_m$ of an article, with X_i being capable of producing $a(X_i) = a_i \geqq 0$. In addition, given n consumers $Y_1, \ldots, Y_n$ of this article, with Y_j having the need $b(Y_j) = b_j \geqq 0$ of this product. Furthermore, let the costs $k(X_i, Y_j) = k_{ij}$ be known for transporting one unit of this article from X_i to Y_j. The problem is to satisfy all needs $b(Y_j)$ such that the overall transportation costs are minimal.

If all data are given, then this problem can be solved using the means made available by the theory of optimization, provided that the cost functions k_{ij} are "decent", i.e., linear or convex.

In this chapter, we want to restrict ourselves to the case the transportation can be realized only via certain given paths. Furthermore, imagine two auxiliary points, *source* Q and *sink* S, being introduced so that only one producer and one consumer are formally present. In addition, some further points may appear playing the role, say, of a distributor. The fact that there are only one producer and one consumer does not essentially restrict generality, since we can also conceive in the transportation network, besides the auxiliary points Q and S, edges to be added (for the undirected case) and arcs (for the directed case) going from Q to the producers X_i and having capacity restrictions which equal the quantity that is producible in point X_i, and we can also conceive arcs added going from the consumers Y_j to sink S with capacity restrictions equalling the need in Y_j.

The problem which we would like to deal with here can thus be formulated as follows:

Given a (undirected) graph $G(\mathfrak{X}, \mathfrak{U})$ with two vertices Q and S being marked. A *capacity* $c(A, B)$ $\big(c(A, B) = c(B, A)\big)$ is assigned to each edge $u = (A, B)$ from $\mathfrak{U}$. Furthermore, let a real number $k(A, B) = k(B, A)$, also called *costs per unit* (in general, we will presume integrality), be assigned to each edge (A, B). The number $c(A, B)$ means that $c(A, B)$ product unities at the most can be transported on the edge (A, B), and this independently of whether from A to B or from B to A. The number $k(A, B)$ indicates that the transportation of one product unit along the edge (A, B), independently of the transportation direction, entails costs amounting to $k(A, B)$. If in addition to this the quantity v of goods to be transported from Q to S is also given, we search among all possibilities of the realization of this transportation (as far as it is actually possible to realize it) for the one entailing minimum overall costs.

For solving the given problem, we think the vertices of the graph to be numbered from 1 to n. We shall also presume the graph to be simple, i.e., without loops and multiple edges. Then, the edges of G can be written in the form (i, j), with i and j being natural numbers between 1 and n.

If we denote by x_{ij} that quantity of the product which is transported from the vertex i to the vertex j along the edge (i, j), as far as this edge exists in G (in general, $x_{ji} = 0$ will be true for the case $x_{ij} > 0$) and with $x_{ii} = 0$ for $i = 1, 2, \ldots, n$ and $x_{ij} = x_{ji} = 0$ for $(i, j) \notin \mathfrak{U}$, our problem takes the following form:

Find a flow $\varphi = (x_{ij})$ on a graph G, with the constraints

$$0 \leqq x_{ij} \leqq c_{ij}, \qquad (i, j) \in \mathfrak{U},$$

$$\sum_{i=1}^{n} x_{ij} - \sum_{k=1}^{n} x_{jk} = \begin{cases} -v & \text{for } j = Q, \\ v & \text{for } j = S, \\ 0 & \text{otherwise} \end{cases}$$

being effective. Then, the objective function

$$Z = \sum_{(i,j)} k_{ij} \cdot x_{ij}$$

will be minimized.

For the existence of a minimum, it is obviously necessary and sufficient that v does not exceed the maximally possible flow going from Q to S.

2.2. THE SOLUTION ACCORDING TO BUSACKER AND GOWEN

Let us try to solve the transportation problem formulated in paragraph 2.1 by using an algorithm found by R.G. Busacker and P. J. Gowen. The proof that the algorithm really yields minimum transportation costs will be established in section 2.4.

Algorithm of Busacker and Gowen

(i) Put $x_{ij} = x_{ji} = 0$ for each edge $(j, i) = (i, j) \in \mathfrak{U}$.

(ii) Define *modified costs*

$$k'_{ij} = \begin{cases} k_{ij} & \text{if} \quad 0 \leqq x_{ij} < c_{ij}, \\ \infty & \text{if} \quad x_{ij} = c_{ij}, \\ -k_{ij} & \text{if} \quad x_{ji} > 0. \end{cases}$$

(iii) Find a least expensive (= shortest) path from Q to S using the modified costs k'_{ij}. Then let so much flow pass through this cheapest simple path that it will no longer be the cheapest one. Either one succeeds in obtaining the required flow of the value v composed of the sum of the original flow going from Q to S and the flow found on the cheapest simple path (in this case, the algorithm terminates and the problem is solved), or the cheapest simple path has been saturated without the desired flow of value v being obtained yet. In the latter case, go back to step (ii).

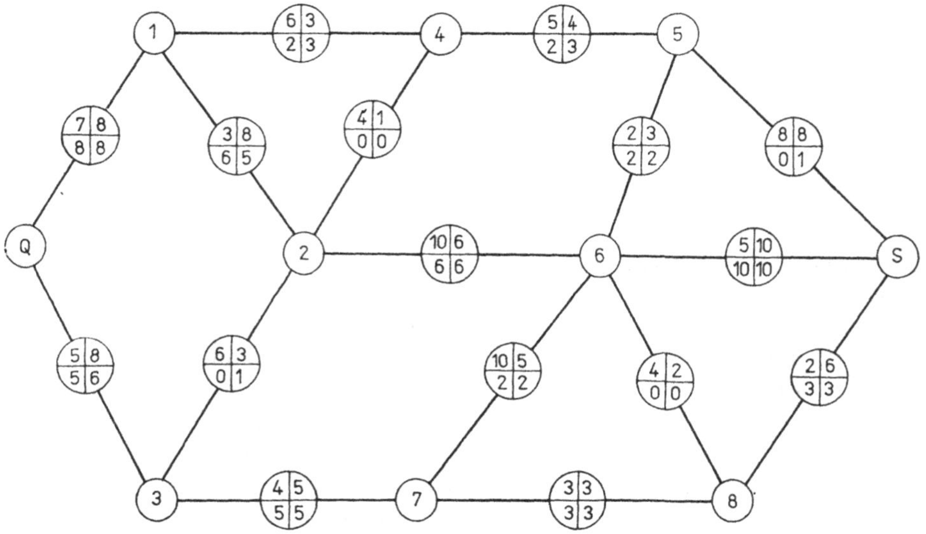

Fig. 2.1

Which effect is achieved by these modified costs?

If a flow is already passing through the edge (i, j), and this from i to j, and if one likes to send an even greater flow quantity in this direction through (i, j), then the costs k_{ij} have to be spent per product unit. If the edge (i, j) is saturated, then no flow can go any longer in the direction of the saturation flow. Therefore, the costs are set infinite for an additional transportation, thus preventing a supplementary quantity of goods from being transported in this direction. But if a flow is going through (i, j) from j to i, then, if this flow direction is not favourable, the flow direction from i to j is favoured by a negative cost set-up.

Let us take an example. Given the graph in Fig. 2.1. At the edges we have left space for four sets of data. The number at the upper right indicates the capacity of the edge, the number at the upper left stands for the transportation costs per unit along this edge. The space at the lower left is reserved as "field of operation", the number at the lower right indicates an optimum flow distribution for a given $v = v_0$. In the space at the lower left, the flow which already goes along this edge (and which changes in the course of the algorithm) can be written, say in pencil, and, if necessary, erased and replaced by another number according to the algorithm.

We have successively applied the algorithm to the graph represented in Fig. 2.1 and we got the least expensive paths with their costs in the order given by the table, as well as the quantities transportable along these paths (it will be easy for the reader to check it himself):

	Least expensive path							Costs per unit	Transportable quantity
1	Q	3	7	8	S			14	3
2	Q	3	7	6	S			24	2
3	Q	1	2	6	S			25	6
4	Q	1	4	5	6	S		25	2
5	Q	3	2	1	4	5	S	27	1
6	Q	3	2	4	5	S		28	1

In the case of path 5, advantage is derived from the fact that modified costs entail a less expensive path from Q to S.

In Fig. 2.1, we have indicated, in the space at the lower left, the product quantity which must flow in order to realize in S a need 13 with minimum expenditure; at the lower right there is indicated the product quantity which must be transported along this edge in order to satisfy a need of $v_0 = 14$ with minimum costs. The flow direction is not represented at the edges, since it can be directly seen when using the node rules. On the edge going from node 1 to node 2 a flow quantity will pass in case of a need of $v_0 = 13$ with minimum costs which is greater than in the case of a need of $v_0 = 14$. This effect is achieved by using modified costs.

In each step of the algorithm given in chapter 1 it is possible to find a least expensive (i.e. a shortest) path.

2.3. THE SOLUTION ACCORDING TO KLEIN

The approach used in this algorithm is in a way opposite to that of the preceding algorithm. In Busacker's and Gowen's algorithm, the essential point was to increase in each step the product flow quantity only by as much as to make the costs minimum, whereas in this new algorithm, any flow with a value v_0 prescribed is sent from Q to S (if this is possible), and the flow direction is then diverted and the costs are stepwise reduced down to the minimum.

Algorithm according to Klein

(i) Find any feasible flow (which is compatible with the capacities) of the value v_0 prescribed going from Q to S (this step can be realized, for example, by means of the algorithm of Ford and Fulkerson described in chapter 1).

(ii) Make up the modified costs

$$k'_{ij} = \begin{cases} k_{ij} & \text{if} \quad 0 \leqq x_{ij} < c_{ij}, \\ \infty & \text{if} \quad x_{ij} = c_{ij}, \\ -k_{ij} & \text{if} \quad x_{ji} > 0. \end{cases}$$

(iii) Find a circuit of negative length by means of the modified costs. If such a circuit does not exist, the flow is optimum (the proof will still be established). If there exists a negative circuit $\boldsymbol{K}$ (i.e. a circuit of negative length, with the modified costs being taken as lengths), then it is possible to reduce the costs by "diverting" the flow along this circuit. The flow quantity which can be diverted along a negative circuit (cf. Fig. 2.2) may be increased so that an edge of $\boldsymbol{K}$ is saturated, which means, its capacity c_{ij} is reached. Then go back to (ii).

The diversion of a flow along a circuit $\boldsymbol{K}$ is illustrated in Fig. 2.2. The edge valuations at the upper right indicate the capacity, at the upper left they indicate the

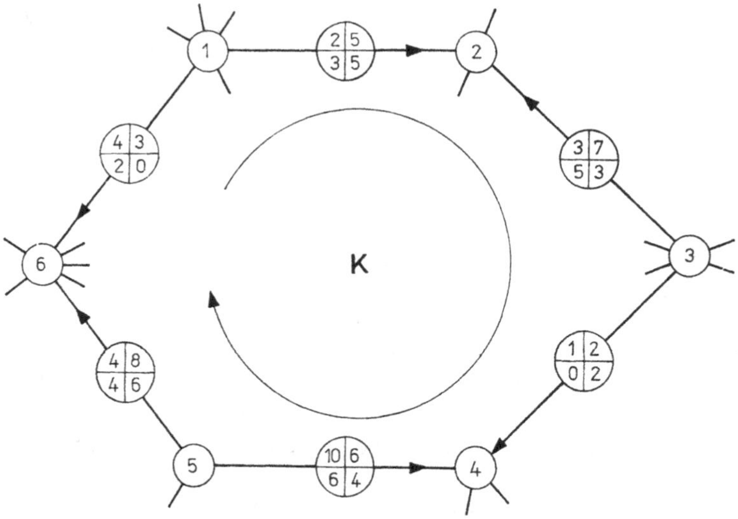

Fig. 2.2

costs per unit, at the lower left a flow going through the edge, namely in the direction fixed by the arrow at the edge, and at the lower right the new flow after "diversion" of two product units in the direction of $\boldsymbol{K}$. Obviously, $\boldsymbol{K}$ is a circuit of negative length, since the lengths (i.e. the costs) in the order of the edges (1, 2), (2, 3), ..., (6, 1) are equal to

$$2, -3, 1, -10, 4, -4,$$

that is, -10 in the sum. Two units may be sent in the direction of $\boldsymbol{K}$, thus entailing a reduction of the costs by 20 units.

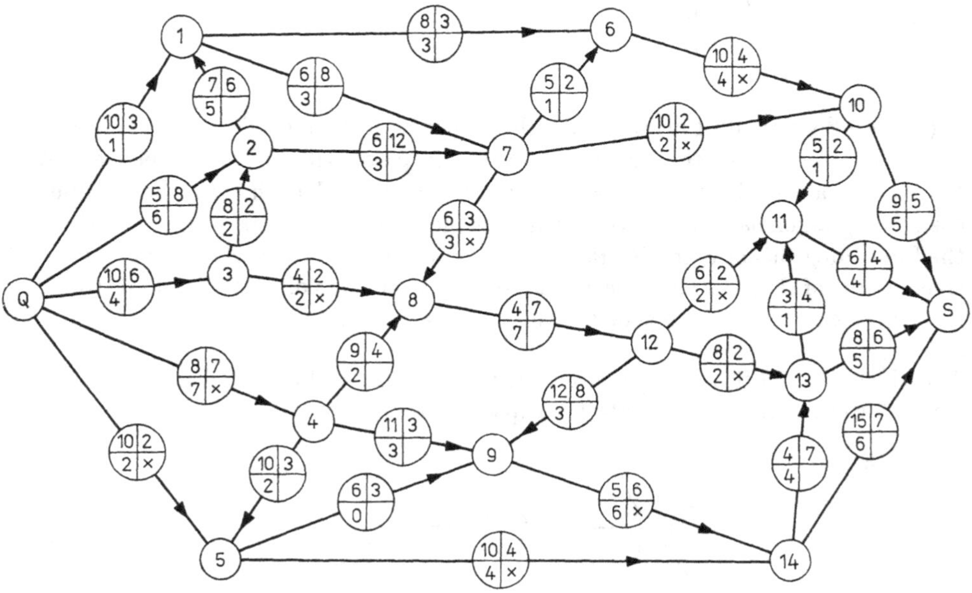

Fig. 2.3

Figure 2.3 serves us for illustrating the algorithm of M. Klein. The vertices have been numbered. The numbers at the upper left stand for the costs per unit, at the upper right they indicate the edge capacities, at the lower left a flow distribution causing a total flow of 20 units from Q to S (it is easy to recognize that 20 units make up a maximum flow), and at the lower right we have drawn crosses in some cases. The corresponding edges (cf. the algorithm of Ford and Fulkerson dealt with in chapter 1) belong to a minimum cut. As can easily be seen, there are two such cuts: one of them is formed by the edges (Q, 5), (Q, 4), (3, 8), (7, 8), (7, 10), (6, 10), the other one by the edges (5, 14), (9, 14), (12, 13), (12, 11), (7, 10), (6, 10).

You can recognize that those edges which belong to a cut of minimum capacity and which are saturated cannot belong to a circuit of negative length. If the flow

v_0 required is smaller than the maximum flow, then edges of a minimum cut may, of course, belong to circuits of negative length. It is also possible for saturated edges not belonging to a minimum cut to lie in circuits of negative length.

How can circuits of negative length be found now? The procedure is essentially described with the algorithm for determining shortest simple paths of negative length (cf. chapter 1) even though we have explicitly excluded negative circuits there. For the method presented in that chapter for finding a shortest simple path it was necessary to exclude circuits of negative length, since otherwise one could not determine the length of a "shortest" simple path by traversing such a circuit sufficiently often, because this length would have become smaller than any negative number. In our case, however, there is not so much to happen, since a circuit of negative length cannot be traversed as often as one likes, because each traversal produces a change in the flow. Any increase of the flow, however, is not feasible because of the capacity restrictions on the edges.

So we first try to find the distances (i.e. the least expensive paths) between Q and each of the vertices. Using the values indicated in Fig. 2.3 for the costs (at the upper left), for the capacities (at the upper right) and the initial flow (at the lower left) and applying the modified costs, we get one after the other the following distances:

Number of the node	Distance	Reasons
2	5	flux $\varphi(Q, 2)$ through arc $(Q, 2)$ $(= 6)$ is smaller than its capacity $c(Q, 2)$ $(= 8)$
1	10	$\varphi(Q, 1) < c(Q, 1)$
3	10	$\varphi(Q, 3) < c(Q, 3)$
7	11	$\varphi(2, 7) < c(2, 7)$
6	16	$\varphi(7, 6) < c(7, 6)$

All other vertices get the distance ∞, since they lie beyond a minimum cut. In such a case, those negative circuits containing vertices of a distance shorter than ∞ cannot contain vertices of a distance ∞ as well.

Above all, we shall remove all circuits of negative length in that part of which the vertices have a finite distance from Q. For this, consider the distance tree (see the doubly drawn arcs in Fig. 2.4) and all edges of which the two end-points are included in the distance tree (in other words, consider the graph which is spanned by all vertices having a finite distance from Q). Then, consider any edge with both of its end-points lying in the distance tree (the edge itself, however, does not belong to it). Together with the distance tree a circuit is generated which we want to examine. We choose the edge say (1, 7). The resulting circuit $(Q, 2, 7, 1, Q)$ is obviously of negative length, since on the simple path $(Q, 1, 7)$ a unit is being transported with the costs $10 + 6 = 16$, although on the cheaper simple path

(Q, 2, 7), which only costs 11 units, goods could still be transported. If one successively carries out these "price reductions", then one obtains for the subgraph of Fig. 2.4 a least expensive transportation, as indicated in the space at the lower right.

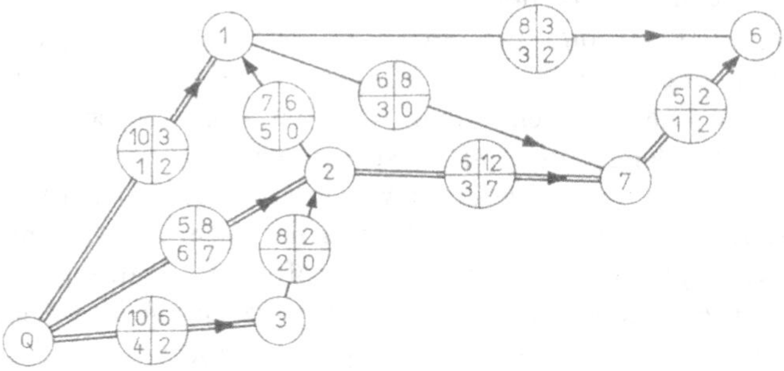

Fig. 2.4

Difficulties may possibly arise in the search for negative circuits, since no distance tree is constructible in the whole graph if the product quantity to be transported is equal to a minimum cut capacity. In such a case, the vertices lying beyond this cut are not "reachable", that is, they have the distance ∞ from Q. But then, one can easily proceed in the following way: Consider the set of those vertices that can be reached from the end-points of the arcs of a minimum cut. As for this graph, we make analogous reflections choosing, for example, the value 0 as node valuations for these end-points (cf. Fig. 2.5). Thus, the vertices 10, 11, 13, 14, which are the end-points of minimum cut arcs, get a distance valuation of 0, the vertex S is provided with the valuation 8, since the arc (13, S) is not saturated and yields, in addition to this, the smallest increase of costs in comparison with all the other non-saturated arcs. We obtain a circuit of negative length, for example, (13, S, 11, 13). Using the modified costs this circuit is of length -1. So we divert

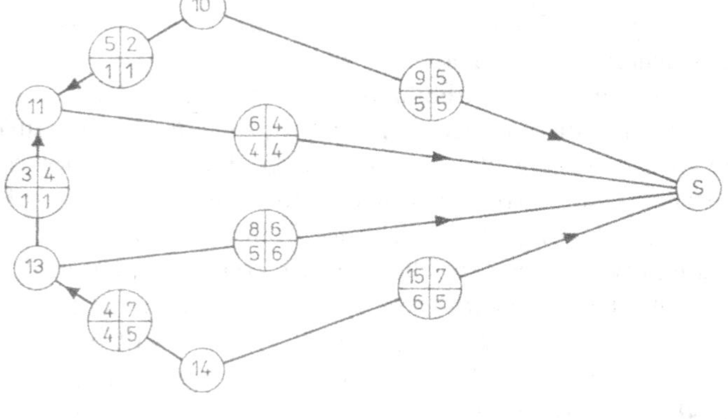

Fig. 2.5

a part of the flow, and that one unit is transported along the arc (13, S) and not along the simple path (13, 11, S). The reduction of the costs amounts to 1. Subsequently to this, another circuit of negative length is found, namely (14, 13, 11, S, 14), etc.

The flow distribution entailing minimum costs is indicated in the space at the lower right.

For reducing the costs, one would have to proceed in an analogous way when several minimum cuts, the capacities of which would be equal to the required product quantity v_0, can be found in the graph. Obviously, it is then possible to effect a diminution of costs separately for each part of the graph as these single parts are separated from each other by minimum cuts. In the example given in Fig. 2.3, no further reduction of the costs can be obtained in that part of the graph lying between the two minimum cuts (cf. Fig. 2.6).

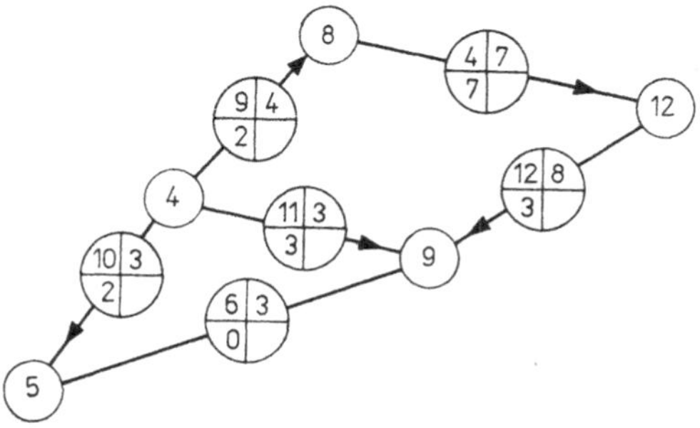

Fig. 2.6

2.4. PROOF OF MINIMALITY

In this section, we want to show that the two algorithms formulated in the preceding expositions really yield a flow with minimum transportation costs. For this, the following theorem will be proved:

THEOREM 2.1. *A flow φ of the value v is optimum if and only if no circuit of negative length exists in the graph $\boldsymbol{G}$ with the costs modified in respect of φ.*

Proof. The necessity of the condition is immediately clear. We shall show its sufficiency. Let φ be a flow of the value v. Assume that in respect of this flow there exists no circuit of negative length in the graph $\boldsymbol{G}_\varphi$. Furthermore, let φ^0 be a flow of the value v entailing minimum (i.e. optimum) costs. The graph $\boldsymbol{G}_{\varphi^0}$, which is provided with the costs modified with respect to φ^0, also has no circuit of negative length, because of optimality. Since no circuits of negative length are present, the following relations hold for any edge (i, j):

$$\begin{aligned} &\text{if} \quad \varphi(i, j) > 0, \quad \text{then} \quad \varphi(j, i) = 0, \\ &\text{if} \quad \varphi^0(i, j) > 0, \quad \text{then} \quad \varphi^0(j, i) = 0. \end{aligned} \tag{1}$$

A directed graph G' is formed from G in the following way: If there exists in G an (undirected) edge $(i, j) = (j, i)$ with the capacity $c_{ij} = c_{ji}$, we form in G' two (directed) arcs $(i, j)'$ and $(j, i)'$ with the capacities $c'(i, j)' = c(i, j)$ and $c'(j, i)' = c(i, j)$, respectively. Because of the relations given in (1), the flux going through $(i, j)'$ and $(j, i)'$ is not greater than $c(i, j)$ independently of the kind φ or φ^0.

If for an arc $(i, j)'$ the relation $\varphi(i, j)' > \varphi^0(i, j)' \geqq 0$ holds, colour $(i, j)'$ *red*. But if for an arc $(i, j)'$ the relation $\varphi^0(i, j)' > \varphi(i, j)' \geqq 0$ holds, then colour $(i, j)'$ *blue*. First, it is possible that, say, $(i, j)'$ is red and $(j, i)'$ is blue if a flow of the type φ is going from i to j, but from j to i a flow of type φ^0. Arcs for which $\varphi(i, j)' = \varphi^0(i, j)'$ is true do not concern us and will thus not be coloured.

Furthermore, we imagine φ^0 (among all cheapest, that means, optimum flows) to be chosen such that the sum of the red and blue arcs is minimum.

Assume that φ is not optimal. Let the costs of the transportation of v units according to φ thus be higher than those of the transportation of v units according to φ^0. Then, φ and φ^0 do not coincide; there exists at least a blue arc $(i_1, i_2)'$. Task! We prove the following lemma.

LEMMA 1. *In G' there exists an elementary cycle μ composed of blue and red arcs, with the blue arcs all being in the same direction as μ and the red ones being oppositely directed to μ.*

For proving the lemma we distinguish between three cases (the task set before was to show that there exists a blue arc $(i_1, i_2)'$):

a) A larger quantity of flow of the type φ^0 than of the type φ is flowing through vertex i_2. Then, there exists at least one arc $(i_2, i_3)'$ starting from i_2 which has more flow of the type φ^0 than of the type φ; so this arc is blue.

b) A larger quantity of flow of the type φ than of the type φ^0 is flowing through vertex i_2; that means that in particular more flow φ than φ^0 flows into i_2. Thus, there exists an arc $(i_3, i_2)'$ which has more flow φ than φ^0 and which therefore is red.

c) The same quantity of flow of the type φ as of the type φ^0 is flowing through i_2. Since the arc $(i_1, i_2)'$ is a blue one, more flow φ^0 than φ is flowing on it, and therefore a larger quantity of flow of the type φ than of the type φ^0 is flowing on the other arcs going to i_2. But then, there is an arc $(i_3, i_2)'$ which transports more flow φ than φ^0 to i_2, and which therefore is a red one.

Naturally, it can be the case that red edges enter a vertex and blue ones leave it, but this is of no importance.

An elementary cycle μ given in the lemma can be found without difficulty. One travels on the blue arc $(i_1, i_2)'$ from i_1 to i_2, goes on travelling from there to i_3 on a blue arc $(i_2, i_3)'$ or on a red arc $(i_3, i_2)'$ etc. When a vertex is reached twice, which will happen because of the finiteness of the graph, then one gets an elementary cycle μ, as stated. It is quite evident that μ does not consist either only of red arcs or only of blue ones, since these form in G_φ or in G_{φ^0} a circuit which would entail negative costs. This, however, was not possible in both cases.

LEMMA 2. *In an elementary cycle μ, as it exists according to lemma 1, it holds: The sum of the costs (per product unit) of all blue arcs of μ equals the sum of the costs (per product unit) of all red arcs of μ.*

Proof. Let μ be an elementary cycle consisting only of blue arcs (having the same direction as μ) and of red ones (being oppositely directed to μ).

Consider μ in G_φ. The blue arcs (since more flow φ^0 than φ is flowing on them) are still capable of absorbing flow of the type φ. The red arcs (since more flow φ than φ^0 is flowing on them) can still absorb flow of the type φ in the sense of the orientation of μ, because the flow of the type φ is flowing in the opposite direction of μ. If we permitted an additional quantity of flow of the type φ to pass in the direction of μ, the costs would not be lower, since no circuits of negative length (i.e. costs) exist in G_φ. Thus, it holds:

The sum of the costs per unit on the blue arcs of μ is at least as high as the sum of the costs per unit on the red arcs of μ, since in case of an increase of φ on the red arcs, costs could be saved along μ, whereas on the blue arcs the costs would go up.

The φ^0 flow may be increased in the direction of $-\mu$ without exceeding the capacities (task!). In doing so, the costs do not decrease (since there exists no circuit of negative length in G).

Since also in G_{φ^0} there exist no circuits of negative length when augmenting the φ^0 flow in the direction of $-\mu$, which would be possible, however, without exceeding the capacities (the reader can easily verify this), i.e., since the costs are not allowed to lower, we get:

The sum of the costs per unit on the red arcs of μ is at least as high as the sum of the costs per unit on the blue arcs of μ.

This proves lemma 2.

TASK 1. In an elementary cycle according to lemma 1, it holds: If $(i, j)'$ is red, then $(j, i)'$ is not blue, and if $(i, j)'$ is blue, then $(j, i)'$ is not red.

Let us now complete the proof of our theorem. Since in an elementary cycle μ according to lemma 1 the red arcs entail costs as high as those of the blue ones, the φ^0 flow in G_{φ^0} can be increased in the direction of $-\mu$ without exceeding the capacities (see proof of lemma 2) and without changing the costs, i.e., the flow $\bar{\varphi}^0$ resulting in this way from φ^0 is also optimal, since φ^0 was optimal as well according to the assumption. Now, increase φ^0 as long as either on an originally blue edge of μ the flow quantity of type φ^0 is as large as that of type φ (because the flow of the type φ^0 diminishes in this case on the blue arcs) or an originally red arc of μ is not red any longer, since the φ^0 flow increases in the direction of this arc. In each case, however, the sum of all blue and red arcs diminishes, but the flow passing through is still optimal. This contradicts the assumption that φ^0 was an optimum flow for which this sum was minimum.

Thus, the theorem is proved.

We have proved at the same time that the algorithm of Klein really yields an optimum flow.

It remains to show that also the algorithm of Busacker and Gowen provides an optimum flow of value v. For this, we prove that at no time of the algorithm are circuits of negative length generated, which verifies all because of the theorem proved just now. Imagine the algorithm to be applied step by step (that is, increasing the flow always by the value 1). Let no circuits of negative length appear in graph $\boldsymbol{G}_i$ $(i = 1, \ldots, w)$ with a flow value i, but let this be for graph $\boldsymbol{G}_{w+1}$ with a flow value of $w + 1 \leqq v$.

Obviously, a circuit $\boldsymbol{K}$ of negative length in $\boldsymbol{G}_{w+1}$ contains an arc which was required in the last step of the algorithm for transporting flow. Consider now this flow path $\boldsymbol{B}$ going from Q to S, which is in $\boldsymbol{G}_w$ a cheapest one according to the rules, and consider also an arc (i, j) on $\boldsymbol{B}$, which is contained in a circuit $\boldsymbol{K}$ of negative length of $\boldsymbol{G}_{w+1}$. If we change the path $\boldsymbol{B}$ such that the arcs lying in $\boldsymbol{K}$ and $\boldsymbol{B}$ are removed and, instead of this, the arcs lying in $\boldsymbol{K}$ but not in $\boldsymbol{B}$ are added, then one obtains a simple path $\boldsymbol{B}'$ of $\boldsymbol{G}_{w+1}$, which is cheaper than $\boldsymbol{B}$ because of the negative value of the costs of $\boldsymbol{K}$. This, however, contradicts the choice of $\boldsymbol{B}$ as a least expensive simple path.

So we have proved as well that the algorithm set up by Busacker and Gowen yields flows of the value v entailing lowest costs.

2.5. CONCLUSIONS

Our originally set problem involved m producers $X_1, \ldots, X_m$ having an output $a(X_i)$ and n consumers $Y_1, \ldots, Y_n$ with the need $b(Y_j)$, whereas in the algorithms only *one* source (and not m pieces) and *one* sink (and not n pieces) could be taken into consideration. At the very beginning we mentioned that this was no essential restriction of generality; you only need to add in the original graph two more auxiliary points Q and S, m auxiliary edges (Q, X_i) and n auxiliary edges (Y_j, S). For these auxiliary edges choose zero as cost functions and $c(Q, X_i) = a(X_i)$ and $c(Y_j, S) = b(Y_j)$, respectively, as capacities. If it is possible to realize a flow of the value v going from the consumers to the producers in the original graph, it will also be feasible in the auxiliary graph. There, this will be in particular a maximum flow, because in the case that $v = \sum_i a(X_i) = \sum_j b(Y_j)$ the set of the edges (Q, X_i) or even the set of the edges (Y_j, S) forms a cut of minimum capacity. It is, however, difficult to decide which of the two algorithms is the more advantageous one. When v is relatively small, the first one would be more suitable, whereas preference is given to the second algorithm for a relatively large v, since a maximum flow may be easily found by applying the algorithm of Ford and Fulkerson. Of course, the following iteration (for finding circuits of negative length) will entail much calculation.

The simple case of a transportation problem involving no capacity restrictions for the arcs is obviously solvable by means of both algorithms, with the second one requiring to be slightly modified, since no maximum flow exists in a graph without capacity restrictions. It will then be possible to use the algorithms if auxiliary points and edges with capacity restrictions on these auxiliary edges are introduced as described above.

The linear transportation problem just described, like the problem of the determination of a maximum flow, represents a special case of the so-called *circulation problem.*

The circulation problem may be solved, for example, by means of the out-of-kilter algorithm (cf. E. Lawler: Combinatorial Optimization, Networks and Matroids; Holt, Rinehart and Winston, 1976).

In chapter 4, we shall discuss some non-linear transportation problems.

2.6. BIBLIOGRAPHY

[1] Busacker, R. G., and P. J. Gowen: A Procedure for Determining a Family of Minimal-Cost Network Flow Patterns, ORO Techn. Report 15, Operations Res. Office, Johns Hopkins Univ., Baltimore 1961.

[2] Busacker, R. G., and T. L. Saaty: Finite Graphs and Networks, An Introduction with Applications, New York 1965 (German: Munich/Vienna 1968; Russian: Moscow 1974).

[3] Ford, L. R., and D. R. Fulkerson: Flows in Networks, Princeton, N.J., 1962 (Russian: Moscow 1966).

[4] Hu, T. C.: Integer Programming and Network Flows, Reading, Mass., 1970 (German: Munich/Vienna 1972; Russian: Moscow 1974).

[5] Klein, M.: A primal method for minimal cost flows, Management Sci. **14** (1967), 205 to 220.

Chapter 3

The cascade algorithm

3.1. FORMULATION OF THE PROBLEM

This chapter is devoted to a problem which we have already discussed in chapter 1. There, we were searching for a shortest path between two given vertices in a (directed or undirected) graph provided with edge or arc valuations, respectively, as lengths (cf. 1.4.). Now, we want to determine at once in a directed graph the distances between two vertices at a time, with an arc valuation being given (to be understood as length).

Given the *arc length matrix* $\boldsymbol{B}(\boldsymbol{G})$ for an arc-valuated directed graph $\boldsymbol{G}(\mathfrak{X}, \mathfrak{U})$, that is, $\boldsymbol{B} = \boldsymbol{B}(\boldsymbol{G}) = (b_{ij})_{i,j=1,\ldots,n}$ if $\boldsymbol{G}$ possesses exactly n vertices. Let

$$b_{ij} = \begin{cases} 0 & \text{for } i = j, \\ l_{ij} & \text{if there exists an arc of length } l_{ij} \\ & \text{going from vertex } i \text{ to vertex } j, \\ \infty & \text{otherwise}. \end{cases}$$

Our aim is to derive from $\boldsymbol{B}(\boldsymbol{G})$ the *distance matrix* $\boldsymbol{D}(\boldsymbol{G}) = (d_{ij})$, with d_{ij} being the length of a shortest path (= simple path) from vertex i to vertex j.

3.2. THE STANDARD METHOD

The method to be described in this section for determining the distance matrix $\boldsymbol{D}(\boldsymbol{G})$ using the arc length matrix $\boldsymbol{B}(\boldsymbol{G})$ of a directed graph $\boldsymbol{G}(\mathfrak{X}, \mathfrak{U})$ is based on an approach by M. Hasse [1].

The distance matrix $\boldsymbol{D}(\boldsymbol{G})$ is calculated by means of the matrix operation described below.

DEFINITION. Given two (n, n) square matrices $\boldsymbol{A} = (a_{ij})$ and $\boldsymbol{B} = (b_{ij})$ with $a_{ij}, b_{ij} \in R^+ \cup \{\infty\}$ (all non-negative real numbers and ∞ are allowed as matrix elements). We understand by the *$\oplus$-product* $\boldsymbol{A} \oplus \boldsymbol{B}$ a matrix $\boldsymbol{C} = (c_{ij})_{i,j=1,\ldots,n}$ with

$$c_{ij} = \min_{1 \leq k \leq n} (a_{ik} + b_{kj}).$$

Here, the rules of operation

$$r + \infty = \infty + r = \infty, \qquad \infty + \infty = \infty$$

will be effective for each $r \in R^+$.

Let us explain the multiplication by means of an example. Given the matrices

$$A = \begin{bmatrix} 1 & 4 & \infty & 3 \\ \infty & 2 & \infty & 1 \\ 0 & 1 & 0 & 5 \\ 3 & 1 & 2 & \infty \end{bmatrix}, \qquad B = \begin{bmatrix} 3 & \infty & \infty & 0 \\ 4 & 2 & \infty & \infty \\ 3 & 5 & 1 & 0 \\ 4 & \infty & \infty & 3 \end{bmatrix}.$$

Determine say c_{11}, and we find

$$c_{11} = \min(1 + 3, 4 + 4, \infty + 3, 3 + 4) = 4.$$

Analogously, find the other c_{ij}, and we get:

$$C = A \oplus B = \begin{bmatrix} 4 & 6 & \infty & 1 \\ 5 & 4 & \infty & 4 \\ 3 & 3 & 1 & 0 \\ 5 & 3 & 3 & 2 \end{bmatrix}.$$

The following two theorems contain simple properties of this matrix operation.

THEOREM 3.1. *For the $\oplus$-product of matrices the associative law is valid, that is,*

$$(R \oplus S) \oplus T = R \oplus (S \oplus T),$$

with R, S, T being (n, n) square matrices.

The reader may prove this theorem by himself.

REMARK. The $\oplus$-product of matrices is *not commutative* as demonstrated by the following example.

$$A = \begin{bmatrix} 2 & 3 & \infty \\ \infty & 1 & 4 \\ 2 & 1 & 5 \end{bmatrix}, \qquad B = \begin{bmatrix} 2 & 4 & 3 \\ \infty & 3 & 1 \\ 4 & \infty & 0 \end{bmatrix}.$$

It results:

$$A \oplus B = \begin{bmatrix} 4 & 6 & 4 \\ 8 & 4 & 2 \\ 4 & 4 & 2 \end{bmatrix}, \qquad B \oplus A = \begin{bmatrix} 4 & 4 & 8 \\ 3 & 2 & 6 \\ 2 & 1 & 5 \end{bmatrix}.$$

DEFINITION. We understand by the *k-th power* A^k of an (n, n) square matrix A with respect to the multiplication $\oplus$

$$A^1 = A, \qquad A^{k+1} = A^k \oplus A.$$

The truth of the following theorem may be verified by the reader himself.

THEOREM 3.2. *Let* $\boldsymbol{A}$ *be a square matrix and* k *and* l *any natural numbers. Then, it holds:*

$$\boldsymbol{A}^{k+l} = \boldsymbol{A}^k \oplus \boldsymbol{A}^l.$$

In particular, the relation

$$\boldsymbol{A}^k \oplus \boldsymbol{A}^l = \boldsymbol{A}^l \oplus \boldsymbol{A}^k$$

results.

Let us now go back to our problem and prove the following theorem.

THEOREM 3.3. *Given the directed graph* $\boldsymbol{G}$ *with the vertices* $X_1, X_2, \ldots, X_n$ *and the arc length matrix* $\boldsymbol{B}$. *If we form* $\boldsymbol{B}^m$ (with respect to the $\oplus$-product), *then for the element* b_{ij}^m *of the matrix* $\boldsymbol{B}^m$ *it holds:*

$$b_{ij}^m = \begin{cases} \textit{length of a shortest simple path} \text{ (of a directed path) } \textit{going from} \\ X_i \textit{ to } X_j \textit{ with } m \textit{ arcs at the most} \text{ (if such a simple path exists)}, \\ 0 \textit{ for } i = j, \\ \infty \textit{ otherwise} \text{ (that is, if there exists no simple path going} \\ \text{from } X_i \text{ to } X_j \text{ with } m \text{ arcs at the most and with } i \neq j). \end{cases}$$

Proof. The truth is verified by means of mathematical induction over m. For $m = 1$ we just employ the definition of the matrix $\boldsymbol{B} = \boldsymbol{B}^1$. Assuming the theorem to be valid for $m \leqq k$, we want to prove it for the case that $m = k + 1$.

Case 1: $i = j$.
It holds $b_{ii}^1 = 0$ (because of the definition of $\boldsymbol{B}$) and $b_{ii}^k = 0$ (because of the induction hypothesis). According to the definition of the $\oplus$-product, and because of the non-negativity of all elements occurring in all of the matrices, it holds $b_{ii}^{k+1} = 0$.

Case 2: $i \neq j$.
Let no simple path exist in $\boldsymbol{G}$ going from X_i to X_j with a number of arcs $\leqq k + 1$. It holds $b_{ij}^{k+1} = \min (b_{il}^k + b_{lj}^1)$, and one can easily see that at least one of the two summands b_{il}^k and b_{lj}^1 is infinite for each l.

Case 3: $i \neq j$, and there exists in $\boldsymbol{G}$ a simple path with a number of arcs $\leqq k + 1$ going from X_i to X_j.
Among all these simple paths choose a simple path $\boldsymbol{R}^0(X_i, X_j)$ of which the length is minimum. Let

$$\boldsymbol{R}^0(X_i, X_j) = (X_i = X_{i_0}, X_{i_1}, \ldots, X_{i_q} = X_j).$$

We get

$$b_{ij}^{k+1} = \min_{1 \leqq l \leqq n} (b_{il}^k + b_{lj}) \leqq b_{ii}^k + b_{ij} = b_{ij}, \tag{1}$$

since according to definition it holds $b_{ii}^k = 0$.

We also distinguish between two other cases.

Case 3.1: $q = 1$ (i.e., the number of arcs of the simple path $\boldsymbol{R}^0 (X_i, X_j)$ equals 1, that is, the simple path consists of one arc only, namely of (X_i, X_j)). Then, the relation $l(\boldsymbol{R}^0) = b_{ij}$ is true for the length $l(\boldsymbol{R}^0)$.

According to the induction hypothesis, b_{il}^k is the length of a shortest simple path going from X_i to X_l $(l = 1, 2, \ldots, n)$ with k arcs at the most. Then, however, $b_{il}^k + b_{lj}$ is the length of a sequence of equally directed arcs from X_i to X_j, and because $\boldsymbol{R}^0$ is the shortest of all simple paths going from X_i to X_j it holds $b_{ij} \leqq b_{il}^k + b_{lj}$ $(l = 1, 2, \ldots, n)$ and also

$$b_{ij} \leqq \min_{1 \leqq l \leqq n} (b_{il}^k + b_{lj}) = b_{ij}^{k+1}.$$

Together with (1) it follows

$$b_{ij} = \min_{1 \leqq l \leqq n} (b_{il}^k + b_{lj}) = b_{ij}^{k+1}.$$

As asserted, b_{ij}^{k+1} is just the length of a shortest simple path going from X_i to X_j with $k + 1$ arcs at the most.

Case 3.2: $q > 1$ (i.e., a shortest simple path from X_i to X_j with $k + 1$ arcs at the most contains at least two arcs).

Then, for the shortest simple path $\boldsymbol{R}^0$ the relation $X_{i_{q-1}} \neq X_i$ is valid. $\boldsymbol{R}^0$ will then split into two partial simple paths $\boldsymbol{R}^0(X_i, X_{i_{q-1}})$ (this is a shortest simple path going from X_i to $X_{i_{q-1}}$ with k arcs at the most) and $\boldsymbol{R}^0(X_{i_{q-1}}, X_j)$ (this is a shortest simple path consisting of an arc of the length $b_{i_{q-1}j}$).

Employing the induction hypothesis we get for the length of this simple path

$$l\big(\boldsymbol{R}^0(X_i, X_j)\big) = l\big(\boldsymbol{R}^0(X_i, X_{i_{q-1}})\big) + l\big(\boldsymbol{R}^0(X_{i_{q-1}}, X_j)\big) = b_{ii_{q-1}}^k + b_{i_{q-1}j}.$$

Obviously, for each $l = 1, \ldots, n$ it holds

$$b_{ii_{q-1}}^k + b_{i_{q-1}j} \leqq b_{il}^k + b_{lj},$$

and thus, also

$$b_{ii_{q-1}}^k + b_{i_{q-1}j} \leqq \min_{1 \leqq l \leqq n} (b_{il}^k + b_{lj}) = b_{ij}^{k+1}.$$

When forming the minimum on the right side, the index $l = i_{q-1}$ had also been taken into account, and therefore it also holds:

$$\min_{1 \leqq l \leqq n} (b_{il}^k + b_{lj}) \leqq b_{ii_{q-1}}^k + b_{i_{q-1}j}.$$

Thus, the equality

$$b_{ij}^{k+1} = b_{ii_{q-1}}^k + b_{i_{q-1}j}$$

holds.

The right side of the relation, however, was just the length of a shortest simple path going from X_i to X_j with $k + 1$ arcs at the most.

Thus, the discussion of all three cases is terminated and theorem 3.3 proved.

COROLLARY. *Since in a graph with n vertices there are no simple paths containing more than $n - 1$ arcs, it holds:*

$$\boldsymbol{B}^n(\boldsymbol{G}) = \boldsymbol{B}^{n-1}(\boldsymbol{G}) = \boldsymbol{D}(\boldsymbol{G}).$$

So we have found a fundamental possibility of determining the distance matrix of a graph.

Furthermore, it can be seen without difficulty that the following theorem is true.

THEOREM 3.4. *Let $\boldsymbol{G}$ be a directed graph with n vertices. If there is a natural number k such that $\boldsymbol{B}^{k+1}(\boldsymbol{G}) = \boldsymbol{B}^k(\boldsymbol{G})$, then*

$$\boldsymbol{D}(\boldsymbol{G}) = \boldsymbol{B}^k(\boldsymbol{G}).$$

Now, we want to calculate the distance matrix of a graph taken as example by means of this standard method (cf. Fig. 3.1):

$$\boldsymbol{B}(\boldsymbol{G}) = \begin{bmatrix} 0 & 2 & \infty & \infty & 3 \\ \infty & 0 & 3 & \infty & 4 \\ 3 & \infty & 0 & \infty & \infty \\ 6 & \infty & 2 & 0 & 3 \\ \infty & 4 & \infty & \infty & 0 \end{bmatrix}, \quad \boldsymbol{B}^2(\boldsymbol{G}) = \begin{bmatrix} 0 & 2 & 5 & \infty & 3 \\ 6 & 0 & 3 & \infty & 4 \\ 3 & 5 & 0 & \infty & 6 \\ 5 & 7 & 2 & 0 & 3 \\ \infty & 4 & 7 & \infty & 0 \end{bmatrix},$$

$$\boldsymbol{B}^3(\boldsymbol{G}) = \begin{bmatrix} 0 & 2 & 5 & \infty & 3 \\ 6 & 0 & 3 & \infty & 4 \\ 3 & 5 & 0 & \infty & 6 \\ 5 & 7 & 2 & 0 & 3 \\ 10 & 4 & 7 & \infty & 0 \end{bmatrix}, \quad \boldsymbol{B}^4(\boldsymbol{G}) = \boldsymbol{B}^3(\boldsymbol{G}).$$

Using the previous theorem, we get $\boldsymbol{B}^3(\boldsymbol{G}) = \boldsymbol{D}(\boldsymbol{G})$. That this result is true may also easily be verified directly with the graph.

REMARK. If $\boldsymbol{G}$ is a graph with the vertices $X_1, X_2, \ldots, X_n$, then there exists for each ordered pair i, j a smallest number k_{ij} ($\leq n$) with $d_{ij} = b_{ij}^{k_{ij}} = b_{ij}^n$. For $d_{ij} \neq \infty$ this means: In $\boldsymbol{G}$ there exists a simple path of minimum length (namely of the length $b_{ij}^{k_{ij}}$) going from X_i to X_j with exactly k_{ij} arcs, and no simple path of the same minimum length with less than k_{ij} arcs. Of course, simple paths of minimum length containing a larger number of arcs may occur.

In our example, the following relations for instance are true:

$$k_{12} = 1; \quad d_{12} = b_{12}^1 = 2,$$

$$k_{51} = 3; \quad d_{51} = b_{51}^3 = 10.$$

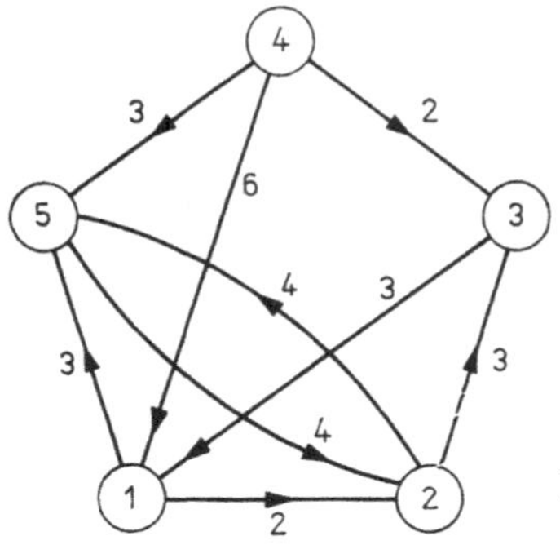

Fig. 3.1

Consider still another one. Given a graph G with the arc length matrix

$$B(G) = \begin{bmatrix} 0 & \infty & \infty & \infty & 2 & \infty & \infty \\ \infty & 0 & \infty & \infty & \infty & \infty & 3 \\ \infty & \infty & 0 & 2 & \infty & 4 & \infty \\ \infty & \infty & 2 & 0 & 4 & \infty & \infty \\ 2 & \infty & \infty & 3 & 0 & \infty & \infty \\ \infty & \infty & 3 & \infty & \infty & 0 & 1 \\ \infty & 2 & \infty & \infty & \infty & 1 & 0 \end{bmatrix}$$

If we establish successively the $\oplus$-powers, then it will finally result:

$$B^6(G) = \begin{bmatrix} 0 & 14 & 7 & 5 & 2 & 11 & 12 \\ 15 & 0 & 7 & 9 & 13 & 4 & 3 \\ 8 & 7 & 0 & 2 & 6 & 4 & 5 \\ 6 & 9 & 2 & 0 & 4 & 6 & 7 \\ 2 & 12 & 5 & 3 & 0 & 9 & 10 \\ 11 & 3 & 3 & 5 & 9 & 0 & 1 \\ 12 & 2 & 4 & 6 & 10 & 1 & 0 \end{bmatrix}$$

and $B^6(G) = D(G)$, but $B^5(G) \neq D(G)$. In this example, we really have to set up $n - 1$ $\oplus$-powers of $B(G)$ to obtain $D(G)$. Of course, this can also be seen directly in graph G itself, since a simple path going from X_1 to X_2, for example, contains six arcs, and there are no shorter ones. Thus, $k_{12} = 6$.

In order to calculate the distance matrix $\boldsymbol{D}(\boldsymbol{G})$ of a graph $\boldsymbol{G}$ with a large number n of vertices a lot of calculation processes will be required. This, however, can be simplified, by using the corollary resulting from theorem 3.3, as follows: The matrix found is squared, and we form one after another the matrices

$$\boldsymbol{B}^1(\boldsymbol{G}) = \boldsymbol{B}(\boldsymbol{G}), \boldsymbol{B}^2(\boldsymbol{G}), \boldsymbol{B}^4(\boldsymbol{G}), \boldsymbol{B}^8(\boldsymbol{G}), \ldots$$

We stop the $\oplus$-exponentiation by $\boldsymbol{B}^{2^k}(\boldsymbol{G})$ either if $2^k \geqq n - 1$ or if $\boldsymbol{B}^{2^{k-1}}(\boldsymbol{G}) = \boldsymbol{B}^{2^k}(\boldsymbol{G})$. Then, obviously $\boldsymbol{D}(\boldsymbol{G}) = \boldsymbol{B}^{2^k}(\boldsymbol{G})$.

In our last example, only three matrix multiplications instead of five would be necessary. We do not want to ignore one disadvantage of this latter method. The first method made it possible to determine exactly the numbers k_{ij} for each couple of natural numbers i, j, whereas with the second one only the bounds of the k_{ij} may be obtained (namely two 2-powers, i.e. $2^r \leqq k_{ij} \leqq 2^{r+1}$, with r depending on the choice of the pair i, j).

3.3. THE REVISED MATRIX ALGORITHM

The calculation processes necessary for determining the distance matrix $\boldsymbol{D}(\boldsymbol{G})$ of a graph $\boldsymbol{G}$ may be reduced in general if instead of the matrix product $\oplus$ introduced above another one is used.

DEFINITION. Let $\boldsymbol{G}$ be a directed graph, $\boldsymbol{B}(\boldsymbol{G})$ its arc length matrix and $X_1, X_2, \ldots, X_n$ its vertices. Put ${}^0\boldsymbol{B}(\boldsymbol{G}) = \boldsymbol{B}(\boldsymbol{G}) = ({}^0b_{ij})$. By the $\oplus_1$-product $\boldsymbol{B}(\boldsymbol{G}) \oplus_1 \boldsymbol{B}(\boldsymbol{G})$ we mean that matrix ${}^1\boldsymbol{B}(\boldsymbol{G}) = ({}^1b_{ij})$ the elements ${}^1b_{ij}$ of which are calculated in the following order (row by row)

$${}^1b_{11}, {}^1b_{12}, \ldots, {}^1b_{1n}, {}^1b_{21}, {}^1b_{22}, \ldots, {}^1b_{2n}, \ldots, {}^1b_{n1}, {}^1b_{n2}, \ldots, {}^1b_{nn}$$

and in the following way:

$${}^1b_{ij} = \min_{1 \leqq l \leqq n} ({}^pb_{il} + {}^qb_{lj}).$$

Here, it holds:

$$p = \begin{cases} 0 & \text{for} \quad l \geqq j, \\ 1 & \text{for} \quad l < j, \end{cases} \qquad q = \begin{cases} 0 & \text{for} \quad l \geqq i, \\ 1 & \text{for} \quad l < i. \end{cases}$$

That is to say, we "square" the matrix $\boldsymbol{B}(\boldsymbol{G})$ with respect to the $\oplus$-product introduced in 3.2 and immediately put each calculated value in the matrix, or in other words: Using the assertion instruction known from programming languages we form in the above-mentioned order

$$b_{ij} := \min_{1 \leqq l \leqq n} (b_{il} + b_{lj}).$$

It is evident that in the case of the use of digital computers storage locations will be saved, since for both the calculation and storage the elements of only one matrix (which is continuously changing) are necessary.

Now, we are able to determine also $\oplus_1$-powers of higher order according to

$$ {}^{k}\boldsymbol{B}(\boldsymbol{G}) = {}^{k-1}\boldsymbol{B}(\boldsymbol{G}) \oplus_1 {}^{k-1}\boldsymbol{B}(\boldsymbol{G}) . $$

Consider the graph represented in Fig. 3.2a. The arc length matrix $\boldsymbol{B}(\boldsymbol{G}) = {}^{0}\boldsymbol{B}(\boldsymbol{G})$ has the following structure:

$$ {}^{0}\boldsymbol{B}(\boldsymbol{G}) = \begin{bmatrix} 0 & 30 & 18 & 6 & \infty \\ \infty & 0 & \infty & \infty & \infty \\ \infty & 10 & 0 & \infty & \infty \\ \infty & \infty & \infty & 0 & 7 \\ \infty & 13 & 2 & \infty & 0 \end{bmatrix} . $$

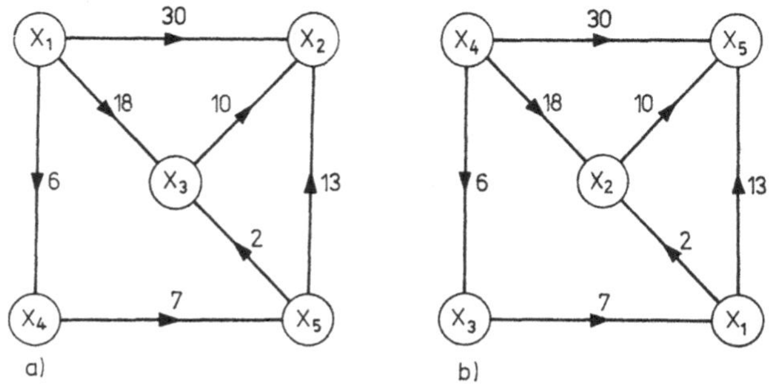

Fig. 3.2

Applying the rules of the $\oplus_1$-multiplication we get successively:

$$ {}^{1}\boldsymbol{B}(\boldsymbol{G}) = \begin{bmatrix} 0 & 28 & 18 & 6 & 13 \\ \infty & 0 & \infty & \infty & \infty \\ \infty & 10 & 0 & \infty & \infty \\ \infty & 20 & 9 & 0 & 7 \\ \infty & 12 & 2 & \infty & 0 \end{bmatrix}, \quad {}^{2}\boldsymbol{B}(\boldsymbol{G}) = \begin{bmatrix} 0 & 25 & 15 & 6 & 13 \\ \infty & 0 & \infty & \infty & \infty \\ \infty & 10 & 0 & \infty & \infty \\ \infty & 19 & 9 & 0 & 7 \\ \infty & 12 & 2 & \infty & 0 \end{bmatrix}, $$

$$ {}^{3}\boldsymbol{B}(\boldsymbol{G}) = \begin{bmatrix} 0 & 25 & 15 & 6 & 13 \\ \infty & 0 & \infty & \infty & \infty \\ \infty & 10 & 0 & \infty & \infty \\ \infty & 19 & 9 & 0 & 7 \\ \infty & 12 & 2 & \infty & 0 \end{bmatrix} . $$

Furthermore, the relation $\boldsymbol{D}(\boldsymbol{G}) = {}^{3}\boldsymbol{B}(\boldsymbol{G})$ results.

If, however, we number the vertices otherwise, as for example, in Fig. 3.2b, then ${}^1\boldsymbol{B}(\boldsymbol{G}) = \boldsymbol{D}(\boldsymbol{G})$ will follow, which the reader can recognize without difficulty.

N.S. Narahari Pandit [3] conjectured that two $\oplus_1$-multiplications would suffice to get the distance matrix for each graph $\boldsymbol{G}$, i.e., that ${}^2\boldsymbol{B}(\boldsymbol{G}) = \boldsymbol{D}(\boldsymbol{G})$ is valid. The graph represented in Fig. 3.3., however, proves (task!) that this is not the case, and certainly not if the vertices are numbered in any way. In paragraph 3.4, we will show that the relation ${}^3\boldsymbol{B}(\boldsymbol{G}) = \boldsymbol{D}(\boldsymbol{G})$ is always true.

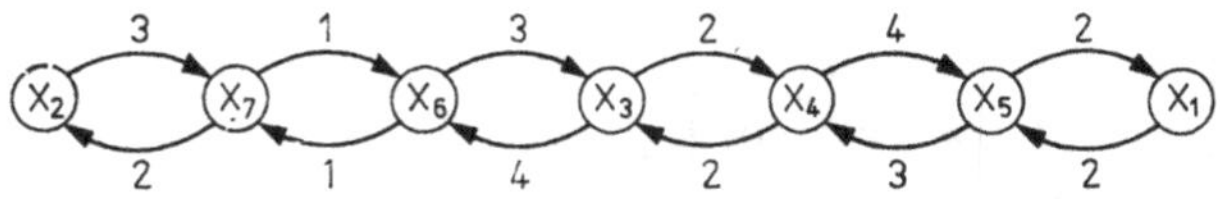

Fig. 3.3

3.4. THE CASCADE ALGORITHM

On the basis of the revised matrix algorithm explained in 3.3, the cascade algorithm for determining the distance matrix $\boldsymbol{D}(\boldsymbol{G})$ of a graph $\boldsymbol{G}$ was developed. In order to understand this algorithm better, we shall describe another matrix operation, namely the $\oplus_2$-product.

DEFINITION. Given a square (n, n)-matrix $\boldsymbol{A} = ({}_0a_{ij})$ with ${}_0a_{ij} \in R^+ \cup \{\infty\}$. The *product* $\boldsymbol{A} \oplus_2 \boldsymbol{A}$ is understood to be a matrix ${}_1\boldsymbol{A} = ({}_1a_{ij})$ whose elements are calculated in the order

$$
\begin{array}{l}
{}_1a_{nn},\ {}_1a_{n,n-1},\ \ldots,\ {}_1a_{n2},\ {}_1a_{n1},\\
{}_1a_{n-1,n},\ {}_1a_{n-1,n-1},\ \ldots,\ {}_1a_{n-1,2},\ {}_1a_{n-1,1}\\
\cdots\cdots\cdots\cdots\cdots\cdots\cdots\\
{}_1a_{1n},\ {}_1a_{1,n-1},\ \ldots,\ {}_1a_{12},\ {}_1a_{11}
\end{array}
$$

according to the rule

$$ {}_1a_{ij} = \min_{1 \leq l \leq n} ({}_pa_{il} + {}_qa_{lj}) $$

with

$$ p = \begin{cases} 0 & \text{for } l \leq j, \\ 1 & \text{for } l > j, \end{cases} \qquad q = \begin{cases} 0 & \text{for } l \leq i, \\ 1 & \text{for } l > i. \end{cases} $$

The $\oplus_2$-product differs from the $\oplus_1$-product only in the fact that the new matrix elements are calculated in exactly the opposite order as in the case of the $\oplus_1$-product. Thus, the newly found matrix elements ${}_1a_{ij}$ are also immediately put in the place (i, j) of the matrix.

The reader may ascertain that the distance matrix is already obtained with $\oplus_2$-multiplication of the arc length matrix of the graph shown in Fig. 3.2a by applying it only once.

It is easy to see that, if the vertices are numbered in another order, one single $\oplus_2$-multiplication is not necessarily sufficient.

Now, we combine the two multiplications $\oplus_1$ and $\oplus_2$ as follows.

DEFINITION. Let $\boldsymbol{G}$ be a directed graph with the vertices $X_1, X_2, \ldots, X_n$ and the arc length matrix $\boldsymbol{B}(\boldsymbol{G})$. The method used to determine the matrix ${}_1({}^1\boldsymbol{B}(\boldsymbol{G}))$ is called *cascade algorithm.*

The matrix ${}^1\boldsymbol{B}(\boldsymbol{G})$ is set up in the first cascade step (forward procedure of the cascade algorithm), the determination of the matrix

$$ {}_1({}^1\boldsymbol{B}(\boldsymbol{G})) = {}^1\boldsymbol{B}(\boldsymbol{G}) \oplus_2 {}^1\boldsymbol{B}(\boldsymbol{G}) $$

follows in the second cascade step (backward procedure of the cascade algorithm).

EXAMPLE. Apply the cascade algorithm to the graph of the example at page 87. Calculate successively

$$ {}^1\boldsymbol{B}(\boldsymbol{G}) = \boldsymbol{B}(\boldsymbol{G}) \oplus_1 \boldsymbol{B}(\boldsymbol{G}) = \begin{bmatrix} 0 & \infty & \infty & 5 & 2 & \infty & \infty \\ \infty & 0 & \infty & \infty & \infty & 4 & 3 \\ \infty & \infty & 0 & 2 & 6 & 4 & 5 \\ 6 & \infty & 2 & 0 & 4 & 6 & 7 \\ 2 & \infty & 5 & 3 & 0 & 9 & 10 \\ \infty & 3 & 3 & 5 & 9 & 0 & 1 \\ \infty & 2 & 4 & 6 & 10 & 1 & 0 \end{bmatrix} $$

in the forward procedure and

$$ {}_1({}^1\boldsymbol{B}(\boldsymbol{G})) = {}^1\boldsymbol{B}(\boldsymbol{G}) \oplus_2 {}^1\boldsymbol{B}(\boldsymbol{G}) = \begin{bmatrix} 0 & 14 & 7 & 5 & 2 & 11 & 12 \\ 15 & 0 & 7 & 9 & 13 & 4 & 3 \\ 8 & 7 & 0 & 2 & 6 & 4 & 5 \\ 6 & 9 & 2 & 0 & 4 & 6 & 7 \\ 2 & 12 & 5 & 3 & 0 & 9 & 10 \\ 11 & 3 & 3 & 5 & 9 & 0 & 1 \\ 12 & 2 & 4 & 6 & 10 & 1 & 0 \end{bmatrix} $$

in the subsequent backward procedure and you will get ${}_1({}^1\boldsymbol{B}(\boldsymbol{G})) = \boldsymbol{D}(\boldsymbol{G})$, that is, the distance matrix, after two matrix operations.

The purpose of our further considerations is to prove the following theorem.

THEOREM 3.5. *If the cascade algorithm is applied to the arc length matrix* $\boldsymbol{B}(\boldsymbol{G})$ *of a graph* $\boldsymbol{G}$*, then one gets the distance matrix* $\boldsymbol{D}(\boldsymbol{G})$ *of* $\boldsymbol{G}$*, that is,*

$$ {}_1({}^1\boldsymbol{B}(\boldsymbol{G})) = \boldsymbol{D}(\boldsymbol{G}). $$

The proof will be based on T. C. Hu [2]. Some preliminary considerations still need to be made.

Let $\boldsymbol{R}(X_i, X_j)$ be a shortest simple path going from X_i to X_j, and let X_k be an interior vertex of this simple path. Then, the subpath $\boldsymbol{R}(X_i, X_k)$ of $\boldsymbol{R}(X_i, X_j)$ is a shortest simple path going from X_i to X_k. Hence it follows that in a directed graph $\boldsymbol{G}$ there exist shortest simple paths connecting two vertices of $\boldsymbol{G}$ and which consist only of one arc. Such arcs shall be called *base arcs*.

DEFINITION. An arc (X_i, X_j) of a graph $\boldsymbol{G}$ is called a *base arc* if a shortest simple path from X_i to X_j is exactly as long as the arc (X_i, X_j), i.e., if $d_{ij} = b_{ij}$ with $\boldsymbol{B} = (b_{ij})$ as arc length matrix of $\boldsymbol{G}$.

If we try to find shortest simple paths going from a fixed vertex of a graph $\boldsymbol{G}$ to all other vertices of $\boldsymbol{G}$ (cf. 1.4.3), and if we are satisfied with one single shortest simple path found (if such simple paths really exist), then it can be shown that the subgraph of $\boldsymbol{G}$ which has been composed of these simple paths is a tree.

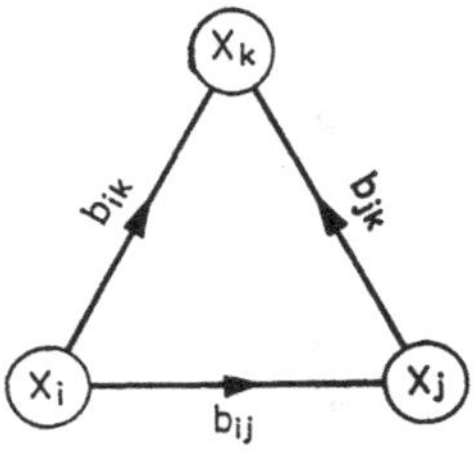

Fig. 3.4

If $\boldsymbol{G}$ contains a subgraph as it is represented in Fig. 3.4, and if we are only interested in shortest simple paths leaving X_i, then one of the two arcs (X_i, X_k) and (X_j, X_k) can surely be removed according as $b_{ik} \geqq b_{ij} + b_{jk}$ or $b_{ij} + b_{jk} \geqq b_{ik}$. But if we like to consider all shortest simple paths (as in the example given in Fig. 3.4, such shortest simple paths going from X_i to X_k and X_j and from X_j to X_k), then it may be that all arcs are required as base arcs, say if $b_{ij} = b_{ik} = b_{jk}$. But if $b_{ik} > b_{ij} + b_{jk}$, then one can substitute $b_{ij} + b_{jk}$ for the valuation b_{ik} of the arc (X_i, X_k) without changing the lengths of the shortest simple paths. This modified arc can then be treated as base arc. For proving theorem 3.5 we shall also introduce new base arcs in this way.

For the following let us explain some further concepts.

DEFINITION. The vertices $X_{i_1}, X_{i_2}, \ldots, X_{i_k}$ of a graph $\boldsymbol{G}$ form

a) an *ascending sequence* if $i_1 < i_2 < \cdots < i_k$,

b) a *descending sequence* if $i_1 > i_2 > \cdots > i_k$,

c) a *down-and-up sequence* if $\min(i_1, i_k) > \max\limits_{2 \leqq j \leqq k-1} i_j$.

If, in particular, $X_{i_1}, X_{i_2}, \ldots, X_{i_k}$ are vertices of a simple path going from X_{i_1} to X_{i_k} (these vertices are in the same order as they are met when traversing the simple path), then we call them an *ascending sequence* or *descending sequence* or *down-and-up sequence of the simple path.*

DEFINITION. An interior vertex X_{i_k} of a simple path

$$\boldsymbol{R}(X_i, X_j) = (X_i = X_{i_1}, X_{i_2}, \ldots, X_{i_k}, \ldots, X_{i_r} = X_j)$$

is called *minimal* (*maximal*) *in the path* $\boldsymbol{R}(X_i, X_j)$ if $i_k < \min(i_{k-1}, i_{k+1})$ (and if $i_k > \max(i_{k-1}, i_{k+1})$, respectively).

DEFINITION. Let $\mathfrak{Q}$ be a sequence of vertices of $\boldsymbol{G}$. A sequence $\mathfrak{P}$ of vertices of $\boldsymbol{G}$ is called a *subsequence* of $\mathfrak{Q}$ if it holds:

a) Each vertex of $\mathfrak{P}$ belongs to $\mathfrak{Q}$.

b) Those vertices which directly succeed one another in $\mathfrak{P}$ are also directly successive in $\mathfrak{Q}$.

In other words: Only those partial sequences in which only "borders" of the original sequence "have been cut away" at the most are called subsequences.

DEFINITION. Given a sequence $\mathfrak{Q}$ of vertices (not necessarily a simple path) of $\boldsymbol{G}$. Delete the vertices in $\mathfrak{Q}$ as follows:

a) Choose (if there is one) a subsequence $\mathfrak{P} = \{X_{i_1}, \ldots, X_{i_k}\}$ of $\mathfrak{Q}$ which is a down-and-up sequence. Then delete all interior vertices X_{i_j} $(j = 2, \ldots, k-1)$ of $\mathfrak{P}$ in $\mathfrak{Q}$.

b) Apply a) to the remaining sequence.

c) If a) cannot be applied any more, the remaining sequence is called a *reduced sequence* of $\mathfrak{Q}$.

EXAMPLE. Let

$$\mathfrak{Q} = \{1, 8, 2, 10, 14, 15, 7, 6, 3, 12, 8, 13, 18, 25, 11, 9\}.$$

The subsequence {7, 6, 3, 12} forms a down-and-up sequence. Applying a) yields {1, 8, 2, 10, 14, 15, 7, 12, 8, 13, 18, 25, 11, 9}. The further application of a) finally leads to a reduced sequence {1, 8, 10, 14, 15, 18, 25, 11, 9}, and that, independently of how the down-and-up sequences are chosen, since the following theorem is true:

THEOREM 3.6. *The reduced sequence* $\mathfrak{R}$ *is well-determined by the basic sequence* $\mathfrak{Q}$.

Another way to find $\mathfrak{R}$ consists in searching for all maximal down-and-up sequences in $\mathfrak{Q}$ (in our example, these are {8, 10, 2}, {15, 7, 6, 3, 12, 8, 13, 18} and in applying a) to them.

Proof. Let $\mathfrak{Q}$ be any sequence of vertices of $\boldsymbol{G}$. We search for all maximal down-and-up sequences in $\mathfrak{Q}$ and denote them by $\mathfrak{P}_1, \ldots, \mathfrak{P}_l$ in the order of the traversing of $\mathfrak{Q}$. If we delete all interior vertices in the $\mathfrak{P}_i$, then a sequence $\mathfrak{P}$ results containing all end-points of the $\mathfrak{P}_i$ and also vertices which possibly exist in $\mathfrak{Q}$ between the down-and-up sequences $\mathfrak{P}_i$ (in our example, the vertices 1, 14, 25, 11, 9 are of this type). $\mathfrak{P}$ is certainly a reduced sequence, since if there still existed a down-and-up sequence in $\mathfrak{P}$, we would not choose the $\mathfrak{P}_i$'s all maximal.

It can also be seen that, if the down-and-up sequences are chosen in another way, the sequence $\mathfrak{P}$ is obtained when breaking off the reduction rule, since each down-and-up sequence is uniquely embedded in a maximal down-and-up sequence (task!).

THEOREM 3.7. *The reduced sequence $\mathfrak{R}$ of a sequence $\mathfrak{Q}$ is either*

a) *an ascending sequence or*

b) *a descending sequence or*

c) *an ascending sequence with subsequent descending sequence.*

In our example, $\mathfrak{R}$ is of type c).

Proof.

1. $\mathfrak{R}$ contains no minimal point (a minimal point is a point whose two neighbouring points in the sequence possess a larger index), since this would then lie in a down-and-up sequence.

2. $\mathfrak{R}$ contains one maximal point at the most, since otherwise a minimal point would exist (namely between those two maximal points).

3. Let $\mathfrak{R}$ contain no maximal point (that is, no point whose two neighbouring points have a smaller index) and no minimal point as well. In this case, however, $\mathfrak{R}$ will be ascending or descending.

4. Let $\mathfrak{R}$ contain exactly one maximal point. Then, $\mathfrak{R}$ consists of an ascending and of a subsequent descending sequence.

Q.E.D.

THEOREM 3.8. *Let $\boldsymbol{R}(X_i, X_j)$ be a shortest simple path going from X_i to X_j. If the vertices of $\boldsymbol{R}$ form an ascending or a descending sequence, then it holds:*

$$^1b_{ij} = d_{ij},$$

that is, one gets the shortest distance between X_j and X_i (the length of a shortest simple path going from X_i to X_j) after having performed the forward procedure of the cascade algorithm.

Proof. Consider a shortest simple path $\boldsymbol{R}(X_i, X_j) = (X_i = X_{i_1}, X_{i_2}, \ldots, X_{i_k} = X_j)$ going from X_i to X_j.

a) Let $X_{i_1}, \ldots, X_{i_k}$ form an ascending sequence, i.e., let $i_1 < i_2 < \cdots < i_k$.

First, note that for any indices r, s always the relation ${}^1b_{rs} \geqq d_{rs}$ is valid, since ${}^1b_{rs}$ is either ∞ (then, there is nothing to show) or results as the length of a simple path from X_r to X_s (i.e., it has never been smaller than the value of a shortest simple path length). Since the arc (X_{i_1}, X_{i_2}) is an arc of a shortest simple path and thus a base arc, it holds:

$$ {}^1b_{i_1 i_2} = {}^0b_{i_1 i_2} = d_{i_1 i_2}. $$

In the first step of the cascade algorithm ${}^1b_{i_1 i_3}$ is calculated as the second of the values ${}^1b_{i_1 r}$ $(r = i_2, \ldots, i_k)$. Since also the arc (X_{i_2}, X_{i_3}) is a base arc, it results:

$$ {}^1b_{i_1 i_3} = {}^1b_{i_1 i_2} + {}^0b_{i_2 i_3} = d_{i_1 i_3}. $$

Proceed in this manner so that you get:

$$ {}^1b_{i_1 i_k} = d_{i_1 i_2} + d_{i_2 i_3} + \cdots + d_{i_{k-1} i_k} = d_{i_1 i_k}. $$

It is naturally possible to calculate other d_{rs} also in the course of the forward step.

b) Let the vertices $X_{i_1}, X_{i_2}, \ldots, X_{i_k}$ of a shortest simple path $\boldsymbol{R}(X_{i_1}, X_{i_k})$ going from X_{i_1} to X_{i_k} form a descending sequence, i.e., let $i_1 > i_2 > \cdots > i_k$.

In the forward procedure of the cascade algorithm the first number of the ${}^1b_{ri_k}$ $(r = i_1, i_2, \ldots, i_{k-1})$ to be calculated will be the value ${}^1b_{i_{k-1} i_k}$. Since the arc $(X_{i_{k-1}}, X_{i_k})$ is an arc of a shortest simple path and thus also a base arc, it holds:

$$ {}^1b_{i_{k-1} i_k} = d_{i_{k-1} i_k}. $$

Analogously to case a) (taking into account the order in which the ${}^1b_{ri_k}$ are to be calculated), we get the proposition of the theorem of which the proof is thus established.

THEOREM 3.9. *Let* $\boldsymbol{R}(X_i, X_j)$ *be a shortest simple path going from* X_i *to* X_j. *If the vertices of* $\boldsymbol{R}(X_i, X_j)$ *form a down-and-up sequence, the relation* ${}^1b_{ij} = d_{ij}$ *is true.*

That means that d_{ij} is already ascertained in the forward step of the cascade algorithm (as also in the case of a descending or an ascending sequence; cf. theorem 3.8).

Proof. Let the vertices $X_i = X_{i_1}, X_{i_2}, \ldots, X_{i_k} = X_j$ of a shortest simple path $\boldsymbol{R}(X_i, X_j)$ form a down-and-up sequence. This down-and-up sequence $\mathfrak{Q}$ possesses at least one minimal point.

1. Let $\mathfrak{Q}$ possess exactly one minimal point. This minimal point shall be X_{i_m}. Then, the subsequence $X_{i_1}, X_{i_2}, ..., X_{i_m}$ is a descending, and the subsequence $X_{i_m}, X_{i_{m+1}}, ..., X_{i_k}$ an ascending sequence. According to theorem 3.8, it holds:

$$^1b_{i_1 i_m} = d_{i_1 i_m} \quad \text{and} \quad ^1b_{i_m i_k} = d_{i_m i_k}.$$

Because of the calculation rule for determining the $^1b_{ij}$, the value $^1b_{i_1 i_m}$ is calculated before $^1b_{i_1 i_k}$ and $^1b_{i_m i_k}$ before $^1b_{i_1 i_k}$. Since our simple path going from X_{i_1} to X_{i_k} is of minimum length, we get

$$d_{i_1 i_m} + d_{i_m i_k} = {}^1b_{i_1 i_m} + {}^1b_{i_m i_k} = {}^1b_{i_1 i_k}.$$

2. Let $\mathfrak{Q}$ have more than one minimal vertex. Delete all those interior points in $\mathfrak{Q}$ which are not maximal points. Let the sequence $\mathfrak{F}$ of vertices develop (if, for example, $\mathfrak{Q} = \{15, 4, 8, 3, 7, 12, 1, 5, 10, 17\}$, then $\mathfrak{F} = \{15, 8, 12, 17\}$).

Write $\mathfrak{F} = \{X_{i_1} = Y_1, Y_2, Y_3, ..., Y_l = X_{i_k}\}$. Here, $Y_2, ..., Y_{l-1}$ are maximal vertices of $\mathfrak{Q}$. Among all couples of indices $\bar{p}, \bar{q}$ with $Y_t = X_{\bar{p}}$ and $Y_{t+1} = X_{\bar{q}}$ consider that one for which the value $^1b_{\bar{p}\bar{q}}$ is calculated first in the forward step of the cascade algorithm. We denote this couple of indices by p, q (in our example we would have $p = 8, q = 12$).. The subsequence $\{X_p, ..., X_q\}$ (in our example, this is $\{8, 3, 7, 12\}$) of $\mathfrak{Q}$ possesses only one minimal point (this is point 3 in our example). Thus, because of the case already discussed, we get the relation $^1b_{pq} = d_{pq}$.

The substitution of $^1b_{pq}$ for $^0b_{pq}$ in the course of the forward step corresponds to the replacement of the corresponding subpath (which is in our case of minimum length) by an arc of this subpath length, and that, by a base arc to which the length d_{pq} is assigned. Now, take $\mathfrak{Q}$ and form a sequence $\mathfrak{Q}'$ of vertices by deleting all intermediate points of the subsequence $\{X_p, ..., X_q\}$. $\mathfrak{Q}'$ has one minimal vertex less than $\mathfrak{Q}$. On the basis of $\mathfrak{Q}$, form a sequence $\mathfrak{F}'$ (this corresponds to the transition from $\mathfrak{Q}$ to $\mathfrak{F}$) by deleting that one of the two vertices X_p or X_q which has the smaller index. Now, consider all successive couples of $\mathfrak{F}'$ of vertices and turn to that couple X_p, X_q for which the value $^1b_{pq}$ is calculated first in the forward step of the cascade algorithm, etc. (mathematical induction!).

This procedure can be carried on as long as the remaining down-and-up sequence possesses still one single minimal point (according to the definition of a down-and-up sequence, none of the two points X_{i_1}, X_{i_k} has been deleted in the course of this process). Thus, case 1 is established, and the proof of theorem 3.9 is obvious.

THEOREM 3.10. *Let* $R(X_i, X_j)$ *be a shortest simple path in* G *from* X_i *to* X_j. *If the vertices of* $R(X_i, X_j)$ *form*

 a) *an ascending sequence or*

 b) *a descending sequence or*

 c) *an ascending with a subsequent descending sequence, then it holds*:

$$^1_1b_{ij} = d_{ij}.$$

Proof.

a) If the vertices $X_i = X_{i_1}, X_{i_2}, \ldots, X_{i_k} = X_j$ form an ascending sequence, i.e., if $i_1 < i_2 < \cdots < i_k$, then the backward procedure yields step by step

$$ {}_1^1b_{i_{k-1}i_k} = {}^0b_{i_{k-1}i_k} = d_{i_{k-1}i_k}, \quad \text{since} \quad (X_{i_{k-1}}, X_{i_k}) \quad \text{is a base arc,} $$

$$ {}_1^1b_{i_{k-2}i_k} = {}^0b_{i_{k-2}i_{k-1}} + {}_1^1b_{i_{k-1}i_k} = d_{i_{k-2}i_k}, \quad \text{since} \quad (X_{i_{k-2}}, X_{i_{k-1}}) \quad \text{is a base arc,} $$

$$ \ldots\ldots\ldots\ldots\ldots\ldots\ldots\ldots\ldots\ldots $$

$$ {}_1^1b_{i_1i_k} = {}^0b_{i_1i_2} + {}_1^1b_{i_2i_k} = d_{i_1i_k}. $$

b) If the vertices $X_i = X_{i_1}, X_{i_2}, \ldots, X_{i_k} = X_j$ form a descending sequence, then $i_1 > i_2 > \cdots > i_k$, and we can proceed as in case a).

c) If the vertices first form an ascending sequence $X_i = X_{i_1}, X_{i_2}, \ldots, X_{i_j}$ and subsequently a descending one $X_{i_j}, X_{i_{j+1}}, \ldots, X_{i_k}$, then we conclude as follows: since $i_k < i_j$ and $i_1 < i_j$, then in the backward step ${}_1^1b_{i_1i_k}$ is calculated after ${}_1^1b_{i_1i_j}$ and after ${}_1^1b_{i_ji_k}$. Using a), however, it follows

$$ {}_1^1b_{i_1i_j} = d_{i_1i_j} \quad \text{and} \quad {}_1^1b_{i_ji_k} = d_{i_ji_k}, $$

and hence it follows

$$ {}_1^1b_{i_1i_k} = {}_1^1b_{i_1i_j} + {}_1^1b_{i_ji_k} = d_{i_1i_j} + d_{i_ji_k} = d_{i_1i_k}. $$

Now, we come to the proof of theorem 3.5.

Let $\boldsymbol{R}(X_{i_1}, X_{i_k})$ be a shortest simple path in $\boldsymbol{G}$ from X_i to X_j. We have to show that ${}_1^1b_{i_1i_k} = d_{i_1i_k}$ is true independently of the type of sequence $\mathfrak{Q} = \{X_{i_1}, X_{i_2}, \ldots, X_{i_k}\}$. For this, consider the reduced sequence $\mathfrak{R}$ of $\mathfrak{Q}$.

Two neighbouring vertices in $\mathfrak{R}$ are either connected by a base arc or they are end-points of a down-and-up sequence in $\mathfrak{Q}$. In both cases, the distances between these two neighbouring points in $\mathfrak{R}$ are calculated in the forward process of the cascade algorithm according to the theorems 3.8 and 3.9. Following theorem 3.7, $\mathfrak{R}$ is either an ascending or descending sequence or an ascending with a subsequent descending sequence, and therefore, it holds according to theorem 3.10:

$$ {}_1^1b_{i_1i_k} = d_{i_1i_k}. $$

This proves theorem 3.5.

COROLLARY 1. If in a directed graph with n vertices only the distance between two vertices Y and Z of $\boldsymbol{G}$ is to be calculated, then change the numeration of the vertices of $\boldsymbol{G}$ such that Y is provided with the index $n - 1$ and Z with the index n. Then, the vertices of any simple path going from Y to Z (thus also of a shortest one) always form a down-and-up sequence, and following theorem 3.9, the distance between these two vertices can already be calculated in the forward step of the cascade algorithm. A comparison between this method and that described in chapter 1, however, shows that no advantage is obtained.

COROLLARY 2. If in a directed graph $\boldsymbol{G}$ only the distance between one fixed vertex and all other vertices of $\boldsymbol{G}$ is to be determined, then give the largest index to this vertex when making a renumbering of the vertices of $\boldsymbol{G}$. Those elements ${}^1_1b_{ij}$ which have been calculated in the backward step are not changed any more, and it holds ${}^1_1b_{ij} = d_{ij}$. Thus, we can stop in the backward step when the last line has been calculated.

COROLLARY 3. If only the shortest distances between the vertices of a subset of all vertices are to be calculated, then assign the highest indices to these vertices. If just p vertices are considered, we can break off the backward algorithm when the last p-lines have been calculated.

THEOREM 3.11. *Given a directed graph $\boldsymbol{G}$ without circuits. Then, the vertices of $\boldsymbol{G}$ can suitably be numbered such that ${}^1\boldsymbol{B}(\boldsymbol{G}) = \boldsymbol{D}(\boldsymbol{G})$ is true.*

Proof. In paragraph 1.5, we have given a vertex numeration (in graphs without circuits) in such a way that arcs only go from vertices with a lower index to vertices with larger indices. Thus, according to our terminology, each simple path is ascending. Using theorem 3.8, the proof is evident.

The proposition set up in 3.3 can also be proved now.

THEOREM 3.12. *Let $\boldsymbol{G}$ be a directed graph containing the vertices $X_1, X_2, \ldots, X_n$ with the arc length matrix $\boldsymbol{B}(\boldsymbol{G})$. Then, the relation ${}^3\boldsymbol{B}(\boldsymbol{G}) = \boldsymbol{D}(\boldsymbol{G})$ holds.*

It is thus asserted that the distance matrix can also be calculated by successively carrying out the forward step of the cascade algorithm three times.

Proof. Let $\boldsymbol{R}(X_i, X_j)$ be a shortest simple path going from X_i to X_j, and let $\mathfrak{Q}$ be the sequence of the vertices of $\boldsymbol{R}(X_i, X_j)$. If we form the reduced sequence $\mathfrak{R}$ pertaining to $\mathfrak{Q}$, then we get all distances between the neighbouring vertices in $\mathfrak{R}$ during the first forward process. According to theorem 3.7, it holds:

Either

a) $\mathfrak{R}$ is ascending or descending. Then, according to theorem 3.8, the value d_{ij} is calculated in the second forward step, i.e., ${}^2b_{ij} = d_{ij}$;

or

b) $\mathfrak{R}$ is ascending and has a subsequent descending sequence, i.e., $\mathfrak{R} = \{X_i, \ldots, X_m, \ldots, X_j\}$ with X_m being maximal. Then, d_{im} and d_{mj} are calculated in the second forward step, but not necessarily d_{ij} (unless $\mathfrak{R}$ consists of only three vertices X_i, X_m, X_j). According to theorem 3.8, however, the relation $d_{ij} = {}^3b_{ij}$ results.

FINAL REMARKS. By using the for-statement known from the programming languages, the cascade algorithm can be described as follows:

$$\left.\begin{array}{l} \text{for } i = 1, 2, \ldots, n \text{ do} \\ \quad \text{for } j = 1, 2, \ldots, n \text{ do} \\ \quad\quad \text{for } r = 1, 2, \ldots, n \text{ do} \\ \quad\quad\quad b_{ij} := \min(b_{ij}, b_{ir} + b_{rj}); \end{array}\right\} \text{forward step}$$

$$\left.\begin{array}{l} \text{for } i = n, n-1, \ldots, 2, 1 \text{ do} \\ \quad \text{for } j = n, n-1, \ldots, 2, 1 \text{ do} \\ \quad\quad \text{for } r = 1, 2, \ldots, n \text{ do} \\ \quad\quad\quad b_{ij} := \min(b_{ij}, b_{ir} + b_{rj}); \end{array}\right\} \text{backward step}$$

$\boldsymbol{B} = (b_{ij})$ is now the distance matrix $\boldsymbol{D}(\boldsymbol{G})$.

R. W. Floyd (Algorithm 97: "Shortest Path", Comm. ACM **5** (1962), 345) suggests a modification of the instruction sequence which makes it possible to reduce the number of the necessary operations by half:

$$\begin{array}{l} \text{for } r = 1, 2, \ldots, n \text{ do} \\ \quad \text{for } i = 1, 2, \ldots, n \text{ do} \\ \quad\quad \text{for } j = 1, 2, \ldots, n \text{ do} \\ \quad\quad\quad b_{ij} := \min(b_{ij}, b_{ir} + b_{rj}); \end{array}$$

$\boldsymbol{B} = (b_{ij})$ is now the distance matrix $\boldsymbol{D}(\boldsymbol{G})$ wanted.

However, this algorithm is not suitable for calculation by hand, as the reader may see without difficulty. Instead of only two matrices ${}^1\boldsymbol{B}$ and ${}^1_1\boldsymbol{B}$, n matrices will have to be written down when applying the algorithm of Floyd.

With this we conclude our treatise on the problem of the determination of shortest simple paths in directed graphs.

3.5. BIBLIOGRAPHY

[1] Hasse, M.: Über die Behandlung graphentheoretischer Probleme unter Verwendung der Matrizenrechnung, Wiss. Z. TU Dresden **10** (1961), 1313–1316.

[2] Hu, T. C.: Revised matrix algorithms for shortest paths, SIAM J. Appl. Math. **15** (1967), 207–218.

[3] Narahari Pandit, N. S.: The shortest route problem, An addendum, Operations Res. **9** (1961), 129–132.

Chapter 4

Nonlinear transportation problems

4.1. FORMULATION OF THE PROBLEM

Chapter 2 was devoted to the linear transportation problem. There we considered a directed arc-valuated graph (capacity restrictions!) with two vertices Q, S marked and one (return) arc (S, Q) marked of unbounded capacity. On the return arc of this graph we searched for a flow of a given value which was compatible with the capacities and minimized the transportation costs, with the costs k_{ij} resulting for the transportation of one product unit along the arc (i, j). The function to be minimized was

$$Z = \sum_{(i,j)} k_{ij} x_{ij}$$

with x_{ij} being the flow quantity compatible with the capacities on the arc (i, j).

In section 4.2, we shall relax the assumptions for the cost functions, that is, we will presuppose the $k_{ij}(x)$ to be only convex. In addition to this, we allow the Kirchhoff's node rule not to be satisfied, in other words: Goods may be produced or consumed in the vertices of the graph.

In paragraph 4.3, we shall turn to a generalization of the classical flow problem treated by Ford and Fulkerson, namely: taking into consideration the capacity restrictions, two different flow types shall be transported from two sources to two sinks in a graph. The sum of the two flows is to be maximized. We do not know a solution of the problem for more than two types of flow.

4.2. A CONVEX TRANSPORTATION PROBLEM

We examine the following problem: Given an undirected graph $G(\mathfrak{X}, \mathfrak{U})$ with n vertices denoted by $1, 2, \ldots, n$. Let an intensity (source strength) d_i be assigned to each vertex i with the relation $\sum_{i=1}^{n} d_i = 0$ being valid.

In other words: the quantity produced in the graph must also be consumed.

Let an edge function (a flow) x_{ij} exist on $\boldsymbol{G}$ for which it holds that

$$\sum_{j\in\mathfrak{X}_i^+} x_{ij} - \sum_{j\in\mathfrak{X}_i^-} x_{ji} = d_i \quad \text{for all} \quad i \in \mathfrak{X}, \tag{1}$$

where $\mathfrak{X}_i^+$ and $\mathfrak{X}_i^-$ denote the set of those vertices of $\boldsymbol{G}$ to which a positive flow goes from i and from which a positive flow goes to i, respectively (the denotation corresponds to that used in chapter 1), since a positive flow going from i to j gives an orientation from i to j to the edge $(i, j) = (j, i)$. Thus, we can take our graph to be directed with a non-negative arc function x_{ij}. (Some remarks on the existence of such a flow will be made later.) Let the transportation of the flow quantity $x_{ij} \geqq 0$ along (i, j) entail the costs $k_{ij}(x_{ij}) \geqq 0$. Find a flow on $\boldsymbol{G}$ which minimizes the total costs, i.e., for which relation (1) is true and which minimizes

$$\sum_{(i,j)\in\mathfrak{U}} k_{ij}(x_{ij}).$$

In the following, we shall assume the *cost functions* $k_{ij}(x)$ to be *convex*. Thus, let

$$k_{ij}\big(\lambda x + (1-\lambda)\, y\big) \leqq \lambda k_{ij}(x) + (1-\lambda)\, k_{ij}(y) \quad \text{for any} \quad \lambda \in [0, 1].$$

As is well known, a convex function possesses a left-hand and a right-hand derivative in each interior point and is continuous there. Furthermore, we presuppose that the $k_{ij}(x)$ are right-hand continuous for $x = 0$. We denote the right-hand and the left-hand derivative of $k_{ij}(x)$ by $k_{ij}^+(x)$ and $k_{ij}^-(x)$, respectively. It holds:

$$k_{ij}^-(x) \leqq k_{ij}^+(x).$$

First, let us give a criterion indicating the point at which a convex function becomes minimum.

LEMMA 1. *The point x_0 minimizes the convex function $g(x)/x \geqq 0$ if and only if $g^+(x_0) \geqq 0$ and $g^-(x_0) \leqq 0$ for $x_0 > 0$ and $g^+(x_0) \geqq 0$ for $x_0 = 0$.*

Before proving the lemma, let us provide some properties of convex functions without proving them.

PROPERTY 1. *Let $g(x)$ be convex and $0 < a < A$. Then, it holds*

$$\frac{g(x+a) - g(x)}{a} \leqq \frac{g(x+A) - g(x)}{A},$$

i.e., the difference quotients of convex functions are monotonically increasing with the steplengths.

PROPERTY 2. *Let $g(x)$ be convex and $0 < a < A$. Then, it holds*

$$\frac{g(x) - g(x-A)}{A} \leqq \frac{g(x) - g(x-a)}{a}.$$

PROPERTY 3. *For any $a > 0$ and convex $g(x)$ it holds*

$$\frac{g(x) - g(x-a)}{a} \leqq \frac{g(x+a) - g(x)}{a}.$$

PROPERTY 4. *For any interior point x of the domain of a convex function $g(x)$ it holds*

$$g^-(x) \leqq g^+(x).$$

PROPERTY 5. *Let $h > 0$, $g(x)$ be convex and x as well as $x - h$ be interior points of the domain of $g(x)$. Then, it holds*

$$hg^+(x-h) \leqq g(x) - g(x-h) \leqq hg^-(x).$$

We recommend the reader to prove the above properties by himself.

Now, we come to the proof of lemma 1.

1. Let x_0 be the minimum point of $g(x)$.

a) Let $x_0 > 0$. Since x_0 is the minimum point, it holds for $h > 0$

$$g(x_0 + h) - g(x_0) \geqq 0,$$

thus

$$\frac{g(x_0 + h) - g(x_0)}{h} \geqq 0$$

and thus $g^+(x_0) \geqq 0$.

Correspondingly, it holds for $h > 0$

$$g(x_0) - g(x_0 - h) \leqq 0,$$

thus

$$\frac{g(x_0) - g(x_0 - h)}{h} \leqq 0$$

and thus, the relation $g^-(x_0) \leqq 0$ holds.

b) Let $x_0 = 0$. Then, $g(0 + h) - g(0) \geqq 0$ is true, and thus

$$\frac{g(0 + h) - g(0)}{h} \geqq 0$$

and thus $g^+(0) \geqq 0$.

2. Let $x_0 > 0$ and $g^+(x_0) \geqq 0$, $g^-(x_0) \leqq 0$. Suppose that x_0 would not be a minimum point. Then, there exists a point x_1 with $g(x_1) < g(x_0)$.

a) Let $x_1 > x_0$. Put $x_1 = x_0 + h$, and hence it follows using property 5:

$$0 \leqq hg^+(x_0) \leqq g(x_1) - g(x_1 - h) < 0,$$

which is certainly wrong.

b) Let $x_1 < x_0$. We put $x_0 = x_1 + h$ and get by using property 5

$$0 \geqq hg^-(x_0) \geqq g(x_0) - g(x_0 - h) > 0,$$

which is wrong as well.

c) Let $x_0 = 0$ and $g^+(x_0) \geqq 0$. Then, with $x_1 = x_0 + h$ being supposed to be the minimum point, it results

$$0 \leqq hg^+(x_0) \leqq g(x_1) - g(x_0) < 0,$$

which is not possible.

Thus, the lemma is completely proved.

Now, we are able to give a criterion for the minimization of the transportation costs realized by a flow which satisfies the modified node condition (2):

THEOREM 4.1. *A flow* $\{x_{ij}\}$ *satisfying the condition*

$$\sum_{j \in \mathfrak{X}_i^+} x_{ij} - \sum_{j \in \mathfrak{X}_i^-} x_{ji} = d_i \quad \text{for all} \quad i \in \mathfrak{X} \tag{2}$$

minimizes the total costs

$$\sum_{(i,j) \in \mathfrak{U}} k_{ij}(x_{ij})$$

in case of convex cost functions $k_{ij}(x) \geqq 0$ *if and only if there exists for each vertex* i *a number* V_i *(called potential) for which it holds:*

$$\left.\begin{aligned} V_j - V_i &\leqq k_{ij}^+(x_{ij}) && \text{if} \quad x_{ij} = 0, \\ k_{ij}^-(x_{ij}) \leqq V_j - V_i &\leqq k_{ij}^+(x_{ij}) && \text{if} \quad x_{ij} > 0. \end{aligned}\right\} \tag{3}$$

Proof.

1. Let the conditions (3) be satisfied for a flow $\{\bar{x}_{ij}\}$. We will show that $\{\bar{x}_{ij}\}$ is optimum, that is, it minimizes the total costs. Since a flow is certainly not optimum if for an edge $x_{ij} > 0$ and $x_{ji} > 0$, we may put $x_{ji} = 0$ if $x_{ij} > 0$. We also want to put $k_{ij}(x_{ij}) = 0$ if the edge (i, j) does not lie in G at all or if $x_{ji} > 0$. Then, the modified node condition (2) assumes the form

$$\sum_{j=1}^{n} x_{ij} - \sum_{j=1}^{n} x_{ji} = d_i \qquad (i = 1, \ldots, n). \tag{2'}$$

Using (2′), the total costs can be written as follows:

$$\sum_{i,j=1}^{n} k_{ij}(x_{ij}) = \sum_{i,j=1}^{n} k_{ij}(x_{ij}) + \sum_{i=1}^{n} V_i \left(\sum_{j=1}^{n} x_{ij} - \sum_{j=1}^{n} x_{ji} - d_i \right)$$

$$= \sum_{i,j=1}^{n} k_{ij}(x_{ij}) + \sum_{i,j=1}^{n} V_i x_{ij} - \sum_{i,j=1}^{n} V_i x_{ji} - \sum_{i=1}^{n} V_i d_i$$

$$= \sum_{i,j=1}^{n} \left(k_{ij}(x_{ij}) + (V_i - V_j)\, x_{ij} \right) - \sum_{i=1}^{n} V_i d_i .$$

Obviously, together with $k_{ij}(x_{ij})$, also $g(x_{ij}) = k_{ij}(x_{ij}) + (V_i - V_j)\, x_{ij}$ is convex.

Applying lemma 1 and taking into consideration the conditions fixed in (3), the minimum value of $g(x_{ij})$ just results in the point $x_{ij} = \bar{x}_{ij}$. The convex function

$$\sum_{i,j=1}^{n} k_{ij}(x_{ij}),$$

however, assumes its minimum value when each summand is minimum. Since the summand $\sum_{i=1}^{n} V_i d_i$ is a constant, and thus without any influence on the situation of the minimum, the total costs become minimum when the conditions (3) of the theorem are fulfilled.

2. We show the necessity of the conditions (3) for the minimality of the total costs. Let $\{\bar{x}_{ij}\}$ be a minimum flow. In order to verify the necessity of (3), we have to define a suitable potential.

First, consider the subgraph $G'(\mathfrak{X}, \mathfrak{U}')$ of $G(\mathfrak{X}, \mathfrak{U})$ which has the same set of vertices as G, but the set $\mathfrak{U}' = \{(i, j) : k_{ij}^{+}(\bar{x}_{ij}) = k_{ij}^{-}(\bar{x}_{ij}) = k'_{ij}(\bar{x}_{ij})\}$ of edges (which depends on the minimum flow). We call G' the *support* of G. On each edge of G' there goes a flow $\neq 0$, since for $\bar{x}_{ij} = 0$ the expression $k_{ij}^{-}(\bar{x}_{ij})$ is not defined. According to our remark made above, the graph G' has assumed an orientation such that on each arc there goes a positive flow.

First, let us presuppose that G' is connected. Let G'' be a spanning tree of G'. Starting from any vertex i_0, to which the potential $\bar{V}_{i_0} = 0$ is assigned, we assign a potential to all other vertices in the following manner:

Let the value $\bar{V}_i$ be already assigned to vertex i. Then, we put

$$\left.\begin{aligned} \bar{V}_j &= \bar{V}_i + k'_{ij}(\bar{x}_{ij}) \quad \text{if} \quad x_{ij} > 0, \\ \bar{V}_j &= \bar{V}_i - k'_{ij}(\bar{x}_{ij}) \quad \text{if} \quad x_{ji} > 0. \end{aligned}\right\} \tag{4}$$

Thus, each vertex has been provided with a potential. We will show that the conditions (3) are fulfilled.

2.1. First, we show that the conditions (3) are true for the arcs of G', i.e., for those arcs of G for which the cost function at the minimum point is differentiable. Suppose that in G' there exists an arc (s, t) with $\bar{x}_{st} > 0$ and $\bar{V}_t - \bar{V}_s \neq k'_{st}$. In the spanning tree G'' the value $\overline{V}_s$ has been coordinated to s along a chain $(i_0, i_1, \ldots, i_m = s)$, and the value $\overline{V}_t$ to t along a chain $(i_0 = i_p, i_{p-1}, \ldots, i_{m+1} = t)$. Together with the arc $(s, t) = (i_m, i_{m+1})$ we get a closed sequence of arcs[1])

$$\mu = (i_0, i_1, \ldots, i_m = s, i_{m+1} = t, \ldots, i_p = i_0).$$

The flow going on each arc of μ is always positive (according to the orientation just given to the edges). Now, we modify the flows on the arcs of μ slightly and show that the new flow can be transported "with lower costs". On the arcs of μ^+ (there are arcs on which the flow goes in the same direction as the orientation of μ) we increase the fluxes by a value h which is still to be verified; on the arcs of μ^- we decrease the fluxes by the value h, and the fluxes on all the other arcs remain unchanged. Obviously, the modified node condition (2) continues to be satisfied when effecting this step. The flow modified in this way is denoted by $\{y_{ij}\}$.

Now, consider the cost difference Δ resulting between the old flow and the new one:

$$\Delta = \sum_{(i,j)\in\mu^+} [k_{ij}(\overline{x}_{ij}) - k_{ij}(\overline{x}_{ij} + h)] + \sum_{(i,j)\in\mu^-} [k_{ij}(\overline{x}_{ij}) - k_{ij}(\overline{x}_{ij} - h)].$$

Using the Taylor series, we get

$$\Delta = h \left[-\sum_{(i,j)\in\mu^+} k'_{ij}(\overline{x}_{ij}) + \sum_{(i,j)\in\mu^-} k'_{ij}(\overline{x}_{ij})\right] + o(h).$$

According to the formation rule for the $\overline{V}_i$, it results

$$\begin{aligned}\Delta &= h\big(\overline{V}_{i_0} - \overline{V}_{i_1} + \overline{V}_{i_1} - + \cdots + - \overline{V}_s - k'_{st}(\overline{x}_{st}) \\ &\quad + \overline{V}_t - + \cdots + - \overline{V}_{i_0}\big) + o(h) \\ &= h\big(\overline{V}_t - \overline{V}_s - k'_{st}(\overline{x}_{st})\big) + o(h)\end{aligned}$$

when traversing μ.

For sufficiently small $h > 0$, however, this expression is certainly greater than zero if $\overline{V}_t - \overline{V}_s > k'_{st}(\overline{x}_{st})$. But if $\overline{V}_t - \overline{V}_s < k'_{st}(\overline{x}_{st})$, we modify the flow as follows: On the arcs of μ^+ we decrease the flow by h, on that of μ^- we increase it by h. The resulting difference Δ may be estimated in an analogous way.

So we have shown that on the arcs of G' the relation $\overline{V}_t - \overline{V}_s = k'_{st}(\overline{x}_{st})$ is true. This is exactly the condition (3) for those points at which the cost function is differentiable.

2.2. In the following, we assume that in G there exists an arc (s, t) (not lying in the support of G') for which $k^+_{st}(\overline{x}_{st}) < \overline{V}_t - \overline{V}_s$ or $k^-_{st}(\overline{x}_{st}) > \overline{V}_t - \overline{V}_s$ for $\overline{x}_{st} > 0$ is valid.

Because G' was presupposed to be connected, there exist in G' chains from i_0 to s and from t to i_0 (cf. part 2.1 of the proof). Thus, we have a closed sequence

[1]) μ possesses an elementary cycle containing the arc (s, t). μ itself may also contain certain arcs twice, but then once in μ^+ and once in μ^-.

of arcs[1])

$$\mu = (i_0, i_1, \ldots, i_m = s, i_{m+1} = t, i_{m+2}, \ldots, i_p = i_0).$$

a) Let $k_{st}^+ < \overline{V}_t - \overline{V}_s$. We transform the minimum cost flow $\{\overline{x}_{ij}\}$ into a flow $\{y_{ij}\}$ along μ as follows:

$$y_{ij} = \begin{cases} \overline{x}_{ij} + h & \text{for} \quad (i, j) \in \mu^+, \\ \overline{x}_{ij} - h & \text{for} \quad (i, j) \in \mu^-, \\ \overline{x}_{ij} & \text{for} \quad (i, j) \notin \mu. \end{cases}$$

The value of h will be fixed later.

Now again, we estimate the cost difference. It holds:

$$\begin{aligned} \Delta &= \sum_{(i,j)\in\mu^+} [k_{ij}(\overline{x}_{ij}) - k_{ij}(\overline{x}_{ij} + h)] + \sum_{(i,j)\in\mu^-} [k_{ij}(\overline{x}_{ij}) - k_{ij}(\overline{x}_{ij} - h)] \\ &= h\Bigg[-\sum_{\substack{(i,j)\in\mu^+ \\ (i,j)\neq(s,t)}} k'_{ij}(\overline{x}_{ij}) + \sum_{(i,j)\in\mu^-} k'_{ij}(\overline{x}_{ij})\Bigg] + k_{st}(\overline{x}_{st}) - k_{st}(\overline{x}_{st} + h) + o(h) \\ &\geqq h\Bigg[-\sum_{\substack{(i,j)\in\mu^+ \\ (i,j)\neq(s,t)}} k'_{ij}(\overline{x}_{ij}) - k_{st}^+(\overline{x}_{st}) + \sum_{(i,j)\in\mu^-} k'_{ij}(\overline{x}_{ij})\Bigg] + o(h) \\ &= h[\overline{V}_{i_0} - \overline{V}_{i_1} + \overline{V}_{i_1} - + \cdots + - \overline{V}_s - k_{st}^+(\overline{x}_{st}) + \overline{V}_t - + \cdots + - \overline{V}_{i_0}] + o(h) \\ &= h[\overline{V}_t - \overline{V}_s - k_{st}^+(\overline{x}_{st})] + o(h) > 0, \end{aligned}$$

for a correspondingly small value of h, which contradicts the optimality of $\{\overline{x}_{ij}\}$. Case

b) where $k_{st}^- > \overline{V}_t - \overline{V}_s$ is to be treated analogously. The transformation of the flow $\overline{x}_{ij}$ into y_{ij} will be made as in case a).

In the case of an edge without flux proceed as in case a).

To conclude the proof of the theorem, we still have to examine the case in which the graph G' (the support of G) is not connected.

Suppose that G' is not connected. We form a graph G'' in the following way: Let the graph G' possess exactly q components. Choose now (any) vertex i_j in each of these components. We still introduce a vertex $n + 1$ and connect it with each of the vertices i_j by an arc $(n + 1, i_j)$. We decrease the intensities of the q vertices i_j by the value ε each and assign the intensity $q\varepsilon$ to the vertex $n + 1$. Furthermore, we imagine $\varepsilon > 0$ to be sufficiently small. We presuppose the costs on these new arcs to vanish identically, and consequently, also the derivatives. All other vertices, edges and costs remain unchanged. Let the graph resulting in this way be G''.

---

[1]) Cf. footnote p. 105.

If $\{\bar{x}_{ij}\}$ is an optimal flow in G, than a flow $\{y_{ij}\}$ is feasible in G'' if we put $y_{ij} = \bar{x}_{ij}$ for all arcs from G and $y_{n+1,i_j} = \varepsilon$ for the newly introduced arcs. The support (i.e., the set of those arcs for which $k^+ = k^- = k'$ is true) is connected in G''. For the continuity of the cost functions, we ensure that, in case of a sufficiently small ε, the costs of an optimal flow in G'' differ as slightly as required from the costs of an optimal flow in G.

Now, we perform these studies with the graph G'', with the potentials $\bar{V}_i$ for the vertices i_j, $n - 1$ being equated to zero. The potentials of the other vertices then result (along the arcs of the support) as described above.

EXAMPLE. First, we shall explain the simpler case by means of an example in which the cost functions are all differentiable. Consider the transportation task of Fig. 4.1. At each vertex there is space for two numbers, with the upper one indicating the strength of the point and the lower one the actual potential value. In addition to this, there are also two values assigned to the edges, with the first one (the upper value) indicating the cost function and the second the quantity of flow actually going through the edge.

In Fig. 4.1a, it can be seen that a connected support (these are, in case of differentiable cost functions, all those edges on which a non-zero flow is going) is obtained with the given feasible initial flow (indicated by the lower number on the edges). In this support, we have chosen a spanning tree (doubly drawn edges) and assigned a potential to the vertices according to the rule (4) along this spanning tree (the potentials are indicated by the lower number at each vertex). It can be recognized that, for example, the edge having the cost function x^2 and the flow quantity 4 does not satisfy the optimality condition (3). The potential difference on this edge should virtually be $k'_{ij}(x_{ij}) = 2 \cdot 4 = 8$, but in fact it is $21 - 3 = 18$. It is thus possible to increase the flow by simultaneously reducing the costs along the cycle μ drawn in. If it is increased by the value h, then the costs

$$3(4 + h) + (4 + h)^2 + 2(5 - h)^2 + (2 - h)$$

will result on the edges of μ.

The costs become minimum for $h = 5/3$. Figure 4.1b shows the improved flow. Here again, the vertices are provided with the new potentials. On the edge with the cost function $4x$ the optimality criterion is violated (although there is no flow going through this edge). The flow can be augmented in the direction of the arising cycle μ. The resulting costs amount to

$$4h + (3 - h)\,6 + (1 + h)^2 + \left(\frac{17}{3} - h\right)^2$$

for the arcs of μ and take their minimum for $h = 17/6$.

The reader may perform the improvements step by step by himself. The optimal transportation is represented in Fig. 4.1c.

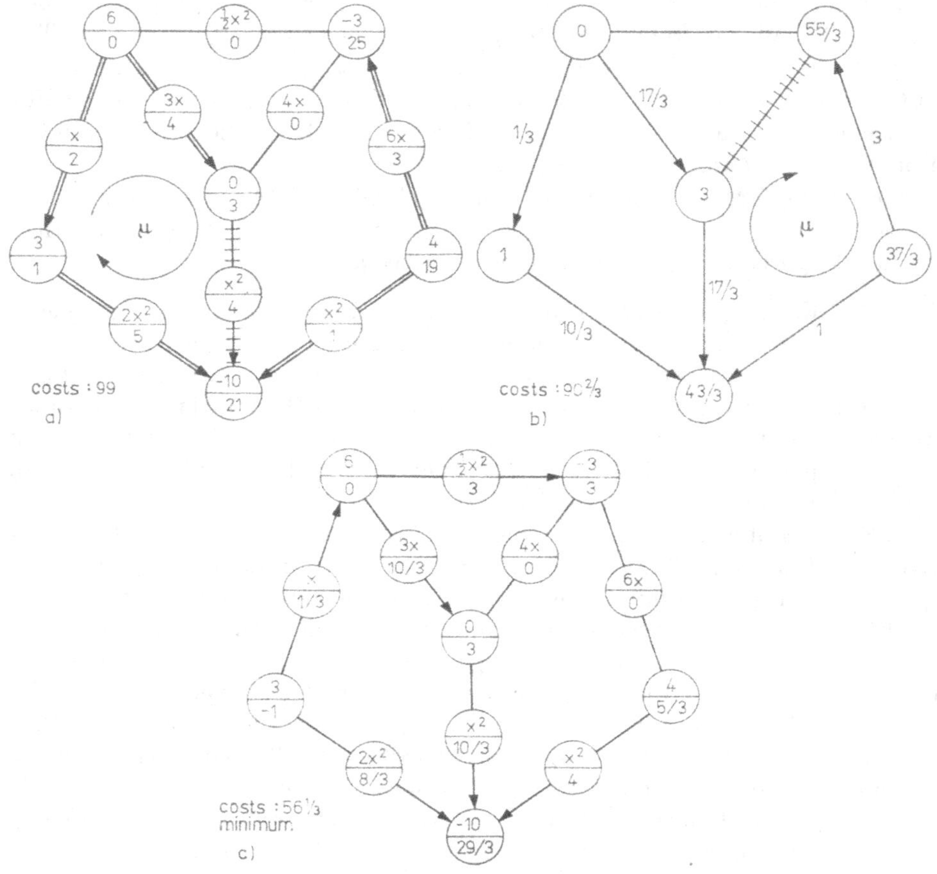

Fig. 4.1

We were not able to decide the question of the finiteness of the algorithm described above. It has been made clear that a relatively large number of steps was necessary, although all cost functions had been presupposed to be differentiable.

Now let us take another example without a connected support of the initial graph.

For this, consider Fig. 4.2a. The strength (need or quantity produced) is put in the vertices of the graph, the cost functions and a feasible initial flow are indicated at the edges. The support of the graph, which, of course; depends on the flow value, is marked by doubly drawn edges. The direction of the flow is marked by the orientation of the edges. It can be seen that the support is not connected, and therefore, we have to introduce an auxiliary point to which we assign a strength 2ε

since the support possesses two components. We connect the auxiliary point with an arbitrarily chosen vertex of each of the components. The strength of each of these two vertices is reduced by ε, and arcs going from the auxiliary point to each of the chosen vertices are introduced, with ε flowing on each of them. The transportation on them is free of charge. The result is represented in Fig. 4.2b. Now, we are able to assign a potential to each vertex along a support (in our case, this is a spanning tree).

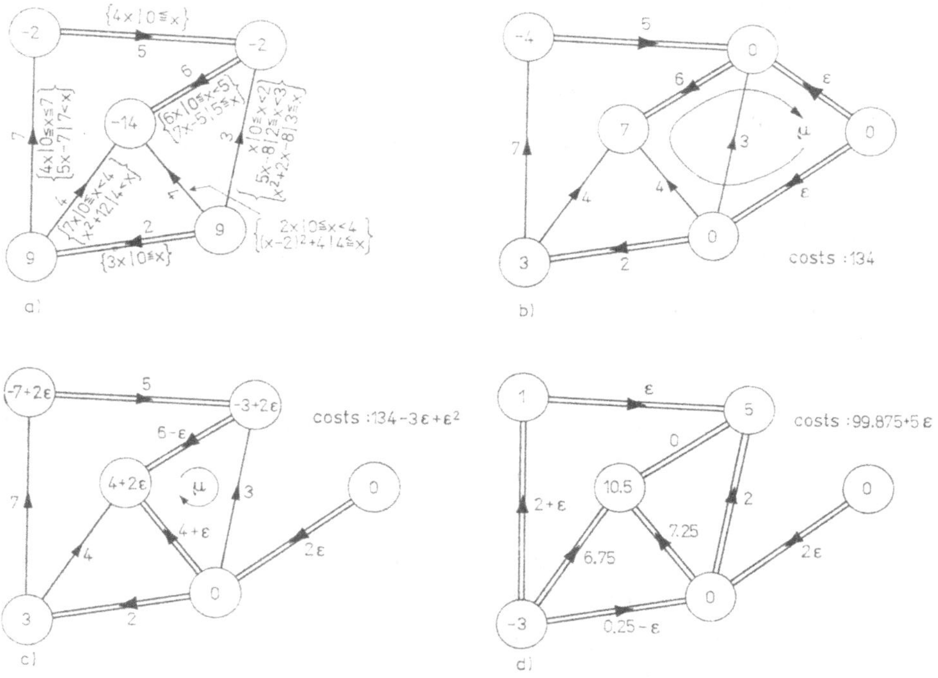

Fig. 4.2

We have drawn in a cycle μ along which the costs may be reduced, since for the arc of μ, which is not contained in the support, the optimality criterion (3) is violated.

The result of the improvement is indicated in Fig. 4.2c.

We recommend the reader to complete the algorithm. Figure 4.2d shows an optimal flow.

The doubly drawn edges in Figs. 4.2c and 4.2d satisfy the optimality criterion each time; the numbers indicated in the vertices stand for the potentials developing along a support of the graph.

In the course of the algorithm, it may happen that unconnected supports develop again. In such a case, an auxiliary point will have to be introduced. Once a connected support is ascertained in the course of the algorithm, than one can also immediately make ε tend to zero in order to omit the auxiliary point. Here, it may occur, however, that the new support is not connected any more. Thus, the support is not connected any more in the optimal graph (after ε has tended to zero), since the vertex with the potential 5 is incident with two edges on which no flow is going. In case of the third edge, the cost function is not differentiable, with the flow value being 2, nevertheless the optimality criterion is fulfilled for all arcs.

With this, we conclude our expositions on transportation problems involving convex cost functions.

If we assume the cost functions to be not only convex, but, in addition to this, also differentiable, the optimality criterion will be simplified as follows:

THEOREM 4.2. *Let the cost functions $k_{ij}(x_{ij})$ be convex and differentiable. A flow $\{x_{ij}\}$ is optimal if and only if there exists a potential $\{V_k\}$ such that*

$$V_j - V_i \leqq k_{ij}^{+}(x_{ij}) \quad \textit{for} \quad x_{ij} = 0$$

and

$$V_j - V_i = k_{ij}'(x_{ij}) \quad \textit{for} \quad x_{ij} > 0.$$

If we also suppose that a capacity restriction is imposed on each edge, i.e., that

$$0 \leqq x_{ij} \leqq c_{ij} \quad \text{for all} \quad (i, j) \in \mathfrak{U},$$

then there results as optimality criterion:

THEOREM 4.3. *Let the cost functions be convex and differentiable. Then, a flow $\{x_{ij}\}$ is optimal if and only if there exists a potential $\{V_k\}$ such that*

$$V_j - V_i \leqq k_{ij}^{+}(x_{ij}) \quad \textit{for} \quad x_{ij} = 0,$$

$$V_j - V_i = k_{ij}'(x_{ij}) \quad \textit{for} \quad 0 < x_{ij} < c_{ij},$$

$$V_j - V_i \geqq k_{ij}'(x_{ij}) \quad \textit{for} \quad x_{ij} = c_{ij}.$$

For the case where the cost functions are convex and piecewise linear, the so-called combined method offers itself. This method makes it possible to transform the nonlinear transportation task into a linear one. The interested reader's attention is drawn to the textbook of Ermolev and Mel'nik [2].

4.3. A MULTI-FLOW PROBLEM

Extending the flow problems discussed in chapter 1 (the algorithm of Ford and Fulkerson will be mentioned here in particular), this section will be devoted to the following:

Given a connected graph $G(\mathfrak{X}, \mathfrak{U})$ thought to be undirected. The vertices are numbered, and the edges are denoted by u_{ij}. If the graph is simple (and this assumption does not restrict our problem to be dealt with in the following), let $u_{ij} = (i, j)$ be true, that is, let u_{ij} be incident with i and j.

The flow problems discussed so far took into account only one single type of flow. The extension made will be in the sense that various types of flows are considered.

Thus, the kth type of flow is shortly called kth flow. For a fixed type of flow, say for the kth flow, we imagine exactly one source N_k in G from which kth flow will be sent to exactly one sink $N_{k'}$. By possibly introducing auxiliary points it can be arranged that for m types of flow there are exactly m sources and exactly m sinks available. By $F(k; k')$ we mean a kth flow which is sent from N_k to $N_{k'}$. The value of the flow $F(k; k')$, i.e., its amount, is denoted by $f(k; k')$. Let a real number $c_{ij} > 0$, i.e., the capacity, be assigned to each edge u_{ij} of G. Let $c_{ij} = c_{ji}$ be true (since G has been supposed to be undirected).

Let x_{ij}^k be the value of the kth flow in the edge u_{ij} going from i to j. Because no multiple edges have been admitted (and since loops are obviously useless), let

$$x_{ij}^k = -x_{ji}^k \quad \text{for all } k$$

(that means that positive kth flow is suitably sent through an edge in one direction only; but positive flow of differing types, however, can pass through an edge in the opposite direction).

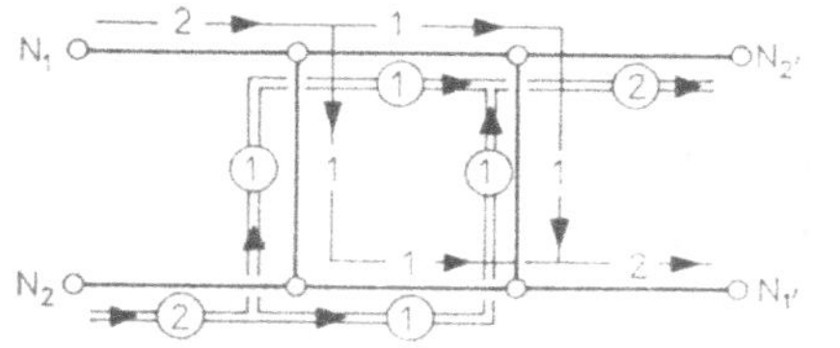

Fig. 4.3

For this consider Fig. 4.3. Let $c_{ij} = 2$ be true for each edge of the graph. The numbers which are circled indicate the respective value of the second flow, the other numbers to be found at the edges stand for the value of the first flow. It can be seen that for all those vertices which are neither sources nor sinks the Kirchhoff's node conditions are fulfilled for each type of flow. Since the source edges only have a capacity of 2 each, no type of flow can be of a value greater than 2. Thus, the flow distribution from Fig. 4.3 already yields a maximum flow, and the following relation holds:

$$f_{\max}(1; 1') = f_{\max}(2; 2') = 2.$$

The exact definition of what we understand by a maximum flow will still be given. As far as our example is concerned, no error can arise, since a maximum value of only 2 of each type can be transported.

Another flow distribution on the graph, however, entails a reduction of at least one of the two flow values, as can easily be seen.

For m types of flow, let the capacity restrictions

$$\sum_{k=1}^{m} |x_{ij}^k| \leqq c_{ij} \quad \text{for all} \quad u_{ij} \in \mathfrak{U}$$

be required.

The first problem which we are faced with is the following:

Given the capacities c_{ij} of the edges and also m non-negative numbers r_i ($i = 1, 2, \ldots, m$). Do then m types of flow $F(k; k')$ exist such that $f(k; k') = r_k$ is true for $k = 1, \ldots, m$ and

$$\sum_{k=1}^{m} |x_{ji}^k| \leqq c_{ij} \quad \text{holds for all} \quad u_{ij} \in \mathfrak{U}?$$

In this section, we shall consider almost exclusively the case $m = 2$. Although it could be shown that some properties also apply to any number m, there are no far-reaching properties of graphs known on which m different types of flow are to be transported, except for the case $m = 1, 2$.

DEFINITION. A set $\{v_{ij}\}$ of edges is called a *cut disconnecting the vertices* $\{N_1, N_2, \ldots, N_m\}$ *from the vertices* $\{N_1, N_2, \ldots, N_{m'}\}$ if

a) after removing the edges of v_{ij} from G, there exists no path from N_k to $N_{k'}$ for any k ($k = 1, 2, \ldots, m$) and

b) no proper subset of $\{v_{ij}\}$ satisfies condition a).

Note that for $i \neq j$, the vertices N_i and $N_{j'}$ may be contained in one component when the edges from $\{v_{ij}\}$ have been removed.

By the *capacity* of such a cut we mean the sum of the capacities of all edges of the cut.

A *minimum cut* is a cut of minimum capacity.

Let us denote a cut by $(1, \ldots, m \mid 1', \ldots, m')$ and its capacity by $c(1, \ldots, m \mid 1', \ldots, m')$.

LEMMA 1. *If a cut* $(1, \ldots, m \mid 1', \ldots, m')$ *is removed from a graph, then the resulting graph possesses* $m + 1$ *components at the most.*

Proof. Imagine the edges of the cut to be successively deleted from the graph G. Because a cut is required to be minimum (b), each edge lies on (at least) one path from N_i to $N_{i'}$ for (at least) one i ($1 \leqq i \leqq m$). If the number of the components increases when deleting, one after the other, the edges of the cut, then at least one pair N_i and $N_{i'}$ of vertices have been disconnected. That means, however, that $m + 1$ components at the most can result, since there exist exactly m pairs of such vertices. This proves lemma 1.

Now we turn to an estimation of the capacity of a minimum cut for the case $m = 2$. For this purpose, consider two particular operations.

Operation a): We identify the vertices N_1 and N_2 to form one point, and the vertices $N_{1'}$ and $N_{2'}$ to form another one. Let the resulting graph be G_a.

Operation b): We identify the vertices N_1 and $N_{2'}$ to form one point, and the vertices N_2 and $N_{1'}$ to form another one. Let the graph resulting in this way be G_b.

In these operations, multiple edges may develop, but on the other hand, edges may also get lost, as for example in a) if N_1 and N_2 or if $N_{1'}$ and $N_{2'}$ are connected by an edge. All other edges, however, remain unchanged like their capacities (cf. Fig. 4.4).

Possibly arising double edges may be replaced by one single edge the capacity of which is put equal to the sum of the capacities of the edges from which this edge has resulted. In Fig. 4.4b, the vertices N_1, N_2 and the vertices $N_{1'}$, $N_{2'}$ are

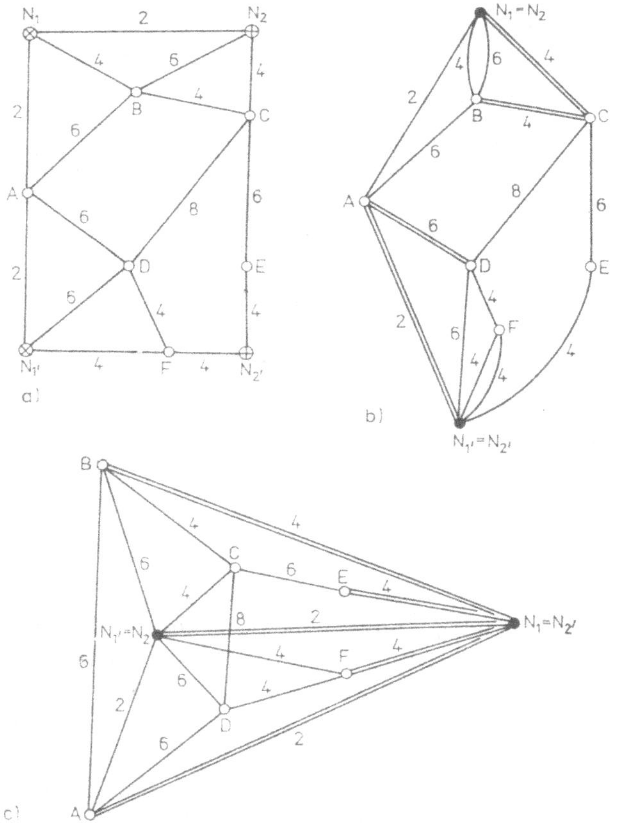

Fig. 4.4

identified (operation a)), whereas the operation b) has been carried out in case of the representations in Fig. 4.4c. In this latter case, a non-planar graph results, as can easily be verified (i.e., however the graph is drawn in the plane, edge crossings will result in every case).

We denote by $(1-2 \mid 1'-2')$ a cut in G_a which disconnects the vertex being formed by identifying N_1 and N_2 from that vertex ensuing from the identification of $N_{1'}$ and $N_{2'}$. Let $(1-2' \mid 1'-2)$ be a corresponding cut in G_b (in Figs. 4.4b and 4.4c, we have marked corresponding cuts by a doubly drawn edge). Now we can make a statement about minimum cuts in G, as far as we know the minimum cuts in G_a and G_b.

LEMMA 2. *Let $(1, 2 \mid 1', 2')$, $(1-2 \mid 1'-2')$, $(1-2' \mid 1'-2)$ be minimum cuts in G, G_a and G_b, respectively. Then, the following relation exists between their capacities:*

$$c(1, 2 \mid 1', 2') = \min [c(1-2 \mid 1'-2'), c(1'-2 \mid 1-2')].$$

Proof. If we remove a minimum cut $(1, 2 \mid 1', 2') = \mathfrak{S}$ from G, then the resulting graph splits into two or three components according to lemma 1.

1. Let $G-\mathfrak{S}$ consist of two components.

a) Let one of the components contain N_1 and N_2, and let the other one contain $N_{1'}$ and $N_{2'}$. Then, there will be contained in $G-\mathfrak{S}$ a path W_{12} going from N_1 to N_2 (since each component is connected), as well as a path W'_{12} (in the other component) connecting $N_{1'}$ with $N_{2'}$. In G_a let the set $\mathfrak{S}_a$ of edges correspond to the set $\mathfrak{S}$ of edges of G. This set $\mathfrak{S}_a$ of edges disconnects the vertex $(1-2)$, which has emerged from the identification of N_1 and N_2, from the vertex $(1'-2')$ resulting from the identification of $N_{1'}$ and $N_{2'}$. That is, $\mathfrak{S}_a$ contains a cut (or it is already the cut itself) disconnecting $(1-2)$ from $(1'-2')$. For the minimum cuts, however, this leads to the relation

$$c(1, 2 \mid 1', 2') \geqq c(1-2 \mid 1'-2').$$

Suppose the relation

$$c(1, 2 \mid 1', 2') > c(1-2 \mid 1'-2') \tag{1}$$

would be true.

Let $\mathfrak{T}_a$ be a minimum cut in G_a, and let $\mathfrak{T}$ be the set of edges corresponding to $\mathfrak{T}_a$ in G. According to (1), $\mathfrak{T}$ is no cut in G, thus, without restriction of generality, there exists in G a path W going from N_1 to $N_{1'}$ which contains no edge of $\mathfrak{T}$. Not all edges of W can have an edge in G_a corresponding to them, since otherwise $\mathfrak{T}_a$ would not be a cut of G_a. Without restriction of generality, W contains the dege (N_1, N_2) and perhaps also $(N_{1'}, N_{2'})$. To all other edges of W there corresponds in G_a a set $\mathfrak{W}_a$ of edges forming in G_a a path W_a connecting $(1-2)$ with $(1'-2')$. W_a, however, has no edge in common with $\mathfrak{T}_a$, which contradicts the property of $\mathfrak{T}_a$ of being a cut. Thus, we have shown that the relation

$$c(1, 2 \mid 1', 2') = c(1-2 \mid 1'-2')$$

is true.

b) Let one of the components contain N_1 and $N_{2'}$ and the other one $N_{1'}$ and N_2. By rewriting $N_2 \leftrightarrow N_{2'}$, we get case a). Then, it results

$$c(1, 2 \mid 1', 2') = c(1-2' \mid 1'-2).$$

That case a) is obtained, indeed, by merely rewriting the vertices becomes clear in the following way. Whether the flow of type 2 is sent from N_2 to $N_{2'}$ or vice versa, is of no importance because of the symmetry (undirected graph!).

2. Let $G-\mathfrak{S}$ have three components.

a) Let N_1 and N_2 be contained in one component, $N_{1'}$ in the second and $N_{2'}$ in the third component (Fig. 4.5). Obviously, the edges $\mathfrak{S}_a$ in G_a corresponding to the edges of $\mathfrak{S}$ disconnect the vertices (1—2) and (1'—2'), i.e., it holds:

$$c(1, 2 \mid 1', 2') \geqq c(1-2' \mid 1'-2).$$

Suppose the following relation would hold:

$$c(1, 2 \mid 1', 2') > c(1-2' \mid 1'-2).$$

Considering a minimum cut $\mathfrak{T}_a$ in G_a, we shall state that the reflections made in 1a) can easily be applied.

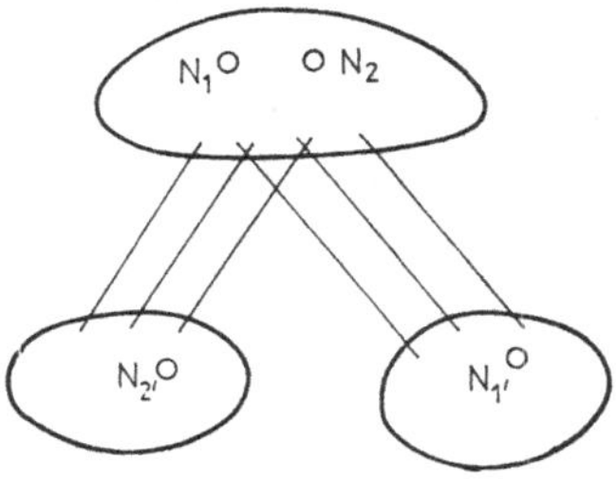

Fig. 4.5

b) Let the first component contain N_1 and $N_{2'}$, the second one N_2 and the third one $N_{1'}$. Rewriting $N_2 \leftrightarrow N_{2'}$ leads to case 2a), and it follows:

$$c(1, 2 \mid 1', 2') = c(1-2' \mid 1'-2).$$

Since in case of three resulting components in $\mathfrak{S}-G$ either of the cases 2a) or 2b) will occur, lemma 2 is finally proved.

Now, let $(1 \mid 1')$ and $(2 \mid 2')$ be minimum cuts in G disconnecting N_1 from N_1 and N_2 from $N_{2'}$, respectively (i.e., the minimum cuts disconnecting the source from the sink in the simple flow problem of Ford and Fulkerson). Then, the following lemma will hold.

LEMMA 3. *The relation* $c(1 \mid 1') + c(2 \mid 2') \geqq c(1, 2 \mid 1', 2')$ *holds.*

Proof. If we remove from G all edges of a cut disconnecting N_1 from $N_{1'}$ and, in addition, all edges of a cut disconnecting N_2 from $N_{2'}$ (here, some edges may be contained in both cuts, but they are counted only once in a cut $(1, 2 \mid 1', 2')$), then, of course, N_1 is disconnected from $N_{1'}$ and also N_2 from $N_{2'}$, which proves lemma 3.

THEOREM 4.4. *Two flows* $F(1;\, 1')$ *and* $F(2;\, 2')$ *can be realized in a graph at the same time if and only if*

a) $f(1;\, 1') \leqq c(1 \mid 1')$,

b) $f(2;\, 2') \leqq c(2 \mid 2')$,

c) $f(1;\, 1') + f(2;\, 2') \leqq c(1, 2 \mid 1', 2')$ *hold.*

Furthermore, the maximum of the sum of both types of flow is equal to the minimum of the capacities of all cuts which disconnect two pairs of vertices at a time (namely N_1 from $N_{1'}$ and N_2 from $N_{2'}$), i.e.,

$$\max \big(f(1;\, 1') + f(2;\, 2')\big) = \min c(1, 2 \mid 1', 2') = \min \big(c(1{-}2 \mid 1'{-}2'),\, c(1{-}2' \mid 1'{-}2)\big). \quad (2)$$

Proof. The necessity of the conditions a) to c) is immediately clear. The sufficiency of the conditions and the truth of the relation (2) are verified by means of a suitable algorithm.

To avoid mistakes, we denote in the following the vertices not only by natural numbers (i.e., by $1, 2, \ldots, i, j, k, 1', 2', \ldots$), but by writing $N_1, N_2, \ldots, N_i, N_j, N_k, N_{1'}, N_{2'}, \ldots$ By a *chain* from N_i to N_j we mean an edge sequence

$$u_{ii_1},\, u_{i_1i_2},\, \ldots,\, u_{i_{r-1}i_r},\, u_{i_rj}.$$

If none of the vertices incident with the edges of a chain occurs more than twice, then the chain is called *simple* or *path* going from N_i to N_j. By the *capacity of a path* W we mean the smallest of the edge capacities of W. A flow of the value x going on a path W from N_i to N_j is shortly called *path flow of* the *value* x. First, we perform labelling procedures similarly to the algorithm of Ford and Fulkerson described in chapter 1.

Define four classes of vertices $\mathfrak{X}_1, \mathfrak{X}_2, \mathfrak{X}_b, \mathfrak{X}_f$ (note that for each edge u_{ij} the relation $|x_{ij}^1| + |x_{ij}^2| \leqq c_{ij}$ is true):

1. $\mathfrak{X}_1$: N_1 belongs to $\mathfrak{X}_1$.

Let $N_i \in \mathfrak{X}_1$ and u_{ij} be an edge with $x_{ij}^1 + |x_{ij}^2| < c_{ij}$. Then also let $N_j \in \mathfrak{X}_1$. Together with N_i, N_j will also belong to $\mathfrak{X}_1$ if an amount of the first flow can still be sent from N_i to N_j without causing a change in the value or in the direction of the second flow.

2. $\mathfrak{X}_2$: N_2 belongs to $\mathfrak{X}_2$.

Let $N_i \in \mathfrak{X}_2$ and u_{ij} be an edge with $|x_{ij}^1| + x_{ij}^2 < c_{ij}$. Then also let $N_j \in \mathfrak{X}_2$. Together with N_i, N_j will also belong to $\mathfrak{X}_2$ if an amount of the second flow could still be sent from N_i to N_j without changing the value or the direction of the first one.

3. $\mathfrak{X}_b$: N_2 belongs to $\mathfrak{X}_b$.

Let $N_i \in \mathfrak{X}_b$ and u_{ij} be an edge with $x_{ij}^1 + x_{ij}^2 < c_{ij}$. Then also let $N_j \in \mathfrak{X}_b$. Together with N_i, N_j will also belong to $\mathfrak{X}_b$ if at least either of the two flows going from N_i to N_j can still be augmented.

4. $\mathfrak{X}_f$: N_2 belongs to $\mathfrak{X}_f$.

Let $N_i \in \mathfrak{X}_f$ and u_{ij} be an edge with $x_{ji}^1 + x_{ij}^2 < c_{ij}$. Then also let $N_j \in \mathfrak{X}_f$. N_j will also belong to $\mathfrak{X}_f$, together with N_i, if either second flow going from N_i to N_j or first flow going from N_j to N_i could be augmented.

It is immediately clear that, together with N_i, N_j also pertains to each (the same) of the four classes of vertices if one edge u_{ij} is not saturated (thus, if $|x_{ij}^1| + |x_{ij}^2| < c_{ij}$). That means that those edges are especially interesting of which the capacity is saturated. In the following, discussion will be of how to divert flows in a suitable manner in order to augment the sum of both flows.

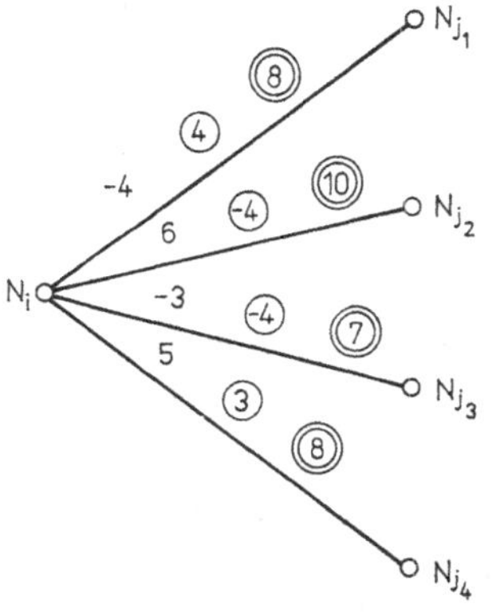

Fig. 4.6

Let us have another look at Fig. 4.6. The reader may prove that the following statement holds: Together with N_i also N_{j_1} and N_{j_3} are contained in $\mathfrak{X}_1$ if the doubly circled number indicates the capacity, the simply circled number is $x_{ij_k}^2$ and the non-circled number is $x_{ij_k}^1$.

Find out which of the vertices N_{j_k} $(k = 1, 2, 3, 4)$ lie in the classes $\mathfrak{X}_2$, $\mathfrak{X}_b$, $\mathfrak{X}_f$ together with N_i.

The reader will recognize without difficulty that the prescription for determining the set $\mathfrak{X}_1$ of vertices represents the labelling algorithm of Ford and Fulkerson (for the first flow) and the prescription for determining $\mathfrak{X}_2$ that for the second flow.

Assume $N_{2'} \in \mathfrak{X}_b$. Then there exists a labelling path $\boldsymbol{W}$ going from N_2 to $N_{2'}$, and let

$$\boldsymbol{W} = (N_2 = N_{i_1}, N_{i_2}, N_{i_3}, \ldots, N_{i_s} = N_{2'}).$$

Imagine each edge $(N_{i_j}, N_{i_{j+1}})$ with $j = 1, \ldots, s-1$ to be oriented from N_{i_j} to $N_{i_{j+1}}$. For each of these edges $u_{i_j i_{j+1}}$ it holds:

$$x^1_{i_j i_{j+1}} + x^2_{i_j i_{j+1}} < c_{i_j i_{j+1}}.$$

Such a $\boldsymbol{W}$ is called a *backward path*. Correspondingly, a path $\boldsymbol{W}$ is called a *forward path* if $N_{2'}$ is contained in $\mathfrak{X}_f$ and if $\boldsymbol{W}$ is a labelling path from N_2 to $N_{2'}$ so that with $\boldsymbol{W} = (N_2 = N_{k_1}, N_{k_2}, \ldots, N_{k_l} = N_{2'})$ for each edge $u_{k_j k_{j+1}}$ $(j = 1, \ldots, l-1)$ of $\boldsymbol{W}$ the relation

$$x^1_{k_{j+1} k_j} + x^2_{k_j k_{j+1}} < c_{k_j k_{j+1}}$$

is valid. If there exist in $\boldsymbol{G}$ both a forward path and a backward path, then we say that $\boldsymbol{G}$ contains a *double path*.

Let $\overline{\mathfrak{X}}_b$ be the set of those vertices of $\boldsymbol{G}$ not contained in $\mathfrak{X}_b$, and let $\overline{\mathfrak{X}}_1$, $\overline{\mathfrak{X}}_2$, $\overline{\mathfrak{X}}_f$, be defined correspondingly.

LEMMA 4. *It holds $N_{2'} \in \overline{\mathfrak{X}}_b$ if and only if there exists no backward path.*

Proof.

1. Let $N_{2'} \in \overline{\mathfrak{X}}_b$ be true. Consider the set $\mathfrak{S}$ of the edges $u_{ij} = (N_i, N_j)$ with $N_i \in \mathfrak{X}_b$ and $N_j \in \overline{\mathfrak{X}}_b$. For each of these edges, it holds

$$x^1_{ij} + x^2_{ij} = c_{ij}.$$

Since $\mathfrak{S}$ disconnects the vertices N_2 and $N_{2'}$ from each other, and since on each path between these two vertices at least one edge of $\mathfrak{S}$ can be found, there is no backward path.

2. Assume there exists no backward path and $N_{2'} \in \mathfrak{X}_b$. Then, there exists a chain and thus a path $\boldsymbol{W} = (N_2 = N_{i_1}, N_{i_2}, \ldots, N_{i_k} = N_{2'})$ going from N_2 to $N_{2'}$, and for each edge $u_{i_j i_{j+1}}$ it holds:

$$x^1_{i_j i_{j+1}} + x^2_{i_j i_{j+1}} < c_{i_j i_{j+1}}.$$

In this case, however, these edges form a backward path, which contradicts the assumption.

This proves lemma 4.

We can prove the following lemma quite analogously.

LEMMA 5. *It holds $N_{2'} \in \mathfrak{X}_f$ if and only if there exists no forward path.*

LEMMA 6. *Let $N_{2'} \in \overline{\mathfrak{X}}_b$ be true and $\mathfrak{S}$ be the set of all edges (X, Y) with $X \in \mathfrak{X}_b$ and $Y \in \overline{\mathfrak{X}}_b$. If there exists an edge $u_{ij} \in \mathfrak{S}$ with $x_{ij}^1 \neq 0$, then it holds $N_1 \in \mathfrak{X}_b$ and $N_{1'} \in \overline{\mathfrak{X}}_b$.*

In other words: *A set $\mathfrak{S}$ of edges through which goes an amount of first flow directed in any way disconnects N_1 and N_2 from $N_{2'}$ and $N_{1'}$.*

Proof.

1. Suppose that it would hold $N_1 \in \overline{\mathfrak{X}}_b$, $N_{1'} \in \mathfrak{X}_b$. Since an amount of first flow goes (from N_1 to $N_{1'}$), there is an edge $u_{ij} \in \mathfrak{S}$ with $N_j \in \overline{\mathfrak{X}}_b$, $N_i \in \mathfrak{X}_b$ and $x_{ji}^1 = -x_{ij}^1 > 0$ (cf. Fig. 4.7). According to the definition of the set $\mathfrak{X}_b$, it holds $x_{ij}^1 + x_{ij}^2 = c_{ij}$, thus

$$x_{ij}^2 = c_{ij} - x_{ij}^1 = c_{ij} + x_{ji}^1 \geqq c_{ij}.$$

But since the relation $x_{ij}^2 \leqq c_{ij}$ is equally true because of the capacity restriction, we get $x_{ij}^2 = c_{ij}$, thus $x_{ij}^1 = 0$, which contradicts the assumption $x_{ij}^1 \neq 0$.

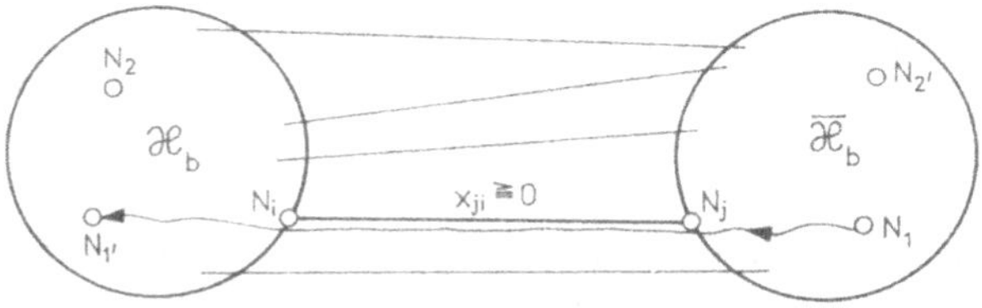

Fig. 4.7

2. Let $N_1, N_{1'} \in \mathfrak{X}_b$ or $N_1, N_{1'} \in \overline{\mathfrak{X}}_b$ be true. Choose any edge $u_{ij} \in \mathfrak{S}$ with $N_i \in \mathfrak{X}_b$ and $N_j \in \overline{\mathfrak{X}}_b$, and let $x_{ji}^1 > 0$ be true without restriction of generality. Then, one gets

$$x_{ij}^2 = c_{ij} - x_{ij}^1 = c_{ij} + x_{ji}^1 > c_{ij}$$

which yields a contradiction to the relation $x_{ij}^2 \leqq c_{ij}$.

Thus, lemma 6 is proved.

The following proposition can be verified quite analogously:

LEMMA 7. *Let $N_{2'} \in \overline{\mathfrak{X}}_f$ be true and let $\mathfrak{S}$ be the set of all edges (X, Y) with $X \in \mathfrak{X}_f$ and $Y \in \overline{\mathfrak{X}}_f$. If there exists an edge u_{ij} with $x_{ij}^1 \neq 0$, then it holds $N_{1'} \in \mathfrak{X}_f$ and $N_1 \in \overline{\mathfrak{X}}_f$.*

In other words: *A set $\mathfrak{S}$ of edges through which goes an amount of first flow directed in any way disconnects $N_{1'}$ and N_2 from N_1 and $N_{2'}$.*

The next lemma follows directly from lemmata 4 and 5.

LEMMA 8. *A double path* (i.e., *both a forward and a backward path from N_2 to $N_{2'}$*) *exists if and only if neither $N_{2'} \in \overline{\mathfrak{X}}_f$ nor $N_{2'} \in \overline{\mathfrak{X}}_b$ is true.*

Now, we give an algorithm which makes it possible to decide whether or not flows $F(1;1')$ and $F(2;2')$ are realizable in a graph. If so, we shall construct these flows. This algorithm also makes it possible to construct such flows that $f(1;1') + f(2;2')$ becomes maximal.

For all c_{ij} we suppose the character of being even-numbered and we are thus able, as we shall see, to assure the integrality of the particular flows. For any real $c_{ij} > 0$, Hu [5] writes that the theorem can be proved as well, but this would be very troublesome. Supposing the character of being even-numbered certainly represents no restriction for machine computation.

We set ourselves the task of constructing flows of the values $r_1 = f(1;1')$ and $r_2 = f(2;2')$, with r_1 and r_2 being assumed to be even numbers.

Algorithm for finding a two-type flow

(i) Using the algorithm of Ford and Fulkerson (chapter 1), starting out from the feasible flow of the value zero, augment successively the value of the first flow up to the value r_1. If $r_1 \leqq c(1 \mid 1')$, this task can be fulfilled, otherwise the condition of theorem 4.4, which we wish to prove, is not satisfied.

If it is required to maximize the sum of the two flows, then maximize first $F(1;1')$. Since all c_{ij} are even numbers, the maximum flow of type 1 will also be even-numbered.

(ii) If possible, construct a flow of type 2 of the value r_2, by using the algorithm of Ford and Fulkerson, on this same graph with the reduced capacities $c'_{ij} = c_{ij} - |x^1_{ij}|$; in case of a maximization of the sum of both flows, construct a maximum flow of type 2. If, however, the flow of type 2 to be constructed in this way does not reach the desired value r_2, then go over to (iii) just as in case of a maximization of the sum of both flows.

(iii) Search for a double path. If none exists (i.e., if no forward and backward path can be found at the same time), the task is unsolvable or the sum of the flows is already maximum.

(iv) If a forward and a backward path exist, then proceed as follows: On the forward path,

flow 1 is decreased by 1 on each edge,

flow 2 is augmented by 1 on each edge.

On the backward path,

flow 1 is augmented by 1 on each edge,

flow 2 is augmented by 1 on each edge.

The steps (ii), (iii), (iv) are repeated as often as either the desired flow value r_2 is reached (the flow value of the first flow remains unchanged!) or the sum of the flows cannot be augmented any more. In the latter case, a flow distribution has been obtained such that the sum of both types is maximum.

Before proving all our assertions set up, we want to explain the algorithm by means of an example.

For this, take the graph represented in Fig. 4.8 (it is the same as in Fig. 4.4). Let the first number on each edge indicate its capacity, the second number will indicate the value of the first flow (obviously, this value is maximum) and the third the value of the second flow. The orientation given in Fig. 4.8a has been chosen so that the first flows are all non-negative. Now, determine a maximum flow of type 2 on the graph with reduced capacities. The result of this procedure is illustrated in Fig. 4.8b. We have now reached a total flow of the value $8 + 4 = 12$. Step (ii) is just finished so that we can go over to step (iii) in searching for a double

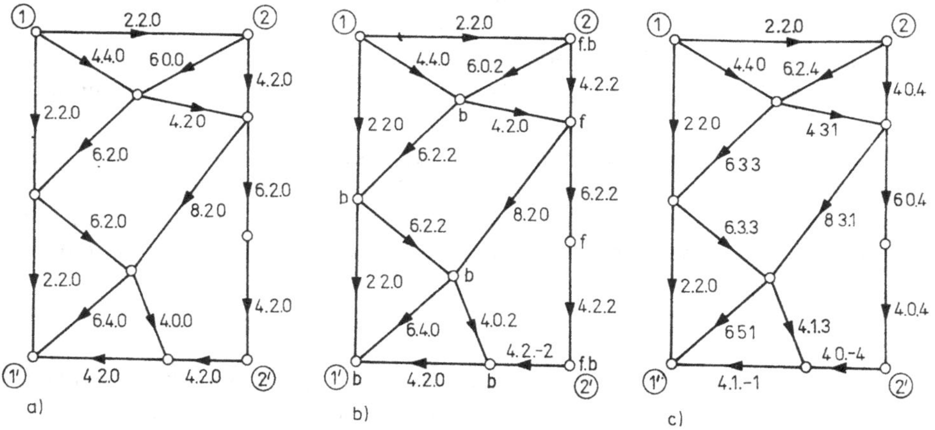

Fig. 4.8

path. In Fig. 4.8b, labellings f and b have been made on certain vertices according to whether they lie in $\mathfrak{X}_f$ or in $\mathfrak{X}_b$, respectively. Of course, it is also possible for vertices to have both labels. Now, step (iv) is to be effected. A repeated application of (iii) and (iv) yields the graph represented in Fig. 4.8c. The sum of the flows amounts to $8 + 8 = 16$ in this case. This sum is maximum, too, as can easily be seen. After step (iv) has been carried out, it may occur that step (ii) has to be performed again.

Our next task is to prove the following assertions:

a) *After step* (iv) *has been performed, the relation*

$$|x_{ij}^1| + |x_{ij}^2| \leqq c_{ij}$$

holds for each edge u_{ij}.

b) *If the algorithm terminates, we have obtained a maximum sum of the flows (or a first flow of the value* r_1 *and a second one of the value* r_2*), or the required values* r_1 *and* r_2 *can never be reached.*

c) *Both flows have integer values on each edge.*

d) *The algorithm is finite.*

First, we prove c). After having carried out the steps (i) and (ii), the statement is obviously satisfied. Imagining step (iv) to be performed, each type of flow is then augmented by 1 on the backward path and thus remains integer, whereas on the forward path, the first flow is decreased by 1 and the second one augmented by 1, both remaining integer in this way.

A special characteristic to be recognized is that, when determining the maximum sum, the sum of both flows on each edge is even-numbered (this is the case even if the character of being even-numbered is required for r_1). The remaining capacity on each edge will also be even-numbered, since the sum of the flows is even (and since also the c_{ij} have been supposed to be even). Thus, in the case of a possible application of step (ii), the flow of type 2 will be modified by the value 2 on each edge. This means, however, that the sum of the flows remains even-numbered.

Proof of d). Since carrying out step (iv) will not change the value of the first flow, but will cause an increase in the second flow by 2, the algorithm terminates in case the capacities are assumed to be finite. If there is no finite cut disconnecting N_1 from $N_{1'}$ and N_2 from $N_{2'}$, then the flow can become as high as required.

Proof of a). As we have seen, the relation

$$|x_{ij}^1| + |x_{ij}^2| \leqq c_{ij} - 2$$

holds (even in the case of r_1 and r_2 being even-numbered) on edges of a forward path or on edges of a backward path on which the capacities are not saturated. Thus, we can change the flow values by 1 in each case without exceeding the capacity. Of special interest therefore are those cases in which the capacities are saturated, i.e., in which $|x_{ij}^1| + |x_{ij}^2| = c_{ij}$ is true.

Because of the relation

$$-x_{ij}^1 + x_{ij}^2 < c_{ij}$$

the following cases may occur on a forward path:

1. $x_{ij}^1 \geqq 0$, $x_{ij}^2 \geqq 0$. Since all flow values are integers, we get after having performed step (iv):

$$|x_{ij}^1 - 1| + |x_{ij}^2 + 1| = c_{ij}.$$

2. $x_{ij}^1 \geqq 0$, $x_{ij}^2 \leqq 0$. Here, in at least one of the two inequalities there is no sign of equality, since no flow would then go through this edge, i.e., the capacity of this edge would not be saturated. After step (iv), it follows

$$|x_{ij}^1 - 1| + |x_{ij}^2 + 1| \leqq |x_{ij}^1| + |x_{ij}^2| = c_{ij}$$

and for the case $x_{ij}^1 > 0$, $x_{ij}^2 < 0$, even an inequality develops leading to step (ii) to be possibly carried out.

3. $x^1_{ij} < 0$, $x^2_{ij} < 0$. Then, it holds

$$|x^1_{ij} - 1| + |x^2_{ij} + 1| = |x^1_{ij}| + 1 + |x^2_{ij}| - 1 = c_{ij}.$$

Because of $|x^1_{ij}| + |x^2_{ij}| = c_{ij}$ and

$$x^1_{ij} + x^2_{ij} < c_{ij}$$

the following three cases are possible on a backward path:

1. $x^1_{ij} \geqq 0$, $x^2_{ij} < 0$. According to (iv), it results

$$|x^1_{ij} + 1| + |x^2_{ij} + 1| = |x^1_{ij}| + 1 + |x^2_{ij}| - 1 = c_{ij}.$$

2. $x^1_{ij} < 0$, $x^2_{ij} \geqq 0$. Then, it holds

$$|x^1_{ij} + 1| + |x^2_{ij} + 1| = |x^1_{ij}| - 1 + |x^2_{ij}| + 1 = c_{ij}.$$

3. $x^1_{ij} < 0$, $x^2_{ij} < 0$. We get

$$|x^1_{ij} + 1| + |x^2_{ij} + 1| = |x^1_{ij}| - 1 + |x^2_{ij}| - 1 = c_{ij} - 2.$$

In this case as well, the edge is not saturated any more, and it may happen that step (ii) has to be carried out.

Proof of b): If we have flow values of r_1 and r_2 for the flows 1 and 2, respectively, when breaking off the algorithm, then nothing is to be shown.

The algorithm terminates if neither (ii) nor (iv) is applicable. Since in case of the applicability of (ii), also (iv) would be applicable (in that case, however, there exists a path from N_2 to $N_{2'}$ on which no edge has a saturated capacity; but this path is both a forward and also a backward path and is thus a double path; therefore (iv) is applicable), we may assume that there exists no double path when terminating the algorithm. Thus, it holds either $N_{2'} \in \overline{\mathfrak{X}}_f$ or $N_{2'} \in \overline{\mathfrak{X}}_b$.

1. $N_{2'} \in \overline{\mathfrak{X}}_b$. Consider the set $\mathfrak{S}$ of all edges u_{ij} with $N_i \in \mathfrak{X}_b$ and $N_j \in \overline{\mathfrak{X}}_b$. It holds

$$x^1_{ij} + x^2_{ij} = c_{ij},$$

i.e., $x^1_{ij} \geqq 0$ and $x^2_{ij} \geqq 0$ (since it is not possible to send through u_{ij} an amount of any type of flow greater then c_{ij}). If $x^1_{ij} > 0$, then, because of lemma 6, it follows that $N_1 \in \mathfrak{X}_b$ and $N_{1'} \in \overline{\mathfrak{X}}_b$. Increasing one of the two types of flow going through the edge u_{ij} (and thus, aiming at increasing the sum of the flows) leads to a diminution of the other type of flow by this very amount (and thus, to no increase in the sum). If $x^1_{ij} = 0$, the second flow will saturate the capacity c_{ij}, and this type of flow may not be augmented on u_{ij} at all (not even at the cost of the first flow). If one has first flow of the value r_1, but not yet second flow of the value r_2, then second flow may be increased only by diminishing the first flow.

That means that the demand made on the values of the two types of flow cannot be satisfied.

2. $N_{2'} \in \overline{\mathfrak{X}}_f$. Consider the set $\mathfrak{S}$ of all edges u_{ij} with $N_i \in \mathfrak{X}_f$ and $N_j \in \overline{\mathfrak{X}}_f$. Then, it holds

$$-x_{ij}^1 + x_{ij}^2 = c_{ij}.$$

Thus, a non-negative flow of type 1 is flowing from N_j to N_i.

If $x_{ji}^1 = 0$, the edge u_{ij} is saturated with second flow, and no greater amount of second flow may pass through this edge. If the value r_2 for the second flow is not yet reached, then it will not be possible to reach it by further using u_{ij}.

But if $x_{ji}^1 > 0$, then, because of lemma 7, it holds $N_{1'} \in \mathfrak{X}_f$ and $N_1 \in \overline{\mathfrak{X}}_f$. Increasing the first flow thus brings about an equal reduction of the second flow, and vice versa. In any case, the sum of the two flows cannot be further increased.

Thereby, we have shown how a flow distribution can be found such that the sum of both flows is maximum.

Using lemma 2 and including the considerations made last, one can also see that the conditions a) to c) of theorem 4.4 are in fact sufficient for assuring flows $F(1; 1')$ and $F(2; 2')$.

The proof of this statement may be left to the reader.

It can be seen without difficulty that, when searching for a maximum flow (i.e., for a flow distribution in the case of which the sum of the two flows is maximum), the second type of flow can be increased, if need be, also by more than one unit. The reader who has to deal with this task will find out, after a little practice, by how much the flow can be augmented. Writing down the formulae and proving them turns out to be rather complicated. We think that the algorithm given by us in this form is also complicated enough.

It can easily be seen that the case in which, say, N_1 coincides with one of the other points, say with N_2, can be treated as a special case by suitably introducing an auxiliary vertex and an edge of a sufficiently high capacity.

CONCLUSIONS. A conjecture that an m-type flow $F(1, \ldots, m; 1', \ldots, m')$ exists if and only if the $2^m - 1$ inequalities of the form

$$f(k; k') \leqq c(k \mid k'), \ldots \qquad \binom{m}{1} \text{ inequalities,}$$

$$f(i; i') + f(j; j') \leqq c(i, j \mid i', j'), \ldots \binom{m}{2} \text{ inequalities,}$$

. . . .

are fulfilled, turns out to be wrong, as is shown by the example of Fig. 4.9. It can

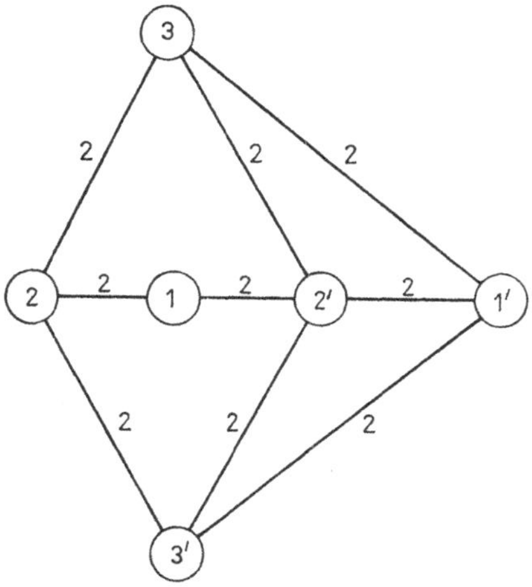

Fig. 4.9

be verified without difficulty that it holds

$$c(1 \mid 1') = 4, \quad c(1, 2 \mid 1', 2') = 6,$$
$$c(2 \mid 2') = 6, \quad c(1, 3 \mid 1', 3') = 6,$$
$$c(3 \mid 3') = 6, \quad c(2, 3 \mid 2', 3') = 8,$$
$$c(1, 2, 3 \mid 1', 2', 3') = 8.$$

A flow of the value

$$f(1; 1') = 4,$$
$$f(2; 2') = 2,$$
$$f(3; 3') = 1$$

satisfies, however, the conditions mentioned above, but cannot be realized.

As far as we know, maximum flows in case of more than two types of flow can only be found by the use of optimization methods, i.e., by means of algorithms in the programs of which only little of graphs can be recognized. These algorithms thus become rather sophisticated.

4.4. BIBLIOGRAPHY

[1] Busacker, R. G., and T. L. Saaty: Finite Graphs and Networks, An Introduction with Applications, New York 1965 (German: Munich/Vienna 1968; Russian: Moscow 1974).

[2] Ermolev, Ju. M., and I. M. Mel'nik: Extremal Problems on Graphs [Russian], Kiev 1968.

[3] Ford, L. R., and D. R. Fulkerson: Flows in Networks, Princeton, N.J., 1962 (Russian: Moscow 1966).

[4] Hu, T. C.: Multi-commodity networks flows, Operations Res. **11** (1963), 344–360.

[5] Hu, T. C.: Integer Programming and Network Flows, Reading, Mass., 1970 (German: Munich/Vienna 1972; Russian: Moscow 1974).

Chapter 5

Communication and supply networks

5.1. FORMULATION OF THE PROBLEM

In contrast with the preceding chapters, where the graphs were fixed, this one is concerned with finding a suitable graph which serves to solve either a supply or a communication problem with costs as low as possible.

Let us first set up a mathematical formulation of the *supply problem.*

Given two finite sets of points of the Euclidean plane, namely the set $\mathfrak{P} = \{X_1, X_2, \ldots, X_m\}$ of the producers (sources) and the set $\mathfrak{C} = \{X_{m+1}, X_{m+2}, \ldots, X_{m+n}\}$ of the consumers (sinks). Both sets form together the set $\mathfrak{X}$ of the pairwise distinct fixed points: $\mathfrak{X} = \mathfrak{P} \cup \mathfrak{C}$. Let the producer X_μ produce exactly a_μ units of a well-defined material per time unit ($a_\mu > 0$, integer; $\mu = 1, 2, \ldots, m$). Here, a_μ will be called the *yield* of the producer X_μ. Let the consumer $X_{m+\nu}$ have a *need* of b_ν units of this very material per time unit ($b_\nu > 0$, integer; $\nu = 1, 2, \ldots, n$). Assume the total demand may be satisfied by the total yield, i.e., let

$$\sum_{\mu=1}^{m} a_\mu \geqq \sum_{\nu=1}^{n} b_\nu .$$

Further, let $k(y)$ be a defined function for $y = 0, 1, 2, \ldots$ the so-called *cost function* with the properties:

$$k(0) = 0, \quad k(y) > 0 \text{ for } y > 0, \quad k(y+1) \geqq k(y),$$

$$k(x+y) \leqq k(x) + k(y) \quad \text{for} \quad x, y = 0, 1, 2, \ldots$$

Let the installation of a supply channel of the length l, along which y material units at the most may be transported per time unit, entail the costs $l \cdot k(y)$. By a *supply network* $\boldsymbol{N} = (\mathfrak{X}^*, \mathfrak{U}; c)$ we mean the geometric realization of a finite directed graph with the set $\mathfrak{X}^*$ of vertices and the set $\mathfrak{U}$ of arcs. To each $u \in \mathfrak{U}$ there is assigned a non-negative integer $y = c(u)$, called capacity of u. We explain $c(u)$ as maximally possible flux on u with the dimension "material units per time unit". In a supply network, each arc u has a positive geometric length $l(u)$.

Consider the set $\mathfrak{N}^*$ of all supply networks assuring the supply of the consumers $X_{m+1}, X_{m+2}, \ldots, X_{m+n}$ (that is, the simultaneous supply of each of the $X_{m+\nu}$ with the required quantity b_ν) by the producers $X_1, X_2, \ldots, X_m$. For these networks of $\mathfrak{N}^*$, especially $\mathfrak{X} \subseteqq \mathfrak{X}^*$ holds.

The total costs caused by installing a supply network $\boldsymbol{N} \in \mathfrak{N}^*$ are denoted by $K = K(\boldsymbol{N})$. Then, it holds

$$K(\boldsymbol{N}) = \sum_{u \in \mathfrak{U}} k\big(c(u)\big) \cdot l(u) \quad \text{with} \quad \boldsymbol{N} = (\mathfrak{X}^*, \mathfrak{U}, c).$$

Among the points of the set $\mathfrak{X}^*$ we distinguish between the *fixed points* (these form the set $\mathfrak{X}$) and those branching points which are not fixed points at the same time. The last-mentioned points are called *Steiner's points* (after Jakob Steiner, 1796—1863) and are put together to form the set $\mathfrak{S}$. Thus, it holds $\mathfrak{X} \cap \mathfrak{S} = \emptyset$ and $\mathfrak{X} \cup \mathfrak{S} = \mathfrak{X}^*$.

Let $\mathfrak{N}$ denote the set of all supply networks $\boldsymbol{N} \in \mathfrak{N}^*$ which possess no Steiner's points (thus, for these networks it holds $\mathfrak{X} = \mathfrak{X}^*$, and no branching point of the supply flow is allowed to occur beyond fixed points). According to whether we allow all networks of $\mathfrak{N}^*$ or only those of $\mathfrak{N}$ to compete, we get the first or the second problem formulated in the following.

PROBLEM A (Unrestricted problem of the cheapest supply network). Among all supply networks $\boldsymbol{N} \in \mathfrak{N}^*$ find those which cause the lowest total costs of installation (let the Steiner's points cause no costs).

PROBLEM B (Restricted problem of the cheapest supply network). Among all supply networks $\boldsymbol{N} \in \mathfrak{N}$ find those which cause the lowest total costs of installation.

We subsequently come to the formulation of the *communication problem.*

Given a finite set $\mathfrak{X} = \{X_1, X_2, \ldots, X_n\}$ of points of the Euclidean plane. These points are to be connected by a (generally connected) network (like, for example, a telephone network). Let the number x_{ij} of the telephone connections be fixed for each couple of points X_i, X_j, with $x_{ij} = x_{ji}$ and $x_{ii} = 0\ (i, j = 1, 2, \ldots, n)$ being supposed. Let $k(y)$ be a defined function (the so-called *cost function*) for $y = 0, 1, 2, \ldots$ with the properties

$$k(0) = 0, \quad k(y) > 0 \text{ for } y > 0, \quad k(y+1) \geqq k(y),$$

$$k(x+y) \leqq k(x) + k(y) \quad \text{for} \quad x, y = 0, 1, 2, \ldots$$

Let the overall laying of x connections (in a ditch) of the length l involve the costs $k(x) \cdot l$. By a *communication network* $\boldsymbol{N} = (\mathfrak{X}^*, \mathfrak{U}, c)$ we mean the geometric realization of a finite undirected graph consisting of the set $\mathfrak{X}^*$ of vertices and the set $\mathfrak{U}$ of edges on the edges u of which a capacity c is defined which assigns to each edge u a non-negative integer $y = c(u)$. Here, $c(u)$ stands for the maximum number of the lines that can be laid along u. In a communication network, each edge is of positive geometric length $l(u)$.

Consider the set $\mathfrak{N}^*$ of all communication networks assuring each of the numbers of connections x_{ij}. For these networks, especially $\mathfrak{X} \subseteq \mathfrak{X}^*$ holds. The total costs caused by the installation of a communication network $\boldsymbol{N} \in \mathfrak{N}^*$ are denoted by $K = K(\boldsymbol{N})$. Then, it holds

$$K(\boldsymbol{N}) = \sum_{u \in \mathfrak{U}} k(c(u)) \cdot l(u) \quad \text{with} \quad \boldsymbol{N} = (\mathfrak{X}^*, \mathfrak{U}, c).$$

Among the points of the set $\mathfrak{X}^*$ we distinguish between the fixed points (these form the set $\mathfrak{X}$) and those branching points which are not fixed points at the same time (as in the case of the supply networks, these points are called *Steiner's points*). These Steiner's points are put together to form the set $\mathfrak{S}$. Thus, it holds $\mathfrak{X} \cap \mathfrak{S} = \emptyset$, $\mathfrak{X} \cup \mathfrak{S} = \mathfrak{X}^*$.

Let $\mathfrak{N}$ denote the set of all communication networks $\boldsymbol{N} \in \mathfrak{N}^*$ which do not contain any Steiner's points (it holds $\mathfrak{X} = \mathfrak{X}^*$, and branchings are allowed only in the fixed points). According to whether we allow all networks of $\mathfrak{N}^*$ or only those of $\mathfrak{N}$ to compete, we distinguish between the two following problems.

PROBLEM A′ (Unrestricted problem of the cheapest communication network). Among all communication networks $\boldsymbol{N} \in \mathfrak{N}^*$ find those causing the lowest total installation costs K (let any Steiner's points cause no costs at all).

PROBLEM B′ (Restricted problem of the cheapest communication network). Among all communication networks $\boldsymbol{N} \in \mathfrak{N}$ find those causing the lowest total installation costs K.

EXAMPLE 1. The four corner points of a square are to be connected by a telephone network. The number of cables to be laid for each couple of points will be given by the following table (cf. Fig. 5.1):

x_{ij}	X_1	X_2	X_3	X_4
X_1	0	1	1	1
X_2	1	0	1	1
X_3	1	1	0	4
X_4	1	1	4	0

In Fig. 5.1, three networks are indicated which establish the required connections. The first two networks ramify only in fixed points, the third one contains a branching point which is not a fixed point and thus a Steiner's point. If we denote by $d(X_i, X_j)$ the Euclidean distance of the points X_i and X_j from each other, then the first network will cause the costs

$$d(X_1, X_2) \left[k(2)\left(1 + \sqrt{2}\right) + k(3) + k(4)\right].$$

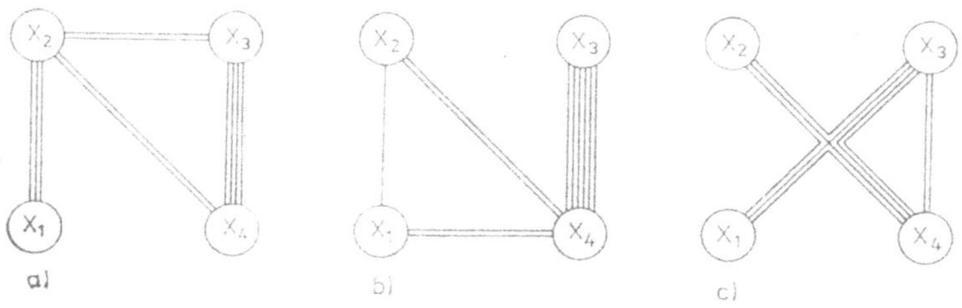

Fig. 5.1

Trying to solve this "minor" task (i.e., to find the minimum network) by calculating the costs of all possible networks, with the restriction, however, that network ramifications are only allowed at fixed points (problem B) and that the connection between two places is realized always by a straight line, would necessitate the application of computers for calculating this whole range of variants.

EXAMPLE 2. In Fig. 5.2a, there are represented two producers and five consumers of a material, with the distances between the fixed points being given by the following table (the producible or the required quantities are indicated at the vertices of Fig. 5.2a):

	X_1	X_2	X_3	X_4	X_5	X_6	X_7
X_1	0	10	6	9	6.5	3	6.5
X_2		0	7	2	4	7	5
X_3			0	5.1	3.2	4.3	7.1
X_4				0	2.5	6	5.4
X_5					0	3.6	4.4
X_6						0	4
X_7							0

Figures 5.2b and 5.2c show two "feasible" networks without any Steiner's points, Fig. 5.2d represents a network containing two Steiner's points, with all demands being fulfilled. Finally, in Fig. 5.2e there is shown a network with one Steiner's point and one fixed point which is at the same time a branching point.

A network as shown in Fig. 5.2f can certainly not be a minimum network, since material is transported in both directions along a certain segment. Of course, it is a network satisfying the needs of the consumers without overtaxing any source, and it is therefore a feasible network.

Let us make some further observations concerning the prerequisites to be fulfilled by the cost function $k(y)$:

The conditions $k(0) = 0$ and $k(y) > 0$ for $y = 1, 2, \ldots$ are immediately clear.

The monotonicity condition $k(y + 1) \geqq k(y)$ for $y = 0, 1, 2, \ldots$ is obvious in so

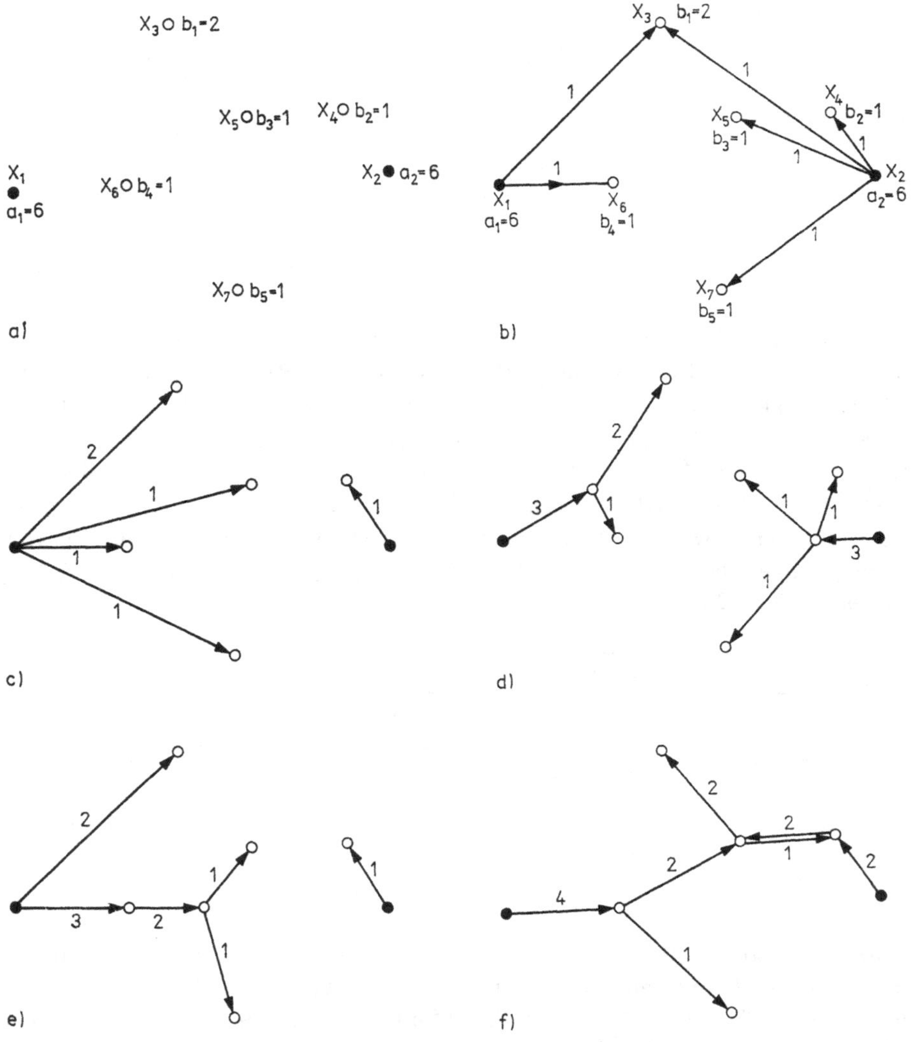

Fig. 5.2

far as the costs will not be lower if a greater quantity of material is sent along a certain segment or if more cables are laid in this same ditch.

The condition $k(x + y) \leqq k(x) + k(y)$, called *subadditivity*, becomes clear as follows: The simultaneous laying of $x + y$ cables (or the simultaneous transportation of $x + y$ material units) should not be more expensive than a separate laying of x cables and of y cables (or than a separate transportation of x material units and of y material units).

5.2. NETWORKS WITHOUT STEINER'S POINTS

For treating the supply problem (problem B) implying the restriction that ramifications are only allowed at the fixed points, we formulate the following algorithm. In case of any (monotonic, subadditive) cost function, however, this algorithm does not yield the minimum network, but in many practical cases an optimum network or also a network that can be regarded as useful. It is largely based on that of Busacker and Gowen formulated in chapter 2:

Algorithm

Let $X_1, X_2, \ldots, X_{m+n}$ be exactly $m + n$ pairwise distinct points of the Euclidean plane, and let $\boldsymbol{K}_{m+n}$ denote the geometric realization of the complete graph with $m + n$ vertices $X_1, X_2, \ldots, X_{m+n}$ (two vertices are connected by a straight line each time). Let $d(X_i, X_j)$ be the Euclidean distance of X_i and X_j from each other with

$$d_{ij} := d(X_i, X_j) \quad \text{with} \quad d_{ij} > 0 \quad \text{for} \quad i \neq j \quad \text{and} \quad d_{ii} = 0$$

being true. We introduce integer auxiliary values $y_{ij}, y_{ji} \geqq 0$ $(i \neq j)$ which have the dimension "material units per time unit" (i.e., the dimension of a flux), which change in the course of the algorithm and which indicate the capacity of a supply network $\boldsymbol{N} \in \mathfrak{N}$ when the algorithm terminates. Then, all needs of the consumers are satisfied without exceeding the source strengths. At least one of the two numbers y_{ij} and y_{ji} is equal to zero (that means: A flow goes through an edge of $\boldsymbol{K}_{m+n}$ in one direction at the most).

Minimum costi growth algorithm

(i) Put $y_{ij} = y_{ji} = 0$ for all $i, j = 1, 2, \ldots, m + n$ $(i \neq j)$.

(ii) Assign to each edge (X_i, X_j) the so-called cost growths k'_{ij} and k'_{ji} as follows:

If $y_{ij} = y_{ji} = 0$, let $k'_{ij} = k'_{ji} = k(1)\, d_{ij}$ be valid;
If $y_{ij} > 0$ (and thus $y_{ji} = 0$), let $k'_{ij} = \big(k(y_{ij} + 1) - k(y_{ij})\big)\, d_{ij}$,
$k'_{ji} = \big(k(y_{ij} - 1) - k(y_{ij})\big)\, d_{ij}$ be true.

The case $y_{ij} = 0$ and $y_{ji} > 0$ is transformed into that just discussed by exchanging the indices.

(iii) Regard the numbers y_{ij} as fluxes and that on the edge (X_i, X_j) with the flow direction from X_i to X_j. Among all directed paths going from a source not yet exhausted to a sink which is not yet satisfied, choose such a path for which the total cost growth is minimum (the total cost growth is the sum of all k'_{ij} along the arcs of this path). Then, increase the flow (by taking the flow direction into account) by the value 1 along this path.

(iv) If not all of the needs are satisfied, then go back to (ii).

(v) Delete all edges (X_i, X_j) for which $y_{ij} = y_{ji} = 0$ holds. For each pair i, j $(i \neq j)$ for which $y_{ij} > 0$ holds (and thus, $y_{ji} = 0$) introduce a straight-line directed canal (arc) going from X_i to X_j with the capacity $c_{ij} = y_{ij}$.

Thus, the algorithm terminates.

Consider the example of Fig. 5.3 relating to two sources and four consumers with the cost function $k(y) = \sqrt{y}$. Let the distance matrix (d_{ij}) be given by

$$\begin{array}{c} \\ X_1 \\ X_2 \\ X_3 \\ X_4 \\ X_5 \\ X_6 \end{array} \begin{array}{c} \begin{array}{cccccc} X_1 & X_2 & X_3 & X_4 & X_5 & X_6 \end{array} \\ \begin{bmatrix} 0 & 4 & \sqrt{5} & \sqrt{8} & 5 & \sqrt{17} \\ & 0 & \sqrt{13} & \sqrt{40} & 3 & 5 \\ & & 0 & 5 & \sqrt{34} & 6 \\ & & & 0 & \sqrt{37} & \sqrt{13} \\ & & & & 0 & \sqrt{10} \\ & & & & & 0 \end{bmatrix} \end{array}$$

The numbers indicated at the vertices of Fig. 5.3a stand for the yields for X_1 and X_2 and indicate the required quantities for X_3, X_4, X_5, X_6.

Applying our algorithm we first choose the arc (X_1, X_3), since this is the shortest which goes from a source to a sink. For sending a flow of the value 1, costs which amount to $\sqrt{1} \cdot d_{13} = \sqrt{5}$ will be caused along this arc.

In order to send another material unit from X_1 to X_3, costs amounting to $(\sqrt{2} - 1)\, d_{13} = (\sqrt{2} - 1)\sqrt{5}$ are necessary. One can easily see that this is also a least expensive path going from a non-exhausted source to an unsatisfied sink.

Now, X_3 is satisfied. The next least expensive path going from a source to a sink turns out to be the arc from X_1 to X_4 entailing the costs $\sqrt{1} \cdot d_{14} = \sqrt{8}$.

Again the arc going from X_1 to X_4 turns out to be the next least expensive path, but now entailing the costs $(\sqrt{2} - 1)\sqrt{8}$.

Now, source X_1 is exhausted, and the demand of X_4 cannot be satisfied directly from X_1.

The next cheapest path going from a source which is not exhausted to a sink which is not satisfied is the arc (X_2, X_5). Two material units can be sent one after the other along this arc. The intermediate situation represented in Fig. 5.3b follows. X_4 and X_6 remain as unsatisfied sinks, and the only non-exhausted source which remains is X_2.

The next cheapest path is (X_2, X_3, X_1, X_4) entailing the costs

$$\sqrt{13} - \sqrt{5}\,(\sqrt{2} - 1) + \sqrt{8}\,(\sqrt{3} - \sqrt{2}).$$

The result is shown in Fig. 5.3c.

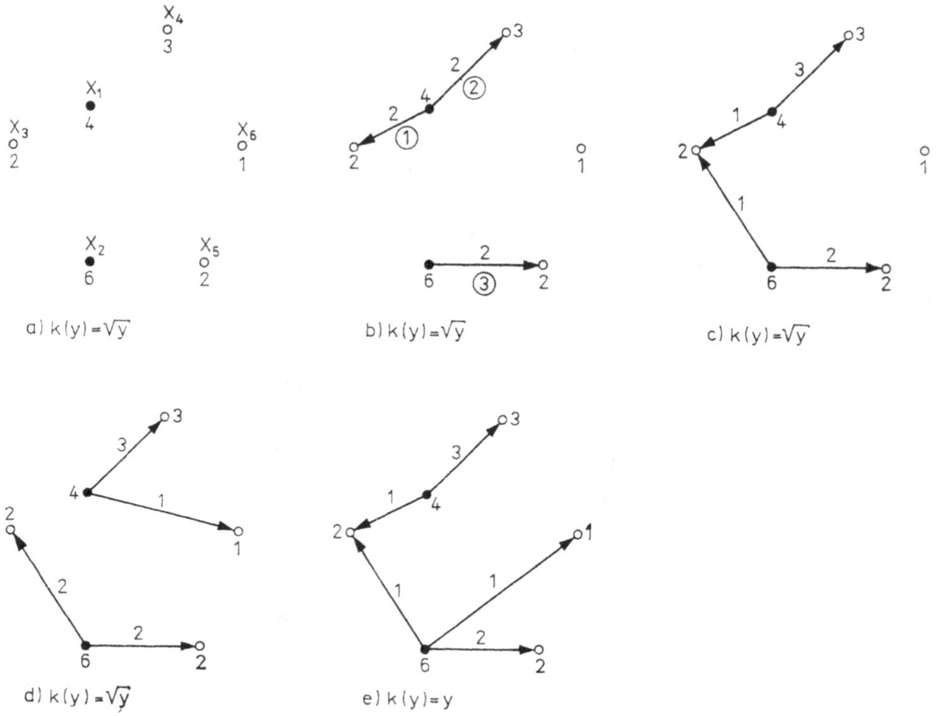

Fig. 5.3

The next cheapest path which results is (X_2, X_3, X_1, X_6) with the costs

$$\sqrt{13}\left(\sqrt{2}-1\right)-\sqrt{5}+\sqrt{17}.$$

Now, all sinks are satisfied, the algorithm terminates (cf. Fig. 5.3d).

The resulting network, however, is not optimal, since it is still possible to send a flow of the value 1 along the path (X_2, X_5, X_6, X_1) from the source X_2 to the source X_1. The cost saving achieved in this case then amounts to exactly

$$\sqrt{17}-\sqrt{10}-3\left(\sqrt{3}-\sqrt{2}\right)>0.$$

From this result we conclude that it is expedient to verify at the end of the algorithm whether there is still a path with negative costs going from a non-exhausted source to an exhausted one. In case there exists such a path, send one material unit along it. Whether an optimal network is always obtained in this way, however, is to be doubted.

If we had chosen $k(y) = y$ as cost function, i.e., a linear function, then the network represented in Fig. 5.3e would have resulted as optimal network. But in practice, linear cost functions are unreal, since the simultaneous laying of supply channels of the capacities x and y would be as costly as the laying of one channel with the total capacity of $x + y$.

For the communication problem also, an algorithm can be found which is similar to the algorithm given for the supply problem. The steps to be performed are all the same, but we must start out from the complete graph with n vertices. When searching for paths causing, if possible, a small increase in costs, it is not necessary for us to search among all paths going from a non-exhausted source to an unsatisfied sink, but among all paths between two vertices each time for which all telephone cables are not yet realized.

EXAMPLE. In the case of example 1 given on page 128, the following network could be generated for a cost function $k(y) = \sqrt{y}$ (the side length of the square is set equal to 1):

— First, say (X_2, X_3) is realized (but not necessarily) with the costs $\sqrt{1} = 1$.

— Then, say (X_1, X_2) is realized with the costs $\sqrt{1} = 1$.

— After this, the connection between X_1 and X_3 is realized along the broken line X_1, X_2, X_3 with the costs $(\sqrt{2} - \sqrt{1}) + (\sqrt{2} - \sqrt{1}) = 0.828 \ldots$

— After this, the direct connection between X_1 and X_4 with the costs 1 could be chosen.

— Then, take the connection from X_2 to X_4, and that, by embedding the broken line X_2, X_1, X_4 with the costs $\sqrt{2} - 1 + \sqrt{3} - \sqrt{2} = 0.732 \ldots$

— Now, we would realize one after the other the last four remaining connections between X_3 and X_4 along the broken line $X_4, X_1, X_2, X_3,$ and the first one of the four broken lines would cause the costs

$$\sqrt{3} - \sqrt{2} + \sqrt{4} - \sqrt{3} + \sqrt{3} - \sqrt{2} = 0.90 \ldots$$

whereas the direct connection from X_3 to X_4 would cause the costs 1.

It is immediately clear that the network resulting in this way is rather expensive. If the four connections from X_3 to X_4 had not been embedded one after the other at the end, but simultaneously, then the direct connection would have been chosen, since this one costs exactly 2 (for the four cable connections) whereas the laying of the four connections via X_1 and X_2 causes the costs

$$2(\sqrt{6} - \sqrt{2}) + \sqrt{7} - \sqrt{3} = 2.98 \ldots$$

From this, it can be seen that, in general, also in the case of communication networks no minimum network will be obtained (by applying the algorithm).

The reader may try to find the minimum network (without Steiner's points) for our example with the cost function $k(y) = \sqrt{y}$.

The following theorems, however, may be proved:

THEOREM 5.1. *If the cost function is linear* (i.e., $k(y) = ay$), *then the algorithm yields a minimum network both for the case of a supply problem and for that of a communication problem. If all selection possibilities left open by the algorithm are realized, then all optimal supply networks will be obtained as well.*

This statement can be proved by means of the considerations made in chapter 2 in connection with the algorithm of R. G. Busacker and P. J. Gowen.

THEOREM 5.2. *If the cost function is quasi-constant* (i.e., $k(0) = 0$, $k(y) = a = \text{const}$ for $y > 0$), *then the algorithm yields a minimum network for the case of a communication problem* (problem B′), *and for that of a supply problem* (problem B) *it yields a minimum network if each of the producers is capable of satisfying the total need, i.e., if*

$$a_\mu \geqq \sum_{\nu=1}^{m} b_\nu \quad \textit{for} \quad \mu = 1, 2, \ldots, m$$

holds.

The minimum network obtained is circuit-free in both cases. As for the supply network, each (connected) component of the minimum network still contains exactly one source.

REMARK. Even if not all of the producers are capable of satisfying the total demand, the algorithm yields a circuit-free network in the single components of which several sources may be found. Whether the supply network obtained is optimal in this case, however, is not known to us.

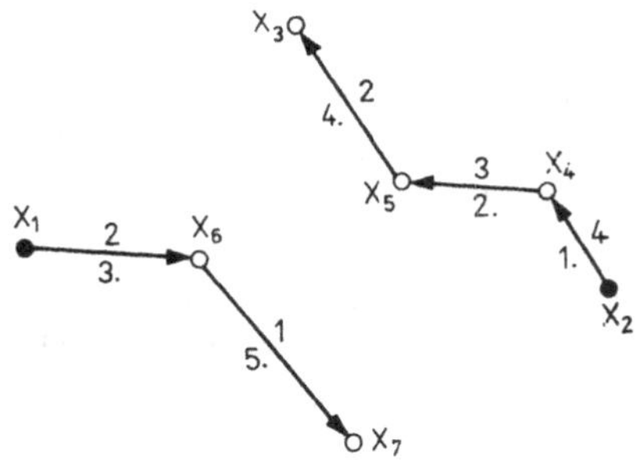

Fig. 5.4

EXAMPLE. Consider the example 2 (Fig. 5.2a) with the distances given in the table on page 129. Let branchings of the network be allowed only in fixed points and the cost function quasi-constant. Then, the network represented in Fig. 5.4 is optimal. The number above each arc indicates the flux on this arc, and the number below shows in what step this arc has been added.

In the case of the supply problem (B) with exactly one source or in the case of the communication problem (B′) with quasi-constant cost function each time, both problems turn into that of the determination of a minimum spanning tree in a length-valuated complete graph $\boldsymbol{G}$. Here, the algorithm becomes considerably simpler. We give an algorithm which yields a sequence $\{\boldsymbol{H}_i\}$ of trees and, when breaking it off, just a minimum spanning tree (cf. H. Sachs [8]).

Algorithm for determining a minimum spanning tree

(i) Let $\boldsymbol{H}$ be a graph consisting of any vertex of $\boldsymbol{G}$ (for example, of the only source X_1 in problem B).

(ii) Among all edges of $\boldsymbol{G}$ with one of its end-points lying in $\boldsymbol{H}$ and the other one lying in $\boldsymbol{G}$, but not in $\boldsymbol{H}$, choose an edge with minimum length; let u be this edge. Add u and that end-point of u which does not lie in $\boldsymbol{H}$ to $\boldsymbol{H}$.

(iii) If $\boldsymbol{H}$ does not contain all vertices of $\boldsymbol{G}$, go back to (ii).

The algorithm terminates.

Still other algorithms have been developed for determining minimum spanning trees (in not necessarily complete graphs) which we shall not discuss here (further algorithms may be found, for example, in [1, 7, 8, 9]). Of course, it is not necessary to start out from a complete graph. Using the above algorithm, minimum spanning trees may also be determined in any edge-valuated connected graphs. In this case, use only a length valuation ∞ instead of edges not existing originally.

The case of quasi-constant costs is of interest in so far as it enables us, by employing the costs for such a network, to estimate the costs for a network when these costs are not quasi-constant.

Estimation of the optimal network costs. Given the problem B or B′ (i.e. without any Steiner's points), with only one source being available in the case of the supply network. Let the cost function be denoted by $k(y)$. Further, let L_0' be the length of a minimum spanning tree $\boldsymbol{N}_0'(\mathfrak{X}, \mathfrak{U}')$ (in a graph with given fixed points). Consider a minimum network with quasi-constant costs $k'(y) = k'(1) = k'$ for $y > 0$. Such a network thus causes the costs $k' \cdot L_0'$. Among all supply networks $\boldsymbol{N}(\mathfrak{X}, \mathfrak{U}, c) \in \mathfrak{N}$ let $\boldsymbol{N}_0 = \boldsymbol{N}_0(\mathfrak{X}, \mathfrak{U}_0, c_0)$ be an optimal network (which, in general, is not known to us and only difficult to calculate in contrast to $\boldsymbol{N}_0'$). Thus, it holds

$$K(\boldsymbol{N}_0) = \sum_{u \in \mathfrak{U}_0} k\big(c_0(u)\big)\, l(u) = \min_{\boldsymbol{N} \in \mathfrak{N}} \sum_{u \in \mathfrak{U}} k\big(c(u)\big)\, l(u).$$

From this, the following two inequalities result:

On the one hand, it holds

$$K(\boldsymbol{N}_0) = \min_{\boldsymbol{N} \in \mathfrak{N}} \sum_{u \in \mathfrak{U}} k\big(c(u)\big)\, l(u) \leqq \sum_{u \in \mathfrak{U}_0} k\big(c(u)\big)\, l(u)$$
$$\leqq k(z) \sum_{u \in \mathfrak{U}_0'} l(u) = k(z)\, L_0'.$$

In the case of the communication problem, it holds $z \leqq \sum_{i,j=1}^{n} x_{ij}$ and in that of the supply problem, it holds $z \leqq \sum_{\nu=1}^{n+1} b_\nu$. The first inequality arises from the fact that the network $\boldsymbol{N}_0'$ must also be taken into consideration when forming the minimum.

On the other hand, it holds

$$k' \cdot L_0' = k' \min_{\boldsymbol{N}\in\mathfrak{N}} \sum_{u\in\mathfrak{U}} l(u) \leqq k' \sum_{u\in\mathfrak{U}_0} l(u) \leqq \sum_{u\in\mathfrak{U}_0} k\big(c(u)\big)\, l(u) = K(\boldsymbol{N}_0).$$

From both inequalities, it follows

$$0 \leqq K(\boldsymbol{N}_0) - k' \cdot L_0' \leqq \big(k(z) - k'\big)\, L_0'.$$

If k' and $k(z)$ do not deviate too much from each other, then the latter inequality yields an estimation of the mistake that can be made at worst if the spanning tree of minimum length is chosen instead of the real cost minimum network (which is, however, only difficult to obtain).

5.3. NETWORKS CONTAINING STEINER'S POINTS

This section is devoted to the problems A and A′, which means that network branchings are allowed to occur also beyond the fixed points. Let us first give a theorem by means of which an algorithm is set up which makes it possible to find the optimal position of the Steiner's points, but this only under the strongly restricting condition that the structure of the network is given in all cases. That means that we have to indicate the adjacency relations (between fixed and fixed points, between Steiner's and Steiner's points and between fixed and Steiner's points), and we are then able to optimize the geometry of the network (in the case of an optimal solution, what are then the distances of the "movable" Steiner's points from each other and from the fixed points adjacent to them?).

We shall discuss in great detail only the case of a road network, that is, the case causing quasi-constant costs. The case with costs which are not quasi-constant can also be treated (quite analogously), but the interested reader's attention will be directed to the bibliography [5, 6].

THEOREM 5.3. *Given the problem* A *with quasi-constant cost function and one source (i.e., the problem of determining a shortest network connecting* $n + 1$ *fixed points with each other, with Steiner's points being allowed to occur). Then, any optimal network* $\boldsymbol{N}_0^*$ *has the following properties:*

1. $\boldsymbol{N}_0^*$ *contains no cycles (circuits).*
2. *Each Steiner's point has the degree* 3.
3. *Two segments meeting in a Steiner's point form an angle of* 120°.

4. *Two segments meeting in a fixed point form an angle of at least* 120° (thus, also a fixed point has the degree 3 at the most).

5. *If* $n + 1$ *is the number of the fixed points of* N_0^*, *then there exist* $n - 1$ *Steiner's points at the most. There exist exactly* $n - 1$ *Steiner's points if and only if each fixed point in* N_0^* *has the degree* 1.

Proof.

1. Obviously, this statement is true since otherwise, it would be possible to reduce the total length by omitting any edge of the circuit.

2. If a Steiner's point S had the degree 1, then it would be possible to omit the edge incident with S, which would also lead to a reduction of the total length.

If a Steiner's point S had the degree 2, that would mean that, because of the minimality of N_0^*, S would lie on that straight line which connects the two neighbouring points of S with each other. In this case, however, S could be omitted without causing a loss.

Suppose S to be in the neighbourhood of at least four vertices. Then, one encounters, during the cyclic traversing of the edges incident with S, two successive segments $\overline{SX}$ and $\overline{SY}$ which form an angle β of 90° at the most.

If $\beta < 90°$, the total length is reduced by modifying N_0^* as follows. In the triangle SXY, the interior angle at S, i.e., β, is maximal, since the segment $\overline{XY}$ is not shorter than each of the two segments $\overline{SX}$ and $\overline{SY}$ because of the minimality of N_0^*. If we draw the perpendicular to the segment $\overline{SY}$ from X and replace the segment $\overline{SX}$ by the perpendicular without changing all else, then a shorter network results, which contradicts the minimality of N_0^*.

We may thus assume that the smallest of all angles at S is exactly 90° (i.e., exactly four vertices are lying in the neighbourhood of S and all angles at S are 90° each). Look at Fig. 5.5. Let X, Y, Z, U be the four neighbours of S, and also

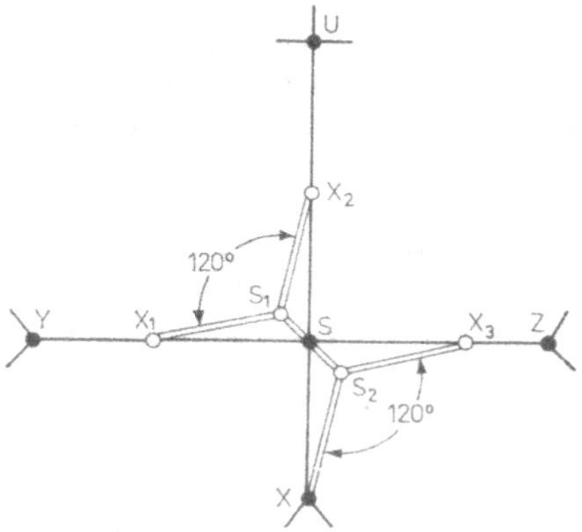

Fig. 5.5

let X be the neighbour lying nearest to S and X_1, X_2, X_3, "new" vertices lying on the segments $\overline{SY}$, $\overline{SU}$, $\overline{SZ}$ which have a distance from S of exactly $\overline{SX}$. Thus, X, X_1, X_2, X_3 form a square. An elementary calculation reveals that, when omitting $\overline{XS}$, $\overline{X_1S}$, $\overline{X_2S}$, $\overline{X_3S}$ and adding $\overline{X_1S_1}$, $\overline{X_2S_1}$, $\overline{S_1S_2}$, $\overline{XS_2}$, $\overline{X_3S_2}$ (with $\overline{X_1S_1}$, $\overline{X_2S_1}$, $\overline{XS_2}$, $\overline{X_3S_2}$ having all the same length), a network is obtained shorter than $\boldsymbol{N}_0^*$ (by exactly $4 - \sqrt{6} - \sqrt{2}$), which contradicts the minimality of $\boldsymbol{N}_0^*$. This proves assertion 2.

3. Let S_0 be a Steiner's point in an optimal network with the three neighbouring points X_1, X_2, X_3. For any point S lying in the interior of the triangle X_1, X_2, X_3, we denote the "costs" of S by $K(S) = \overline{SX_1} + \overline{SX_2} + \overline{SX_3}$. Obviously, $K(S)$ becomes minimum exactly for $S = S_0$. If we denote by (x_i, y_i) the co-ordinates of X_i $(i = 1, 2, 3)$ in a Cartesian system of co-ordinates, then we get as necessary minimum condition for the co-ordinates x, y of the Steiner's point:

$$\sum_{i=1}^{3} \frac{x - x_i}{\sqrt{(x - x_i)^2 + (y - y_i)^2}} = 0, \qquad \sum_{i=1}^{3} \frac{y - y_i}{\sqrt{(x - x_i)^2 + (y - y_i)^2}} = 0.$$

An elementary calculation, though involving a lot of writing, proves assertion 3.

4. This assertion can be proved analogously to the procedure described for assertions 2 and 3.

5. If t is the number of the Steiner's points ($n + 1$ was the number of the fixed points), then there results for the number of the edges on the one hand:

$$\text{number of the edges} = n + t$$

since the network is a circuit-free connected graph (i.e., a tree), and on the other hand:

$$2 \times \text{number of the edges} = \sum_{i=1}^{n+1} v(X_i) + \sum_{i=1}^{t} v(S_i)$$

if we denote the fixed points by X_i and the Steiner's points by S_i, as happens when counting the degrees of all vertices. From both equations it results

$$n + t = \frac{1}{2} \sum_{i=1}^{n+1} v(X_i) + \frac{1}{2} \sum_{i=1}^{t} v(S_i) \geqq \frac{1}{2}(n + 1) + \frac{3}{2} t,$$

since the fixed points have at least the degree 1 and the Steiner's points exactly the degree 3. Hence it follows directly $t \leq n - 1$ (and $t = n - 1$ if there is a sign of equality in the above estimation, i.e. if each fixed point has the degree 1). This proves theorem 5.3.

REMARK (cf. Fig. 5.6). If the cost function is not quasi-constant any more and a cheapest network is to be found connecting the three points X_1, X_2, X_3 with each other (here, let the laying of the segment $\overline{SX_i}$ cause just the costs $k_i d(S, X_i)$), then a theorem similar to 5.3 holds. The properties 1 and 2 are valid, but at one Steiner's point the 120° relation does not necessarily occur, and it holds there:

$$\cos \alpha_1 = \frac{k_1^2 - k_2^2 - k_3^2}{2k_2k_3}. \tag{0}$$

This relation is true correspondingly for α_2 and α_3 by cyclic exchange. An additional condition (which is always met in practice) has to be fulfilled by the cost function, namely $k^2(x + y) < k^2(x) + k^2(y)$ (a linear cost function does not meet this requirement!), otherwise it may happen that the degree of a Steiner's point is higher than 3.

Now, the question arises of how the Steiner's point can be found (in the easiest way possible) in order to construct a shortest network that connects three fixed points with each other. Since the procedure to be described for the case with three fixed points may also be applied to the case with more than three fixed points, we shall explain this in detail.

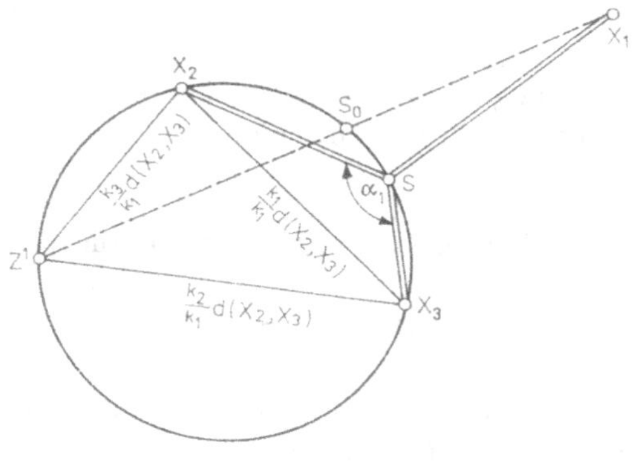

Fig. 5.6

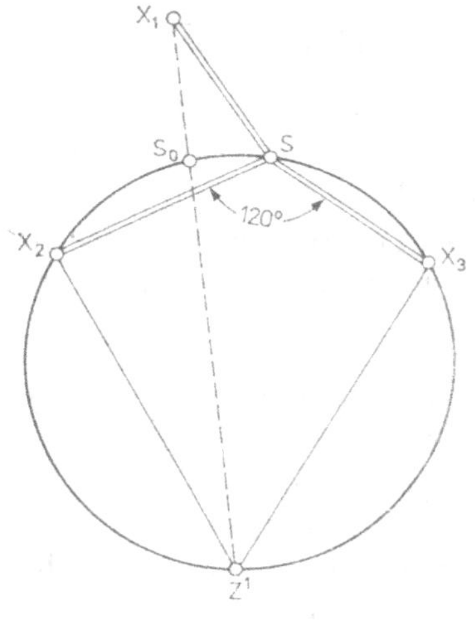

Fig. 5.7

Construction of the Steiner's point (cf. Fig. 5.7)

1. The (optimal) Steiner's point S lies on a circular arc which is a part of the circumcircle of the equilateral triangle X_2, X_3, Z^1. That this assertion is true follows directly from the theorem of the angle at the circumference since just in this case, the interior angle S takes on a value of 120° (of course, S lies also on a circular arc which is a part of the circumcircle of the equilateral triangle X_1, X_2, Z^3 and X_3, X_1, Z^2, respectively).

2. The (optimal) Steiner's point S lies on the connecting line going through X_1 and Z^1 (and, of course, also on the straight line going through X_2, Z^2 and X_3, Z^3).

The validity of this assertion follows directly from the Ptolemy's theorem: In an inscribed tetragon, the sum of the products of the opposite sides equals the product of the diagonals. Thus, for our case it holds

$$d(S, X_3)\, d(X_2, Z^1) + d(S, X_2)\, d(X_3, Z^1) = d(X_2, X_3)\, d(S, Z^1).$$

Because of the relation $d(X_2, X_3) = d(X_2, Z^1) = d(X_3, Z^1)$ it follows

$$d(S, X_3) + d(S, X_2) = d(S, Z^1).$$

Obviously, the sum $d(S, X_1) + d(S, X_2) + d(S, X_3)$ becomes minimum if and only if S lies — as asserted — on the connecting line between X_1 and Z^1.

Moreover, it is seen that in the minimum case, the network spanned by the points X_1, X_2, X_3, S_0 has the same length as the segment $\overline{X_1Z^1}$. We are thus provided with the basis for determining the exact position of the Steiner's points in the case of any number of fixed points (with the topological structure of the network being given).

REMARK. The construction given above may also be applied to the case of not quasi-constant cost functions, but in the place of step 1, do not construct an equilateral triangle X_2, X_3, Z^1, but a triangle which just realizes the relation (0) because of the theorem of the angle at the circumference. This can be achieved by choosing Z^1 such that the segments $\overline{X_2X_3}$, $\overline{X_3Z^1}$, $\overline{Z^1X_2}$ are to each other as $k_1 : k_2 : k_3$ (cf. Fig. 5.6). Moreover, it can be shown that the optimal point S_0 also lies on the connecting line between X_1 and Z^1 and that the network costs $k_1d(S, X_1) + k_2d(S, X_2) + k_3d(S, X_3)$ are equal to $k_1d(X_1, Z^1)$. The interested reader is recommended to verify these properties.

We shall now become acquainted with an algorithm which makes it possible, in the case of a given structure of the network, to determine the position of the Steiner's points so that a minimum network arises among all networks having this structure.

A problem not yet solved is how to decide from what structure one has to start out in order to solve problem A. To find, among the great variety of all possible structures, such one which solves problem A, would not be possible without performing a great deal of work.

In [2], a model of analogy for treating the *road network problem* by way of experiment is proposed by R. Courant and H. Robbins. This model (cf. Fig. 5.8) consists of two parallel transparent plates which are connected with each other by bars perpendicular to them. Each bar corresponds to a fixed point, the ratios of the distances between the bars conform with the ratios of the distances between the respective fixed points. If this model is plunged into a soap solution and then taken out, the soap film forms a connected system of planes perpendicular to the plates. Seen from the top, there will develop a network that can easily be measured after being photographed. A repetition of this experiment should reveal a certain structure turning up most frequently. Then, an approximated solution (often called "suboptimal") is found on the basis of this predominant structure.

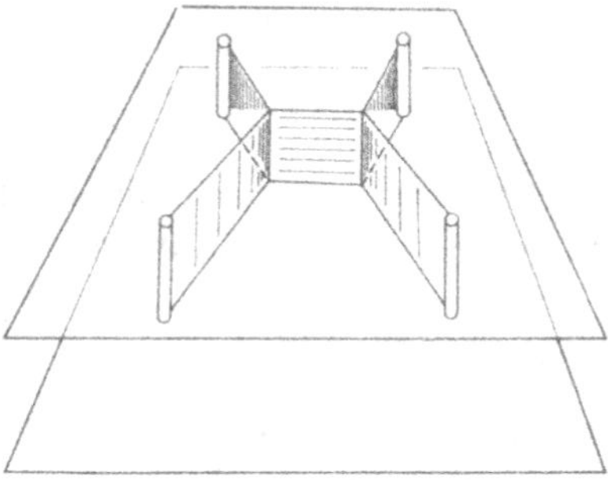

Fig. 5.8

Algorithm of Z. A. Melzak (for optimizing the position of the Steiner's points in case of a given network structure)

Let us explain the algorithm by means of an example. The size of the example chosen should suffice to illustrate how any other case is to be treated. The exact algorithm can be read in E. N. Gilbert and H. O. Pollak [4].

Consider the graph of Fig. 5.9a. The fixed points $X_1, \ldots, X_6$ form the corner points of a regular hexagon, the fixed points X_6, X_7, X_8 as well as X_5, X_6, X_7 form the corner points of an equilateral triangle each time. Find among all networks having the structure indicated in Fig. 5.9a a network of minimum total length (by determining the optimal position of the Steiner's points).

This task can be decomposed into two minor ones (C) and (D) by splitting up the vertex X_5. Combining the solutions of (C) and (D) leads to a solution of the original task (Fig. 5.9b).

In the following construction, we use — without proving it — the fact that if there exists a solution of task (C), then the Steiner's points S_1, S_2, S_3 belong to the convex hull of the fixed points $X_1, \ldots, X_5$.

First, find such a Steiner's point which is connected with two fixed points, say S_1.

If we knew the exact position of the Steiner's point S_2, we would be able to determine the exact (that is, the optimal) position of S_1 by applying the construction rule given on page 140. However, we can certainly construct the point Z^1 as third point of an equilateral triangle together with the other two corner points X_3 and X_4 (cf. Fig. 5.9c). Because the sum of the lengths of the three segments $\overline{S_2S_1}$, $\overline{X_3S_1}$, $\overline{X_4S_1}$ equals the length of the segment $\overline{S_2Z^1}$, as already stated, Z^1 stands vicariously for the two vertices X_3 and X_4, and the optimal network of task (C) is exactly as long as that of task (C_1).

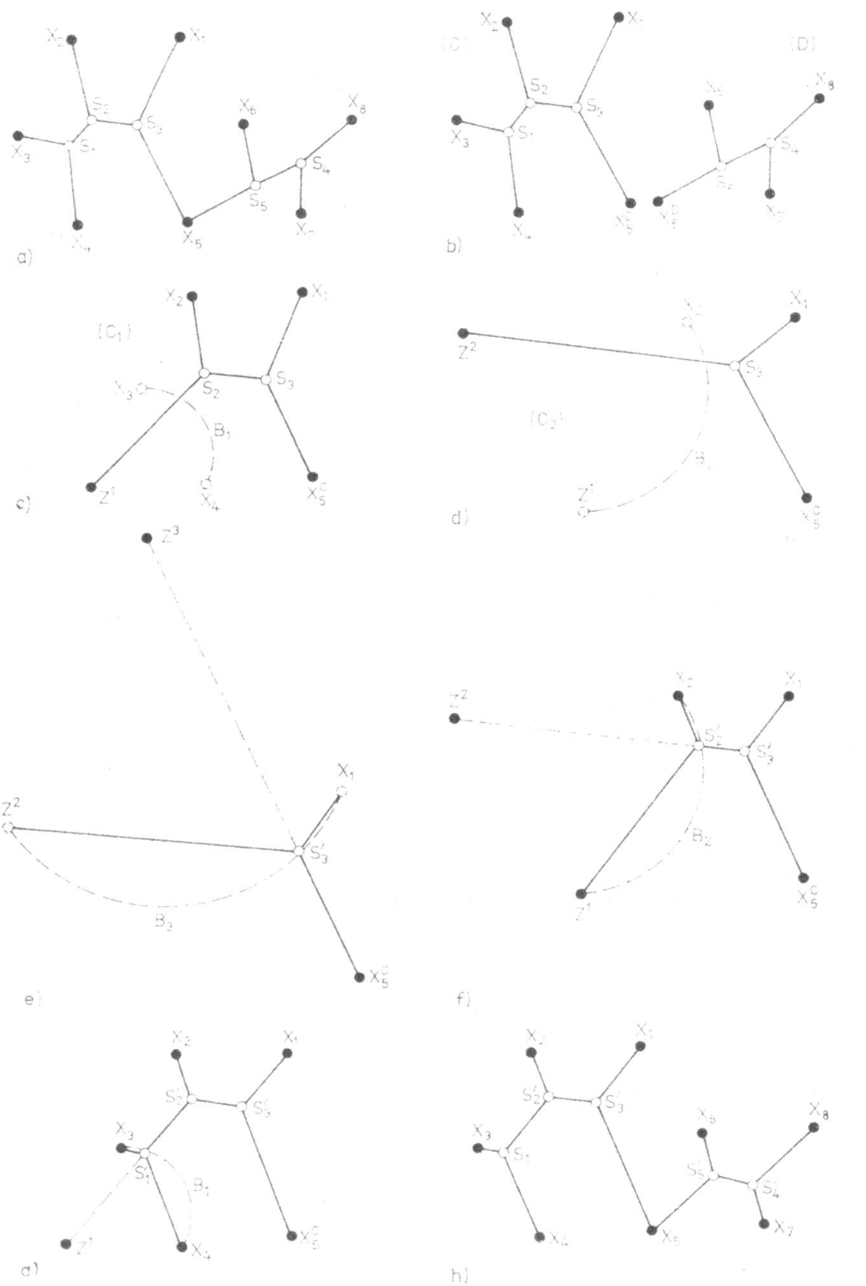

Fig. 5.9

Now, we consider Z^1 to be a fixed point and search for a Steiner's point again lying in the neighbourhood of two fixed points. This is, say, S^2 with the two neighbouring points Z^1 and X_2. Construct Z^2 as third corner point of the equilateral triangle with the other two corner points Z^1 and X_2 (cf. Fig. 5.9d). If we knew the exact position of S_3, we would be able to construct S_2. If we consider Z^2 vicariously for the fixed points Z^1 and X_2, then it remains to determine the optimal position of the Steiner's point S_3 with the three fixed points X_1, $X_5{}^C$ and Z^2 as its neighbours. According to the construction rule for determining the optimal position of a Steiner's point lying in the neighbourhood of the three fixed points (cf. Fig. 5.7), we find S_3' (the optimal point for S_3) on the arc segment $\widehat{X_1Z^2}$ of the circumcircle of the equilateral triangle with the corner points Z^3, Z^2, X_1 and also on the connecting line between Z^3 and $X_5{}^C$ (Fig. 5.9e).

Now, we find S_2' as intersection point of the arc segment $\widehat{X_2Z^1}$ of the circumcircle of the equilateral triangle with the corner points X_2, Z^1, Z^2 and the straight line going through the points Z^2 and S_3' (cf. Fig. 5.9f). Finally, construct S_1' analogously (Fig. 5.9g). Thus, task (C) is solved.

The reader is recommended to solve task (D) by himself. Figure 5.9h illustrates the result of the combination of the solutions obtained for task (C) and task (D).

Although the network thus found is certainly not the shortest one among all networks connecting the fixed points $X_1, \ldots, X_8$ (as the segments $\overline{X_5S_3'}$ and $\overline{X_5S_5'}$ form an angle of less than 120°, which is in contradiction to theorem 5.3, property 4), it is, nevertheless, minimum on condition that the structure indicated in Fig. 5.9a is realized.

It is not always possible to realize a given structure in an optimal way. This holds when the arc segment $\widehat{X_2X_3}$ of the circumcircle of the equilateral triangle with the corner points Z^1, X_2, X_3 does not intersect the straight line going through Z^1 and X_1 (cf. Fig. 5.7).

If a given structure cannot be optimally realized, then this will be revealed in the course of the algorithm.

We have already mentioned that the algorithm discussed just now may be extended to the case of not quasi-constant costs, but this, however, is on condition that the relation

$$k^2(x + y) < k^2(x) + k^2(y) \quad \text{for} \quad x, y > 0$$

holds. This condition assures that exactly three edges meet together in a Steiner's point (task!).

The auxiliary vertex Z^i which is to be constructed in each case is then not the third corner point of an equilateral triangle, but of a triangle which resembles a triangle with the side proportions of $k_1 : k_2 : k_3$, when k_1, k_2, k_3 are the costs per unit on the segments $\overline{S'X_1}$ and $\overline{S'X_2}$ and $\overline{S'X_3}$, resp. (cf. Fig. 5.6 and also p. 140).

5.4. INFLUENCE EXERTED BY THE COST FUNCTION ON THE STRUCTURE OF THE OPTIMAL NETWORK

Besides the general monotonicity and subadditivity conditions for the cost function, we have treated so far, in the essence, only the case of a quasi-constant cost function and the case in which a triangular inequality is valid for the squares of the cost function, namely

$$k^2(x + y) < k^2(x) + k^2(y) \quad \text{for any} \quad x, y > 0.$$

In most cases, this latter condition turned out to be satisfied in the practical examples concerning heat supply networks and also cable networks which we have dealt with.

Now, let us make a few more statements for the case where, in addition to the general conditions to be met by the cost function, we require this function to be *concave*, thus

$$k(x + 1) - k(x) \leqq k(x) - k(x - 1) \quad \text{for} \quad x = 1, 2, \ldots$$

holds. This condition applies in particular also to linear cost functions although, as already mentioned above, linear cost functions do not really occur in practice. However, the case of *affine-linear functions*, which are more likely in reality, is included, i.e.

$$k(x) = a + bx \quad \text{for} \quad x > 0 \; (a, b \text{ constants}).$$

For example, in the case of heat supply problems we have (usefully approximated) cost functions of the form

$$k(x) = ax^b + c \quad \text{for} \quad x > 0 \left(a, c > 0;\, 0 < b < \frac{1}{2}\right)$$

which even meet the requirement $k^2(x + y) < k^2(x) + k^2(y)$. That, however, means that we can use for them all considerations made in the preceding section.

Now, the following theorem may be proved:

THEOREM 5.4. *Let the cost function be concave.*

1. *In the case of either problem* A *or* B, *there exists an optimal network* $\mathbf{N}_0$ *which is a forest (and thus cycle-free). If moreover each of the producers is able to satisfy the total demand of all consumers (i.e., with only one source available), then each component of the forest contains one producer.*

2. *In the case of either problem* A′ *or* B′, *there exists an optimal network having the following property:*

If we consider only the partial graph $\mathbf{G}_i$ *which realizes the* $z_i = \sum_{j=1}^{n} x_{ij}$ *connections* *of the fixed point X_i with the other fixed points, then $\boldsymbol{G}_i$ constitutes a cycle-free subnetwork of the optimal network.*

Proof.

1. Consider an optimal supply network $\boldsymbol{N}_0 = (\mathfrak{X}, \mathfrak{U}_0)$ containing a minimum number s of cycles. For $s = 0$, there is nothing to prove.

In the following, assume $\boldsymbol{N}_0$ to be chosen such that s is minimum with $s \leqq 1$. Orient the edges of $\boldsymbol{N}_0$ so that the orientation corresponds with the flowing direction on this edge (note that an edge exists only if there is a non-vanishing flux on it).

We denote by μ any cycle of $\boldsymbol{N}_0$. Certainly, μ is not a circuit, i.e., no cycle of which the arcs are oriented in the same direction, since otherwise it would be possible to lower the flow along μ and thus also the costs. This, however, would be in contradiction to the cost minimality of $\boldsymbol{N}_0$. Let μ^+ and μ^- be the sets of arcs of μ which are equally and oppositely directed to μ, respectively, and let S^+ and S^- be the minima of the flux quantities going through the arcs of μ^+ and μ^-, respectively.

We transform $\boldsymbol{N}_0$ to $\boldsymbol{N}_t$ in the following way: In each of the arcs of μ^+, the flux is increased by t, in each of the arcs of μ^-, it is reduced by t, and on all other arcs the flux remains unchanged. In doing so, we confine t to the range $-S^+ \leqq t \leqq S^-$. The costs $K(\boldsymbol{N}_t)$ for the network formed in this way amount to

$$k(\boldsymbol{N}_t) = \sum_{y \in \mathfrak{U}_0 \setminus \mu} k(x_y)\, l(y) + \sum_{y \in \mu^+} k(x_y + t)\, l(y) + \sum_{y \in \mu^-} k(x_y - t)\, l(y)\,.$$

Here, we denote by x_y the flux on the edge y in the direction of the orientation of u. Since $k(x)$ is concave and $l(u) > 0$, also $K(\boldsymbol{N}_t)$ is a concave function with respect to t (task!). Therefore, because of $-S^+ \leqq t \leqq S^-$, $K(\boldsymbol{N}_t)$ assumes its minimum at a limit point t', i.e., either for $t = -S^+$ or for $t = S^-$. Consider the network $\boldsymbol{N}_{t'}$: In $\boldsymbol{N}_{t'}$, we delete all arcs on which no flux exists any more. For the choice of t', such an arc exists at least on the cycle μ.

A network results satisfying all demands, and it contains one cycle less than $\boldsymbol{N}_0$ and causes costs not higher than those of $\boldsymbol{N}_0$. This, however, contradicts the assumption that $\boldsymbol{N}_0$ had been chosen as optimal network with a minimum number $l \geqq 1$ of cycles.

The second part of statement 1 may be proved in a similar way. Consider instead of a cycle μ a sequence of arcs (i.e., a chain) between two producers contained in one component. Thus, the proof of statement 1 is complete.

To prove statement 2 of theorem 5.4 may be left to the reader.

If we suppose strict concavity (which is, however, not fulfilled in case of affine functions), then the proof makes clear that for the case of either problem A or B, each minimum network is cycle-free.

We give another example which shows that if the cost function is not concave, the optimal network is, in general, not circuit-free.

EXAMPLE. Let X_1, X_2, X_3 be the corner points of an equilateral triangle, let the cost function $k(x)$ be given by $k(x) = \left[\frac{x+2}{3}\right]$. Here, let $[z]$ mean the largest integer which is not larger than z. Let the number of the telephone connections to be realized be given by the following table:

	X_1	X_1	X_3
X_1	0	1	1
X_2	1	0	14
X_3	1	14	0

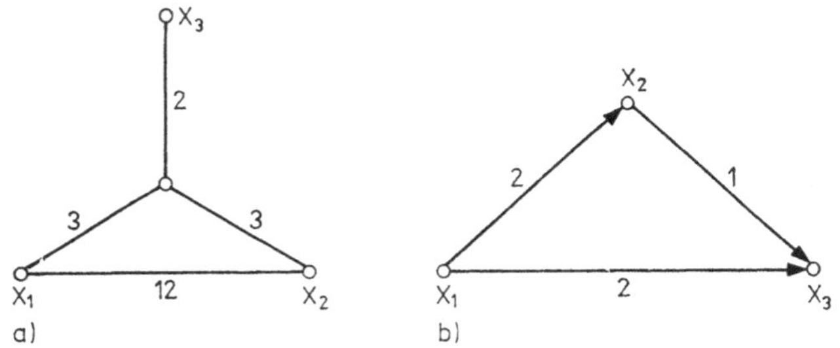

Fig. 5.10

Then, Fig. 5.10a represents the only optimal network. In addition, it is not cycle-free.

The following example shows that even if no Steiner's points are allowed it cannot be ensured that the optimal network is circuit-free.

EXAMPLE. The producer X_1 with the yield 4 will provide for two consumers X_2 and X_3 having a need of 1 and 3, respectively. Let branchings of the network be allowed only at fixed points. Let the distances of the fixed points from each other be $d(X_1, X_2) = d(X_2, X_3) = 1, d(X_1, X_3) = 1.5$. Let the cost function $k(x)$ be given by $k(1) = k(2) = 1$, $k(x) = 2$ for $x = 3, 4, \ldots$ The reader may show that the network represented in Fig. 5.10b is the only optimal network.

We thus conclude this chapter with these considerations. The reader wishing to become acquainted with further results in more detail is referred to the works in [3, 4, 5, 6].

5.5. BIBLIOGRAPHY

[1] Berge, C., und A. Ghouila-Houri: Programme, Spiele, Transportnetze, Leipzig 1969. (Translated from French.)

[2] Courant, R., und H. Robbins: Was ist Mathematik?, 3. edn., Berlin/Heidelberg/New York 1973 (Translated from the Englisch).

[3] Gilbert, E. N.: Minimum Cost Communication Networks, Bell Teleph. Lab. 1966.
[4] Gilbert, E. N., and H. O. Pollak: Steiner minimal trees, SIAM J. Appl. Math. **16** (1968), 1–29.
[5] Hutschenreuther, H.: Eine Lösung des Steiner-Weber-Problems und ihre Anwendung in der Praxis, Diss. A TH Ilmenau, 1972.
[6] Melzak, Z. A.: On the problem of Steiner, Canad. Math. Bull. **4** (1961), 143–148.
[7] Noltemeier, H.: Graphentheorie mit Algorithmen und Anwendungen, Berlin/New York 1976.
[8] Sachs, H.: Einführung in die Theorie der endlichen Graphen, Teil I, Leipzig 1970.
[9] Tinhofer, G.: Methoden der angewandten Graphentheorie, Vienna/New York 1976.

Chapter 6

The assignment and the travelling salesman problems

In this chapter, we wish to treat two discrete optimization problems. However, we shall not conceal that only a small insight into the solution methods can be given, since today there already exist such a variety of different methods and algorithms based on them for solving especially the travelling salesman problem that it is impossible at present to make a final evaluation.

It is well known that a large number of discrete optimization problems exist. Of course, there was good reason for us to choose just the assignment and the travelling salesman problems. For the solution of the assignment problem as we shall describe it (using the so-called Hungarian method), theorems from the field of graph theory (theorem of Ford and Fulkerson, cf. chapter 1; theorem of König, Ore and Hall, cf. Sachs [11]) will be applied, and the travelling salesman problem is a perfectly classical problem of graph theory, as we shall see.

Paragraph 6.1 will be devoted to the assignment problem, section 6.2 to the travelling salesman problem. We wish to present there two solution methods, namely a branch-and-bound algorithm and a heuristic one. In section 6.3, we only want to give the interested reader a short survey of the literature used.

6.1. THE ASSIGNMENT PROBLEM

6.1.1. Formulation of the problem

Let us explain the problem by means of an example. For the fulfilment of n diverse tasks A_i there are available n machines M_j each of which can be employed to fulfil each of the tasks. For each pair (i, j) of natural numbers it is known that the expenditure $d_{ij} \geqq 0$, $i, j = 1, 2, \ldots, n$ (e.g., in money or time units) is required for the machine M_j to fulfil the task A_i. Now, find such an assignment of the machines to the tasks (here, each machine will fulfil exactly one task) that the total expenditure becomes minimum.

Using the terminology of graph theory, the following problem is to be solved:

Given a directed completely bipartite graph $\boldsymbol{G} = \boldsymbol{G}(\mathfrak{X} \cup \mathfrak{Y}; \mathfrak{U})$ with the n vertices $X_1, \ldots, X_n$ of one class and the n vertices $Y_1, \ldots, Y_n$ of the other class, as well as with the n^2 arcs (X_i, Y_j), $i, j = 1, 2, \ldots, n$. Further, let an arc function d be given which assigns to each arc $u \in \mathfrak{U}$ a non-negative integer $d(u)$. Find a set of n independent arcs (i.e., a *linear factor*), thus a set of n arcs so that each of the $2n$ vertices is incident with exactly one arc, with the sum of the valuations $d(u)$ of these n arcs being minimum.

Being an integer optimization task the assignment problem takes on the form

$$\sum_{i,j=1}^{n} d_{ij}x_{ij} \to \min!$$

with the constraints

$$\sum_{j=1}^{n} x_{ij} = 1, \quad i = 1, \ldots n.$$

$$\sum_{i=1}^{n} x_{ij} = 1, \quad j = 1, \ldots, n,$$

$$x_{ij} \in \{0, 1\}, \quad i, j = 1, \ldots, n.$$

The constraints assure that in each row and in each column of the solution matrix $\boldsymbol{X} = (x_{ij})_{i,j=1,\ldots,n}$ there is exactly one 1, with the other numbers all being zeros, that is, $\boldsymbol{X}$ is a permutation matrix.

In principle, the matrix $\boldsymbol{D} = (d_{ij})_{i,j=1,\ldots,n}$ would be allowed to have some elements of the "value" of infinity. For the problem set at the beginning, this would mean $d_{rs} = \infty$: the task A_r cannot be treated by the machine M_s. But if $\boldsymbol{D}$ has a too large number of elements of the value of infinity, then the finiteness of the minimum value of the objective function cannot be guaranteed any more.

6.1.2. *A solution algorithm for the assignment problem*

Before being able to explain the algorithm, we have to give a few definitions.

DEFINITIONS. Two elements c_{ij} and c_{kl} of a matrix

$$\boldsymbol{C} = (c_{ij})_{i,j=1,\ldots,n}$$

are called *independent* if $i \neq k$ and $j \neq l$. A set $\mathfrak{Q}$ of elements of $\boldsymbol{C}$ is called *independent* if two elements of $\mathfrak{Q}$ at a time are independent. A set $\mathfrak{Q}$ of elements of $\boldsymbol{C}$ is called *independent zero-set* if it is independent and if each element of $\mathfrak{Q}$ is of the value zero. $\mathfrak{Q}$ is called *maximum zero-set* if it is an independent zero-set with maximum number of elements.

A set $\mathfrak{R} = \mathfrak{Z} \cup \mathfrak{S}$ consisting of a set of rows $\mathfrak{Z} = \{Z_{i_1}, Z_{i_2}, \ldots, Z_{i_z}\}$ of z rows, and of a set of columns $\mathfrak{S} = \{S_{j_1}, S_{j_2}, \ldots, S_{j_s}\}$ of s columns of a matrix $\boldsymbol{C}$ is called *zero-covering* of $\boldsymbol{C}$ if, after having removed the rows from $\mathfrak{Z}$ and the columns from $\mathfrak{S}$, no zero-element remains any more in $\boldsymbol{C}$. $\mathfrak{R}$ is called *minimum* if among all zero-coverings $z + s$ is minimum.

When solving the assignment problem by means of the Hungarian method, the following three partial tasks have to be solved one after the other:

a) transformation of a matrix to reduce the value of the objective function and to increase the number of the zeros;

b) determination of a maximum zero-set in the transformed matrix;

c) determination of a minimum zero-covering for the maximum zero-set determined in b);

In order to substantiate theoretically the step from b) to c), we prove the following theorem.

THEOREM 6.1. *In a matrix* $\boldsymbol{C} = (c_{ij})_{i,j=1,\ldots,n}$, *the maximum number of independent zeros equals the minimum number of rows and colums after the removal of which the matrix contains no zero-elements any more.*

Proof. Consider the bipartite graph $\boldsymbol{G} = (\mathfrak{Z}, \mathfrak{S}, \mathfrak{U})$. Here, the set $\mathfrak{Z}$ of vertices of the one class of the bipartite graph $\boldsymbol{G}$ consists of n "row vertices" $Z_1, \ldots, Z_n$, and the set $\mathfrak{S}$ of vertices of the other class of $\boldsymbol{G}$ is composed of n "column vertices" $S_1, \ldots, S_n$. An edge $u = (Z_i, S_j)$ is added to $\mathfrak{U}$ if and only if the element c_{ij} of the matrix $\boldsymbol{C}$ is a zero-element, i.e., if $c_{ij} = 0$ holds.

Obviously, a set of pairwise non-adjacent edges of $\boldsymbol{G}$ corresponds to any independent zero-set in $\boldsymbol{C}$, and vice versa. Thus, a maximum set $\mathfrak{V} \subseteq \mathfrak{U}$ with $\mathfrak{V} = \{(Z_{i_1}, S_{j_1}), \ldots, (Z_{i_t}, S_{j_t})\}$ of pairwise non-adjacent edges of $\boldsymbol{G}$ corresponds to a maximum set $\mathfrak{Q} = \{q_{i_1j_1}, \ldots, q_{i_tj_t}\}$ of independent zeros, and vice versa.

Let us say that a set $\mathfrak{M}$ of vertices of $\boldsymbol{G}$ *disconnects the sets* $\mathfrak{Z}$ *and* $\mathfrak{S}$ of vertices from each other if, after having removed the vertices of $\mathfrak{M}$ from $\boldsymbol{G}$, the residual graph contains no more edges. It can directly be seen that a zero-covering $\mathfrak{N}$ of $\boldsymbol{C}$ corresponds to any set disconnecting the sets $\mathfrak{Z}$ and $\mathfrak{S}$ of vertices from each other, and vice versa and that, in particular, a minimum zero-covering $\mathfrak{N}$ of $\boldsymbol{C}$ corresponds to a minimum set $\mathfrak{M}$ of vertices disconnecting $\mathfrak{Z}$ and $\mathfrak{S}$ from each other, and vice versa. But now, the following theorem of D. König holds the proof of which may be found in [11].

THEOREM 6.2. *In a bipartite graph* $\boldsymbol{G} = (\mathfrak{X}, \mathfrak{Y}, \mathfrak{U})$, *the maximum number of pairwise non-adjacent edges equals the number of vertices of a minimum set of vertices disconnecting* $\mathfrak{X}$ *and* $\mathfrak{Y}$ *from each other.*

Figure 6.1 gives some illustration of this latter theorem.

By applying it, one will also be able to prove theorem 6.1. The reader must make it clear to himself in detail.

First, we solve the three partial tasks a), b) and c).

a$_1$) First, we have to carry out a transformation T of the matrix (which is other than the forthcoming ones), thus transforming the matrix $\boldsymbol{D} = (d_{ij})$ underlying the assignment problem into a matrix $\boldsymbol{D}^0 = (d_{ij}^0)_{i,j=1,2,\ldots,n}$. In principle, this

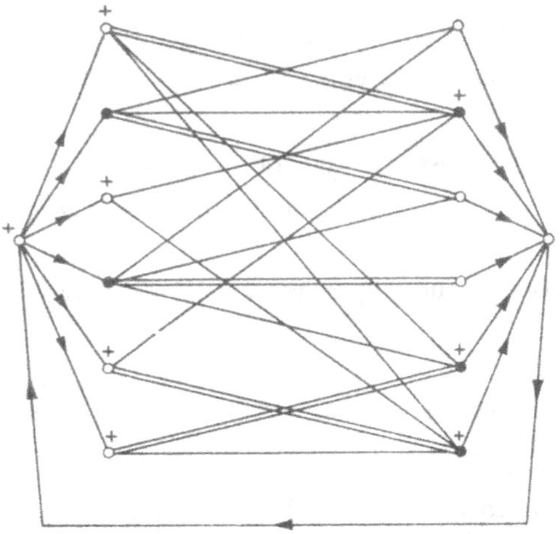

Fig. 6.1

transformation could be omitted, nevertheless one achieves, as a rule, a greater reduction of the minimum of the objective function (compared to the other transformations). The transformation T has the following form:

$$d_{ij}^0 := d_{ij} - u_i^0 - v_j^0$$

with

$$u_i^0 := \min_j d_{ij}, \qquad v_j^0 := \min_i (d_{ij} - u_i^0) \quad \text{for} \quad i, j = 1, 2, \ldots, n.$$

The objective function changes according to

$$\begin{aligned}\sum_{i,j=1}^{n} d_{ij}x_{ij} &= \sum_{i,j=1}^{n} [d_{ij}^0 + u_i^0 + v_j^0]\, x_{ij} \\ &= \sum_{i,j=1}^{n} d_{ij}^0 x_{ij} + \sum_{i=1}^{n} u_i^0 \sum_{j=1}^{n} x_{ij} + \sum_{j=1}^{n} v_j^0 \sum_{i=1}^{n} x_{ij} \\ &= \sum_{i,j=1}^{n} d_{ij}^0 x_{ij} + \sum_{i=1}^{n} u_i^0 + \sum_{j=1}^{n} v_j^0.\end{aligned}$$

Thus, the value of the objective function $\sum_{i,j=1}^{n} d_{ij}x_{ij}$ differs from that of the objective function $\sum_{i,j=1}^{n} d_{ij}^0 x_{ij}$ only by the sum of the transformation constants u_i^0, v_j^0. That is, after the transformation we have in principle the same optimization problem (of course, with a modified matrix) the only difference being that the value of the objective function is smaller by exactly

$$\sum_{i=1}^{n} u_i^0 + \sum_{j=1}^{n} v_j^0.$$

The transformations to be carried out later will have the same effect. They are carried on as long as the minimum value of the objective function becomes zero.

An example of this type of transformation can be found later as a part of the case treated at the end of section 6.1.2.

By this transformation T we have thus assured that each row and each column of the current matrix contain at least one zero (i.e. that the maximum number of independent zeros is $\geqq 2$ if $n \geqq 2$).

b) *Determination of a maximum set of independent zeros.* In the following, consider an n-rowed square matrix $\boldsymbol{D}^k$ (which we have already constructed for $k = 0$). For determining a maximum set of independent zeros we use the algorithm of Ford and Fulkerson described in the first chapter. Consider the following transportation network $\boldsymbol{N}^k = (\mathfrak{X}, \mathfrak{U}^k, c)$: The set $\mathfrak{X}$ consists of n "row vertices" $Z_1, \ldots, Z_n$, n "column vertices" $S_1, \ldots, S_n$, the source Q and the sink S. The set $\mathfrak{U}^k$ of arcs is composed of the n arcs (Q, Z_i), the n arcs (S_j, S), the return arc (S, Q) and of all those arcs (Z_i, S_j) for which $d_{ij}^k = 0$ holds.

With the exception of the return arc (S, Q), for which $c\,(S, Q) = \infty$ is true, the capacity bound c is always 1, i.e., $c(u) = 1$ for all $u \in \mathfrak{U}$ with $u \neq (S, Q)$.

Now, we determine a maximum flow $\varphi_0{}^k$ in $\boldsymbol{N}^k$.

ASSERTION. *A maximum set* $\mathfrak{Q}^k$ *of independent zeros in* $\boldsymbol{D}^k$ *corresponds to the set of arcs* $\{(Z_i, S_j) : \varphi_0{}^k(Z_i, S_j) = 1\}$, *and it holds*:

$$\mathfrak{Q}^k = \{d_{ij}^k : \varphi_0{}^k(Z_i, S_j) = 1\}.$$

Proof. If $\varphi_0{}^k(S, Q) = n$, there is nothing to be proved, since in this case each of the arcs (S_j, S) and thus also each of the arcs (Q, Z_i) is traversed by a maximum flux (of the value $\varphi_0{}^k(S_j, S) = \varphi_0{}^k(Q, Z_i) = 1$). But then, there will also exist n arcs (Z_{i_r}, S_{j_r}), $r = 1, \ldots, n$, with $\varphi_0{}^k(Z_{i_r}, S_{j_r}) = 1$. Since these n arcs are pairwise non-adjacent (task!), there corresponds to them a set of n independent zeros $d_{i_r j_r}^k$ in $\boldsymbol{D}^k$. But since there exist in $\boldsymbol{D}^k$ n independent zeros at the most, the proof of the assertion is obvious in this case.

Let us assume that the maximum number m_k of independent zeros in $\boldsymbol{D}^k$ satisfies the relation $m_k > \varphi_0{}^k(S, Q)$. Since a set of m_k pairwise non-adjacent arcs (Z_{i_r}, S_{j_r}) $r = 1, 2, \ldots, m_k$, corresponds to the m_k independent zeros of $\boldsymbol{D}^k$, a flow of the value $m_k > \varphi_0{}^k(S, Q)$ can be sent through the transportation network $\boldsymbol{N}^k$, which contradicts the maximality of $\varphi_0{}^k$.

The reader may find an example for the application of the algorithm of Ford and Fulkerson for ascertaining a maximum flow in chapter 1, page 36.

c) *Determination of a minimum zero-covering.* As we have seen in theorem 6.1, the number of the rows and columns of a minimum zero-covering is equal to the maximum number of independent zeros of a matrix.

Having applied the algorithm described for determining a maximum zero set, it is now very easy to determine a minimum zero-covering. For this, we prove the following

ASSERTION. *Let* $\mathfrak{Z}^+$ *and* $\mathfrak{Z}^-$ *be the sets of the row vertices marked and not marked, respectively* (after breaking off the algorithm of Ford and Fulkerson for the determination of a maximum set of pairwise non-adjacent edges). *Correspondingly, let* $\mathfrak{S}^+$ *and* $\mathfrak{S}^-$ *be the sets of the column vertices marked and not marked, respectively. Then,* $\mathfrak{Z}^- \cup \mathfrak{S}^+$ *is a set of vertices disconnecting the sets* $\mathfrak{Z}$ *and* $\mathfrak{S}$. *That is, the rows which correspond to the non-marked row vertices in* $\boldsymbol{D}^k$ *form together with the columns of* $\boldsymbol{D}^k$, *which correspond to the marked column vertices, a minimum zero-covering* $\mathfrak{N}$.

Proof. First, we show that $\mathfrak{Z}^- \cup \mathfrak{S}^+$ is a set disconnecting the sets of vertices $\mathfrak{Z}$ and $\mathfrak{S}$.

Let $\mathfrak{M}$ denote the set of the arcs (Z_i, S_j) of $\mathfrak{U}$ on which a flow of the value 1 is flowing after breaking off the algorithm of Ford and Fulkerson, thus the maximum set of pairwise non-adjacent arcs (Z_i, S_j) found. Assume there exists an arc $u = (Z^+, S^-) \in \mathfrak{U}$ with $Z^+ \in \mathfrak{Z}^+$ and $S^- \in \mathfrak{S}^-$. Then, $u \in \mathfrak{M}$ is true since otherwise, S^- would be marked according to the algorithm. But then, a flow of the value 1 goes on the arc (Q, Z^+), that is, Z^+ has been labelled backwards, which means that there exists an $S_1^+ \in \mathfrak{S}^+$ with $\varphi_0^k(Z^+, S_1^+) = 1$. In this case, however, Z^+ is incident with two arcs of $\mathfrak{M}$, which is in contradiction to the circumstance that the value of the flow going through a row vertex equals 1 at the most.

Now, we show that to each vertex $P \in \mathfrak{Z}^- \cup \mathfrak{S}^+$ there exists an arc $u \in \mathfrak{M}$ incident with P.

Obviously, this assertion is true for $P \in \mathfrak{S}^+$ since otherwise, φ_0^k would not be maximum (thát is to say, it would be possible, according to the algorithm, to label the sink S via the arc (P, S) and to increase the flow in this way). Now, let $P \in \mathfrak{Z}^-$. If there existed no arc of $\mathfrak{M}$ incident with P, one could label P via the arc (Q, P), which contradicts $P \in \mathfrak{Z}^-$.

Finally, we show that there exists only one vertex of $\mathfrak{Z}^- \cup \mathfrak{S}^+$ which is incident with any arc $u \in \mathfrak{M}$.

Assume it would hold $u = (Z^-, S^+) \in \mathfrak{M}$ with $Z^- \in \mathfrak{Z}^-$ and $S^+ \in \mathfrak{S}^+$. In this case, using the algorithm of Ford and Fulkerson, Z^- would have to be labelled, since S^+ is labelled and a flow of the value 1 goes through (Z^-, S^+). But this, however, is in contradiction to the choice of $u = (Z^-, S^+)$. The assertion is thus proved.

If we have determined a maximum zero-set, i.e., if the algorithm for determining a maximum set of pairwise disjoint arcs breaks off, then the unlabelled vertices of $\mathfrak{Z}$ and the labelled vertices of $\mathfrak{S}$ represent a minimum zero-covering.

a$_2$) *The transformation* T_k: $\boldsymbol{D}^k \to \boldsymbol{D}^{k+1}$, $k = 0, 1, 2, \ldots$ We imagine the matrix $\boldsymbol{D}^k = (d_{ij}^k)_{i,j=1,\ldots,n}$ to be given for $k \geqq 0$ and form the matrix $\boldsymbol{D}^{k+1}$ by means of

$$d_{ij}^{k+1} := d_{ij}^k - u_i^{k+1} - v_j^{k+1}, \quad i, j = 1, 2, \ldots, n.$$

Here, let

$$u_i^{k+1} := \begin{cases} 0 & \text{if the } i\text{th row is covered, i.e., if } Z_i \in \mathfrak{N}, \\ e_k & \text{if the } i\text{th row is not covered, i.e., if } Z_i \notin \mathfrak{N}, \end{cases}$$

$$v_j^{k+1} := \begin{cases} -e_k & \text{if the } j\text{th column is covered, i.e., if } S_j \in \mathfrak{N}, \\ 0 & \text{if the } j\text{th column is not covered, i.e., if } S_j \notin \mathfrak{N}, \end{cases}$$

with

$$e_k := \left\{\min_{i,j} d_{ij}^k\text{: the index pair } (i, j) \text{ is not covered}\right\}.$$

By this transformation it is assured (task!) that $d_{ij}^{k+1} \geqq 0$ holds for all $i, j = 1, 2, \ldots, n$ (taking into account that no negative elements occurred either in $\boldsymbol{D}$ or in $\boldsymbol{D}^0$).

If we consider the change of the objective function, then, analogously to the transition from $\boldsymbol{D}$ to $\boldsymbol{D}^0$, it results:

$$\sum_{i,j=1}^n d_{ij}^k x_{ij} = \sum_{i,j=1}^n [d_{ij}^{k+1} + u_i^{k+1} + v_j^{k+1}]\, x_{ij}$$

$$= \sum_{i,j=1}^n d_{ij}^{k+1} x_{ij} + \sum_{i=1}^n u_i^{k+1} + \sum_{j=1}^n v_j^{k+1},$$

thus

$$\sum_{i,j=1}^n d_{ij} x_{ij} = \sum_{i,j=1}^n d_{ij}^k x_{ij} + \sum_{s=0}^k \left[\sum_{i=1}^n u_i^s + \sum_{j=1}^n v_j^s\right] \quad \text{for} \quad k = 0, 1, 2, \ldots$$

Further, we can determine the transformation constant $\sum_{i=1}^n u_i^{k+1} + \sum_{j=1}^n v_j^{k+1}$. Let z_k be the number of the rows and s_k the number of the columns of the zero-covering $\mathfrak{N}$. Then, it follows

$$\sum_{i=1}^n u_i^{k+1} + \sum_{j=1}^n v_j^{k+1} = (n - z_k)\, e_k - s_k e_k = (n - z_k - s_k)\, e_k \geqq 1,$$

since $e_k \geqq 1$ (for the definition of e_k and the integrality of the $d_{ij}^k \geqq 0$) and since $n - z_k - s_k \geqq 1$ as far as the zero-covering $\mathfrak{N}$ consists of less than n elements (i.e., if the maximum number of independent zeros is still smaller than n).

In the course of each cycle traversed when solving the assignment problem, $\sum_{i,j=1}^n d_{ij}^k x_{ij}$ decreases by at least 1 if the x_{ij} are fixed. Because the x_{ij} and the d_{ij}^k are non-negative, there exists an index t such that

$$\min_{(x_{ij})} \sum_{i,j=1}^n d_{ij}^t x_{ij} = 0$$

holds. This, however, is achieved if there is a set of exactly n independent zeros in $\boldsymbol{D}^t$. Then, the minimum value of the original objective function is

$$\sum_{s=0}^{t}\left[\sum_{i=1}^{n} u_i^s + \sum_{j=1}^{n} v_j^s\right].$$

The optimal solution, i.e., the solution of the assignment problem, is yielded by a minimum set of independent zeros in $\boldsymbol{D}^t$.

To conclude this section, let us calculate a little example.

$$\begin{array}{r|rrrrrr|r}
 & & & & & & & u_i^0\\ \hline
 & 9 & 7 & 8 & 8 & 6 & 2 & 2\\
 & 10 & 6 & 8 & 10 & 8 & 3 & 3\\
\boldsymbol{D} = & 12 & 8 & 9 & 10 & 7 & 2 & 2\\
 & 6 & 9 & 12 & 12 & 10 & 1 & 1\\
 & 10 & 11 & 9 & 8 & 10 & 1 & 1\\
 & 1 & 2 & 1 & 3 & 4 & 10 & 1\\ \hline
v_j^0 & 0 & 1 & 0 & 2 & 3 & 0 &
\end{array}$$

$$\sum_{i=1}^{6} u_i^0 + \sum_{j=1}^{6} v_j^0 = 16$$

$$\begin{array}{r|rrrrrr|r}
 & & & & & & \downarrow & u_i^1\\ \hline
 & 7 & 4 & 6 & 4 & 1 & 0 & 1\\
 & 7 & 2 & 5 & 5 & 2 & 0 & 1\\
\boldsymbol{D}^0 = & 10 & 5 & 7 & 6 & 2 & 0 & 1\\
 & 5 & 7 & 11 & 9 & 6 & 0' & 1\\
 & 9 & 9 & 8 & 5 & 6 & 0 & 1\\
\rightarrow & 0 & 0 & 0' & 0 & 0 & 9 & 0\\ \hline
v_j^1 & 0 & 0 & 0 & 0 & 0 & -1 &
\end{array}$$

By labelling the zeros with a prime, we have indicated a maximum zero set in $\boldsymbol{D}^0$. The arrows indicate the rows and columns which belong to the minimum zero-covering $\mathfrak{N}$. Since $(1, 5)$ is not covered and $d_{15} = 1$, it holds $e_0 = 1$. It results

$$\sum_{i=1}^{n} u_i^1 + \sum_{j=1}^{n} v_j^1 = (n - z_0 - s_0)\, e_0 = 4.$$

$$\begin{array}{r|rrrrrr|r}
 & & & & & \downarrow & \downarrow & u_i^2\\ \hline
 & 6 & 3 & 5 & 3 & 0' & 0 & 1\\
 & 6 & 1 & 4 & 4 & 1 & 0 & 1\\
\boldsymbol{D}^1 = & 9 & 4 & 6 & 5 & 1 & 0 & 1\\
 & 4 & 6 & 10 & 8 & 5 & 0 & 1\\
 & 8 & 8 & 7 & 4 & 5 & 0' & 1\\
\rightarrow & 0 & 0 & 0' & 0 & 0 & 10 & 0\\ \hline
v_j^2 & 0 & 0 & 0 & 0 & -1 & -1 &
\end{array}$$

Since $d_{22}^1 = 1$ is not covered, it holds $e_1 = 1$. We get

$$\sum_{i=1}^{n} u_i^2 + \sum_{j=1}^{n} v_j^2 = 3.$$

$$
\begin{array}{r}
\\
\rightarrow \\ \rightarrow \\ \boldsymbol{D}^2 = \\ \\ \\ \rightarrow \\ v_j^3
\end{array}
\begin{array}{cccccc|c}
 & & & & & \downarrow & u_i^3 \\
5 & 2 & 4 & 2 & 0' & 0 & 0 \\
5 & 0' & 3 & 3 & 1 & 0 & 0 \\
8 & 3 & 5 & 4 & 1 & 0 & 1 \\
3 & 5 & 9 & 7 & 5 & 0' & 1 \\
7 & 7 & 6 & 3 & 5 & 0 & 1 \\
0 & 0 & 0 & 0' & 1 & 11 & 0 \\
0 & 0 & 0 & 0 & 0 & -1 &
\end{array}
$$

We get $e_2 = 1$ and

$$\sum_{i=1}^{n} u_i^3 + \sum_{j=1}^{n} v_j^3 = 2.$$

If $\mathfrak{R} = \{Z_6, S_2, S_5, S_6\}$ had been chosen, then it would even hold $e_2 = 2$.

$$
\begin{array}{r}
\\
\\ \\ \\ \boldsymbol{D}^3 = \\ \\ \rightarrow \\ v_j^4
\end{array}
\begin{array}{cccccc|c}
 & \downarrow & & & \downarrow & \downarrow & u_i^4 \\
5 & 2 & 4 & 2 & 0' & 1 & 2 \\
5 & 0' & 3 & 3 & 1 & 1 & 2 \\
7 & 2 & 4 & 3 & 0 & 0 & 2 \\
2 & 4 & 8 & 6 & 4 & 0' & 2 \\
6 & 6 & 5 & 2 & 4 & 0 & 2 \\
0' & 0 & 0 & 0 & 1 & 12 & 0 \\
0 & -2 & 0 & 0 & -2 & -2 &
\end{array}
$$

It results $e_3 = 2$ and

$$\sum_{i=1}^{n} u_i^4 + \sum_{j=1}^{n} v_j^4 = 4.$$

$$
\boldsymbol{D}^4 = \begin{bmatrix}
3 & 2 & 2 & 0' & 0 & 1 \\
3 & 0' & 1 & 1 & 1 & 1 \\
5 & 2 & 2 & 1 & 0' & 0 \\
0' & 4 & 6 & 4 & 4 & 0 \\
4 & 6 & 3 & 0 & 4 & 0' \\
0 & 2 & 0' & 0 & 3 & 14
\end{bmatrix}
$$

Now, the number of the elements of a maximum set of independent zeros is equal to 6.

We find two maximum zero-sets:

$$\mathfrak{Q}_1 = \{(1, 5), (2, 2), (3, 6), (4, 1), (5, 4), (6, 3)\},$$

$$\mathfrak{Q}^2 = \{(1, 4), (2, 2), (3, 5), (4, 1), (5, 6), (6, 3)\}.$$

Both sets solve the assignment problem. The minimum value of the objective function is $d_{15} + d_{22} + d_{36} + d_{41} + d_{54} + d_{63} = \sum_{s=0}^{4} \left[\sum_{i=1}^{6} u_i^s + \sum_{j=1}^{6} v_j^s \right] = 29.$

6.2. THE TRAVELLING SALESMAN PROBLEM

6.2.1. Formulation of the problem

The travelling salesman problem was named after the following practical task:

Given n places. The traveller starting from one of them, each place must be visited exactly once. In doing so, the place of departure must be reached again at the end of the travel, and the total distance covered during the travel must be minimum.

Even the following problem can be regarded as a travelling salesman problem:

In a rolling-mill, a number of n various profiles must be produced on a rolling train. When going over from the production of one profile to another, a certain number of roll stands have to be altered each time. Here, the number of the roll stands to be altered depends on from what profile one goes to which other one. Find now an order for the production of the types of profiles which entails a minimum number of all roll stands to be rebuilt.

By introducing another (fictitious) profile, this task can be modelled as a travelling salesman problem.

J. Piehler and E. Seiffart [10, 13] showed that some certain scheduling problems can be regarded as a travelling salesman problem.

Using the terminology of graph theory, the following task can be formulated: Given a simple directed arc-valuated graph $G = G(\mathfrak{X}, \mathfrak{U})$ with n vertices $X_1, X_2, \ldots, X_n$. Let a non-negative integer $d(i, j) = d_{ij}$, called *arc length*, be assigned to each arc $(X_i, X_j) = (i, j)$. Consider the arc length matrix $D = (d_{ij})$, with $d_{ij} = \infty$ being set if $(i, j) \notin \mathfrak{U}$, especially also $d_{ii} = \infty$ being set for $i = 1, 2, \ldots, n$.

Find a *hamiltonian circuit* (i.e., a circuit which contains each vertex of G exactly once) $H = (X_{i_1}, X_{i_2}, \ldots, X_{i_n}, X_{i_{n+1}} = X_{i_1})$ with minimum total length, thus

$$\sum_{j=1}^{n} d_{i_j i_{j+1}} \to \min!$$

When treated as a linear integer optimization problem, the travelling salesman problem takes the following form:

$$\sum_{i,j=1}^{n} d_{ij} z_{ij} \to \min!,$$

$$\sum_{j=1}^{n} z_{ij} = 1 \quad \text{for} \quad i = 1, 2, \ldots, n, \tag{1}$$

$$\sum_{i=1}^{n} z_{ij} = 1 \quad \text{for} \quad j = 1, 2, \ldots, n. \tag{2}$$

If s is any natural number with $1 \leqq s \leqq \left[\frac{n}{2}\right]$ and if $i_1, i_2, \ldots, i_s \in \{1, 2, \ldots, n\}$ are s pairwise distinct of the first n natural numbers, i.e., $i_k \neq i_l$ if $k \neq l$, then let it be true that

$$z_{i_1 i_2} + z_{i_2 i_3} + \cdots + z_{i_{s-1} i_s} + z_{i_s i_1} \leqq s - 1, \tag{3}$$

$$z_{ij} \in \{0, 1\} \quad \text{for} \quad i, j = 1, 2, \ldots, n. \tag{4}$$

The constraints (1), (2) and (4) classify $\boldsymbol{Z} = (z_{ij})_{i,j=1,2,\ldots,n}$ as a permutation matrix (analogously to the matrix $\boldsymbol{X}$ in the case of the assignment problem). The constraints (3) assure that a feasible selection $z_{i_1}, z_{i_2}, \ldots, z_{i_n}$ is constituted such that the sequence of arcs

$$\{(X_{i_1}, X_{i_2}), (X_{i_2}, X_{i_3}), \ldots, (X_{i_{n-1}}, X_{i_n}), (X_{i_n}, X_{i_1})\}$$

forms a hamiltonian circuit (that there occur no circuits of a length $< n$).

If the arc length matrix $\boldsymbol{D}$ is regarded as a matrix corresponding to an assignment problem, then one could try to solve it hoping that the solution of the assignment problem yields a solution of the travelling salesman problem. However, this will generally not be the case. For this, consider the example dealt with in the preceding section where the set $\mathfrak{Q}_1$ of arcs is decomposed into the three circuits (1, 5, 4), (2), (3, 6) (when interpreting it as a travelling salesman problem), whereas the set $\mathfrak{Q}_2$ of arcs is decomposed into the three circuits (1, 4), (2), (3, 5, 6).

We observe that the number of the constraints (3) considerably increases with the number n of vertices (say like $n^{n/2}$). Therefore, it is certainly not expedient to use the classical solution methods of linear optimization for solving the travelling salesman problem.

In what follows, we shall discuss two of the various solution procedures which have been developed so far. The first of these two solution algorithms treated is to be owed to J. D. C. Little, K. G. Murty, D. W. Sweeney and C. Karel [7]. It is a procedure based on the so-called *branch and bound method*. It will not be withheld from the reader that this method is, in principle, nothing else but a checking up (though rather ingenious) of all possibilities while some certain cases can already be excluded according to the state of the procedure. After this, we shall indicate a heuristic method devised by M. Held and R. M. Karp.

6.2.2. *A branch-and-bound solution algorithm for the travelling salesman problem*

An example will serve us to explain the algorithm, since this might be the best way to comprehend it. Given the arc length matrix

$$\boldsymbol{D} = \begin{bmatrix} \infty & 6 & 6 & 8 & 5 & 7 \\ 3 & \infty & 5 & 12 & 4 & 10 \\ 4 & 5 & \infty & 5 & 10 & 7 \\ 6 & 7 & 3 & \infty & 6 & 3 \\ 8 & 6 & 8 & 10 & \infty & 8 \\ 9 & 4 & 2 & 7 & 5 & \infty \end{bmatrix}.$$

Find a round trip of minimum total length leading through the six points.

As is well known, in a complete graph with n vertices there are $(n-1)!$ various hamiltonian circuits. Thus, checking up all possibilities would involve a great deal of work, especially in the case where the number n of vertices is not small.

First, carry out the transformation T which has already been applied in the solution of the assignment problem:

$$d_{ij}^0 := d_{ij} - u_i^0 - v_j^0$$

$$\text{with } u_i^0 := \min_j d_{ij},\ v_j^0 := \min_i (d_{ij} - u_i^0) \text{ for } i, j = 1, 2, \ldots, n.$$

In the course of this transformation, the matrix $\boldsymbol{D}$ is transformed into the matrix $\boldsymbol{D}^0$ with

$$\boldsymbol{D}^0 = \begin{bmatrix} \infty & 1 & 1 & 2 & 0 & 2 \\ 0 & \infty & 2 & 8 & 1 & 7 \\ 0 & 1 & \infty & 0 & 6 & 3 \\ 3 & 4 & 0 & \infty & 3 & 0 \\ 2 & 0 & 2 & 3 & \infty & 2 \\ 7 & 2 & 0 & 4 & 3 & \infty \end{bmatrix}.$$

Here, it holds

$$\sum_{i=1}^{6} u_i^0 + \sum_{j=1}^{6} v_j^0 = 24.$$

In principle, we thus still have the former problem which is to be solved, with the difference that the length of a minimum round trip having the matrix $\boldsymbol{D}^0$ is shorter than that having the matrix $\boldsymbol{D}$, namely by 24 (the proof is already given in 6.1.2). Thus, a round trip has a length of at least 24.

In the following, let $\mathfrak{Z}_0$ denote the set of all hamiltonian circuits in $\boldsymbol{G}$ and $c(\mathfrak{Z}_0)$ a lower bound for the length of any hamiltonian circuit of $\mathfrak{Z}_0$. With the present state, we can say that $c(\mathfrak{Z}_0) = 24$ holds.

The idea of the procedure is the following: The set $\mathfrak{Z}_0$ of all hamiltonian circuits is decomposed into two disjoint subsets $\mathfrak{Z}_0'$ and $\mathfrak{Z}_0''$ (branch). For this, we choose (in principle) any arc $(X_i, X_j) = (i, j)$, which is done, however, according to a purposeful rule. Then, we group all those hamiltonian circuits to $\mathfrak{Z}_0'$ which do not contain the arc (i, j), and all those hamiltonian circuits containing the arc (i, j) are grouped to the subset $\mathfrak{Z}_0''$.

It is easy to recognize that of all $(n - 1)!$ hamiltonian circuits from $\mathfrak{Z}_0$ exactly $(n - 2)!$ lie in $\mathfrak{Z}_0''$ and $(n - 2)\,(n - 2)!$ in $\mathfrak{Z}_0'$. As in the case of an increasing n, $\mathfrak{Z}_0'$ contains considerably more hamiltonian circuits than $\mathfrak{Z}_0''$, the arc (i, j) is chosen such that the lower bound for the length of a hamiltonian circuit of $\mathfrak{Z}_0'$ becomes as large as possible. The matrix $\boldsymbol{D}_0'$ assigned to the set $\mathfrak{Z}_0'$ has ∞ in the position i, j, since the arc (i, j) is contained in no hamiltonian circuit. (Now, i, j is chosen so that after having carried out the transformation T, the reduction constant $\sum_{i=1}^{n} u_i + \sum_{j=1}^{n} v_j$ becomes maximum when applied to $\boldsymbol{D}_0'$. This is just what assures the maximum increase of the lower bound $c(\mathfrak{Z}_0')$ for the length of any hamiltonian circuit of $\mathfrak{Z}_0'$).

For determining the arc (i, j) precisely, we proceed as follows: To each element $d_{rs} \in \boldsymbol{D}^0$ with $d_{rs} = 0$ form

$$w_{rs} := \min_{\substack{k \\ (k \neq s)}} d_{rk} + \min_{\substack{k \\ (k \neq r)}} d_{ks}$$

and then

$$w := \max_{\substack{r,s \\ (d_{rs}=0)}} w_{rs}$$

Now, determine a pair of indices (i, j) for which $w_{ij} = w$ holds. In our example, this means

$$w_{15} = 2,\ w_{21} = 1,\ w_{31} = 0,\ w_{46} = 2,\ w_{52} = 3,\ w_{43} = 0,\ w_{63} = 2,\ w_{34} = 2.$$

Since $w_{52} = \max w_{rs}$, we choose the arc (5, 2). The following two matrices $\boldsymbol{D}_0'$ and $\boldsymbol{D}_0''$ correspond to the two sets $\mathfrak{Z}_0'$ and $\mathfrak{Z}_0''$, respectively (the abbreviations used will probably be clear without any commentaries):

$\{\overline{(5, 2)}\} \subseteq \mathfrak{Z}_0'$

$$\boldsymbol{D}_0' = \begin{bmatrix} \infty & 1 & 1 & 2 & 0 & 2 \\ 0 & \infty & 2 & 8 & 1 & 7 \\ 0 & 1 & \infty & 0 & 6 & 3 \\ 3 & 4 & 0 & \infty & 3 & 0 \\ 2 & \infty & 2 & 3 & \infty & 2 \\ 7 & 2 & 0 & 4 & 3 & \infty \end{bmatrix}$$

We apply the transformation T to $\boldsymbol{D}_0'$ and get (with $u_5 = 2, v_2 = 1$ and $u_i = v_j = 0$ otherwise)

$$\boldsymbol{D}_0' = \begin{bmatrix} \infty & 0 & 1 & 2 & 0 & 2 \\ 0 & \infty & 2 & 8 & 1 & 7 \\ 0 & 0 & \infty & 0 & 6 & 3 \\ 3 & 3 & 0 & \infty & 3 & 0 \\ 0 & \infty & 0 & 1 & \infty & 0 \\ 7 & 1 & 0 & 4 & 3 & \infty \end{bmatrix}$$

In doing so, we have continued to use the designation $\boldsymbol{D}_0'$ although the transformation T is carried out. We get

$$c(\mathfrak{Z}_0') = 24 + 3 = 27$$

$\{(5, 2)\} \subseteq \mathfrak{Z}_0''$

$$\boldsymbol{D}_0'' = \begin{bmatrix} \infty & \infty & 1 & 2 & 0 & 2 \\ 0 & \infty & 2 & 8 & \infty & 7 \\ 0 & \infty & \infty & 0 & 6 & 3 \\ 3 & \infty & 0 & \infty & 3 & 0 \\ \infty & 0 & \infty & \infty & \infty & \infty \\ 7 & \infty & 0 & 4 & 3 & \infty \end{bmatrix}$$

The ∞-values of the fifth row are obtained, since one must go from the vertex X_5 to X_2 by all means. The ∞-values of the second column are obtained analogously. It is clear that the element at the position (2, 5) is equal to ∞ since otherwise, there could occur circuits of length 2.

Obviously, we can omit the second column and the fifth row without undergoing a loss of information if we only remember that they have been omitted. The designations used for the reduced matrix are probably easy to understand.

for the lower bound of the length of any hamiltonian circuit which does not contain the arc (5, 2).

$$\boldsymbol{D}_0'' = \left[\begin{array}{ccccc} \infty & 1 & 2 & 0 & 2 \\ 0 & 2 & 8 & \infty & 7 \\ 0 & \infty & 0 & 6 & 3 \\ 3 & 0 & \infty & 3 & 0 \\ 7 & 0 & 4 & 3 & \infty \end{array}\right] \begin{array}{c} 1 \\ 2 \\ 3 \\ 4 \\ 6 \end{array}$$
$$\begin{array}{ccccc} 1 & 3 & 4 & 5 & 6 \end{array}$$

Although we have omitted one row and one column, we want to stick to the designation $\boldsymbol{D}_0''$. Applying a transformation T to $\boldsymbol{D}_0''$ entails no increase of the lower bound, thus

$c(\mathfrak{Z}_0'') = 24$ holds.

Now, we seek an element $\mathfrak{B} \in \{\mathfrak{Z}_0', \mathfrak{Z}_0''\}$ so that

$$c(\mathfrak{B}) = \min \{c(\mathfrak{Z}_0'), c(\mathfrak{Z}_0'')\}$$

holds. Thus, we choose $\mathfrak{B} = \mathfrak{Z}_0''$ for our example.

We set $\mathfrak{Z}_1 := \mathfrak{B} := \mathfrak{Z}_0''$, and using the matrix $\boldsymbol{D}_1 := \boldsymbol{D}_0''$, we determine an arc (i, j) such that $w_{ij} := \max w_{rs}$ with $d_{rs} = 0$ and $d_{rs} \in \boldsymbol{D}_0''$ holds (cf. p. 161). In our example, we ascertain the arc (1, 5) with $w_{15} = 4$. We get:

$\{(5, 2), (\overline{1, 5})\} \subseteq \mathfrak{Z}_1'$

$$\boldsymbol{D}_1' = \left[\begin{array}{ccccc} \infty & 1 & 2 & \infty & 2 \\ 0 & 2 & 8 & \infty & 7 \\ 0 & \infty & 0 & 6 & 3 \\ 3 & 0 & \infty & 3 & 0 \\ 7 & 0 & 4 & 3 & \infty \end{array}\right] \begin{array}{c} 1 \\ 2 \\ 3 \\ 4 \\ 6 \end{array}$$
$$\begin{array}{ccccc} 1 & 3 & 4 & 5 & 6 \end{array}$$

$\{(5, 2), (1, 5)\} \subseteq \mathfrak{Z}_1''$

$$\boldsymbol{D}_1'' = \left[\begin{array}{cccc} \infty & 2 & 8 & 7 \\ 0 & \infty & 0 & 3 \\ 3 & 0 & \infty & 0 \\ 7 & 0 & 4 & \infty \end{array}\right] \begin{array}{c} 2 \\ 3 \\ 4 \\ 6 \end{array}$$
$$\begin{array}{cccc} 1 & 3 & 4 & 6 \end{array}$$

Applying the transformation T to the matrices $\boldsymbol{D}_1'$, $\boldsymbol{D}_1''$ leads to

$$\boldsymbol{D}_1' := \left[\begin{array}{ccccc} \infty & 0 & 1 & \infty & 1 \\ 0 & 2 & 8 & \infty & 7 \\ 0 & \infty & 0 & 3 & 3 \\ 3 & 0 & \infty & 0 & 0 \\ 7 & 0 & 4 & 0 & \infty \end{array}\right] \begin{array}{c} 1 \\ 2 \\ 3 \\ 4 \\ 6 \end{array}$$
$$\begin{array}{ccccc} 1 & 3 & 4 & 5 & 6 \end{array}$$

with $c(\mathfrak{Z}_1') := c(\mathfrak{Z}_1) + 4 = 28$.

$$\boldsymbol{D}_1'' := \left[\begin{array}{cccc} \infty & 0 & 6 & 5 \\ 0 & \infty & 0 & 3 \\ 3 & 0 & \infty & 0 \\ 7 & 0 & 4 & \infty \end{array}\right] \begin{array}{c} 2 \\ 3 \\ 4 \\ 6 \end{array}$$
$$\begin{array}{cccc} 1 & 3 & 4 & 6 \end{array}$$

with $c(\mathfrak{Z}_1'') := c(\mathfrak{Z}_1) + 2 = 26$.

Now, we seek the element $\mathfrak{B} \in \{\mathfrak{Z}_0', \mathfrak{Z}_1', \mathfrak{Z}_1''\}$ with the minimum $c(\mathfrak{B})$, i.e., for our case $\mathfrak{Z}_1''$ with $c(\mathfrak{Z}_1'') = 26$. We set $\mathfrak{Z}_2 := \mathfrak{Z}_1''$ and determine the arc (2, 3) with the aid of the branching criterion for $w_{23} = \max w_{rs} = 5$. We continue this branching process relative to that arc, and after having carried out the transformation T, we get:

$\{(5, 2), (1, 5), \overline{(2, 3)}\} \subseteqq \mathfrak{Z}_2'$

$$D_2' = \begin{array}{cccc|c} \infty & \infty & 1 & 0 & 2 \\ 0 & \infty & 0 & 3 & 3 \\ 3 & 0 & \infty & 0 & 4 \\ 7 & 0 & 4 & \infty & 6 \\ \hline 1 & 3 & 4 & 6 & \end{array}$$

with $u_1 = 5$ and $u_i = v_j = 0$ otherwise, thus

$c(\mathfrak{Z}_2') = c(\mathfrak{Z}_1'') + 5 = 31.$

$\{(5, 2), (1, 5), (2, 3)\} \subseteqq \mathfrak{Z}_2''$

$$D_2'' = \begin{array}{ccc|c} \infty & 0 & 3 & 3 \\ 0 & \infty & 0 & 4 \\ 0 & 0 & \infty & 6 \\ \hline 1 & 4 & 6 & \end{array}$$

with $u_3 = 4$, $v_1 = 3$ and $u_i = v_j = 0$ otherwise, thus

$c(\mathfrak{Z}_2'') = c(\mathfrak{Z}_1'') + 7 = 33.$

The lower bound c which is now the smallest one is $c(\mathfrak{Z}_0') = 27$. We set $\mathfrak{Z}_3 := \mathfrak{Z}_0'$, determine for example the arc (1, 5) with the aid of the branching criterion (it could have been also one of the other three arcs (2, 1), (3, 4), (6, 3) with $w_{rs} = 1$) and get:

$\{\overline{(5, 2)}, \overline{(1, 5)}\} \subseteqq \mathfrak{Z}_3'$

$$D_3' := \begin{bmatrix} \infty & 0 & 1 & 2 & \infty & 2 \\ 0 & \infty & 2 & 8 & 0 & 7 \\ 0 & 0 & \infty & 0 & 5 & 3 \\ 3 & 3 & 0 & \infty & 2 & 0 \\ 0 & \infty & 0 & 1 & \infty & 0 \\ 7 & 1 & 0 & 4 & 2 & \infty \end{bmatrix}$$

with $c(\mathfrak{Z}_3') = c(\mathfrak{Z}_3) + 1 = 28.$

$\{\overline{(5, 2)}, (1, 5)\} \subseteqq \mathfrak{Z}_3''$

$$D_3'' = \begin{array}{ccccc|c} 0 & \infty & 2 & 8 & 7 & 2 \\ 0 & 0 & \infty & 0 & 3 & 3 \\ 3 & 3 & 0 & \infty & 0 & 4 \\ \infty & \infty & 0 & 1 & 0 & 5 \\ 7 & 1 & 0 & 4 & \infty & 6 \\ \hline 1 & 2 & 3 & 4 & 6 & \end{array}$$

with $c(\mathfrak{Z}_3'') = c(\mathfrak{Z}_3) = 27.$

Then, we branch $\mathfrak{Z}_3''$ and choose the arc (2, 1). With $\mathfrak{Z}_4 := \mathfrak{Z}_3''$ we get:

$\{\overline{(5, 2)}, (1, 5), \overline{(2, 1)}\} \subseteqq \mathfrak{Z}_4'$

$$D_4' = \begin{array}{ccccc|c} \infty & \infty & 2 & 8 & 7 & 2 \\ 0 & 0 & \infty & 0 & 3 & 3 \\ 3 & 3 & 0 & \infty & 0 & 4 \\ \infty & \infty & 0 & 1 & 0 & 5 \\ 7 & 1 & 0 & 4 & \infty & 6 \\ \hline 1 & 2 & 3 & 4 & 6 & \end{array}$$

with $c(\mathfrak{Z}_4') := c(\mathfrak{Z}_4) + 2 = 29.$

$\{\overline{(5, 2)}, (1, 5), (2, 1)\} \subseteqq \mathfrak{Z}_4''$

$$D_4'' = \begin{array}{cccc|c} 0 & \infty & 0 & 3 & 3 \\ 3 & 0 & \infty & 0 & 4 \\ \infty & 0 & 1 & 0 & 5 \\ 1 & 0 & 4 & \infty & 6 \\ \hline 2 & 3 & 4 & 6 & \end{array}$$

with $c(\mathfrak{Z}_4'') := c(\mathfrak{Z}_4) = 27.$

We continue to branch $\mathfrak{Z}_5 := \mathfrak{Z}_4''$ and can choose for example the arc (3, 4) for branching. We get

$$\{\overline{(5, 2)}, (1, 5), (2, 1), \overline{(3, 4)}\} \subseteq \mathfrak{Z}_5'$$

$$\boldsymbol{D}_5' = \begin{bmatrix} 0 & \infty & \infty & 3 \\ 3 & 0 & \infty & 0 \\ \infty & 0 & 0 & 0 \\ 1 & 0 & 3 & \infty \end{bmatrix} \begin{matrix} 3 \\ 4 \\ 5 \\ 6 \end{matrix} \qquad \text{columns: } 2 \; 3 \; 4 \; 6$$

with $c(\mathfrak{Z}_5') = c(\mathfrak{Z}_5) + 1 = 28$.

$$\{\overline{(5, 2)}, (1, 5), (2, 1), (3, 4)\} \subseteq \mathfrak{Z}_5''$$

$$\boldsymbol{D}_5'' = \begin{bmatrix} 2 & \infty & 0 \\ \infty & 0 & 0 \\ 0 & 0 & \infty \end{bmatrix} \begin{matrix} 4 \\ 5 \\ 6 \end{matrix} \qquad \text{columns: } 2 \; 3 \; 6$$

with $c(\mathfrak{Z}_5'') = c(\mathfrak{Z}_5) + 1 = 28$.

We thus have four sets with the same lower bound $c = 28$ at our disposal, namely $\mathfrak{Z}_1'$, $\mathfrak{Z}_3'$, $\mathfrak{Z}_5'$ and $\mathfrak{Z}_5''$, to continue the branching process. It is advisable to carry on branching in that set which contains as few elements as possible. In our case, we would choose $\mathfrak{Z}_6 := \mathfrak{Z}_5''$ and get (when branching say the arc (6, 2))

$$\{\overline{(5, 2)}, (1, 5), (2, 1), (3, 4), \overline{(6, 2)}\} \subseteq \mathfrak{Z}_6'$$

$$\boldsymbol{D}_6' = \begin{bmatrix} 0 & \infty & 0 \\ \infty & 0 & 0 \\ \infty & 0 & \infty \end{bmatrix} \begin{matrix} 4 \\ 5 \\ 6 \end{matrix} \qquad \text{columns: } 2 \; 3 \; 6$$

with $c(\mathfrak{Z}_6') = c(\mathfrak{Z}_6) + 2 = 30$.

$$\{\overline{(5, 2)}, (1, 5), (2, 1), (3, 4), (6, 2)\} \subseteq \mathfrak{Z}_6''$$

$$\boldsymbol{D}_6'' = \begin{bmatrix} \infty & 0 \\ 0 & \infty \end{bmatrix} \begin{matrix} 4 \\ 5 \end{matrix} \qquad \text{columns: } 3 \; 6$$

with $c(\mathfrak{Z}_6'') = c(\mathfrak{Z}_6) + 0 = 28$.

Thus, we have found a hamiltonian circuit of minimum length, namely

$$\{(1, 5), (5, 3), (3, 4), (4, 6), (6, 2), (2, 1)\}$$

of the length

$$5 \quad +8 \quad +5 \quad +3 \quad +4 \quad +3 \quad = 28.$$

The procedure given above can be described by means of a valuated rooted tree. Figure 6.2 shows the *rooted tree* for our example.

If one wishes to find all hamiltonian circuits of minimum length, then branching has to be carried on wherever the lower bound has not yet exceeded the value 28. The reader may find out by himself that there exist still other hamiltonian circuits of length 28.

In an analogous way one can make out all hamiltonian circuits which do not exceed a well-defined value in their lengths.

In an unfavourable situation, however, this procedure may involve a great deal of work. From the point of view of computation, it is less advantageous if a set $\mathfrak{Z}_i'$ or $\mathfrak{Z}_i''$ which has been rejected for the present might perhaps be required again, after a certain time. The number of the sets to be branched potentially increases by 1 with each branching.

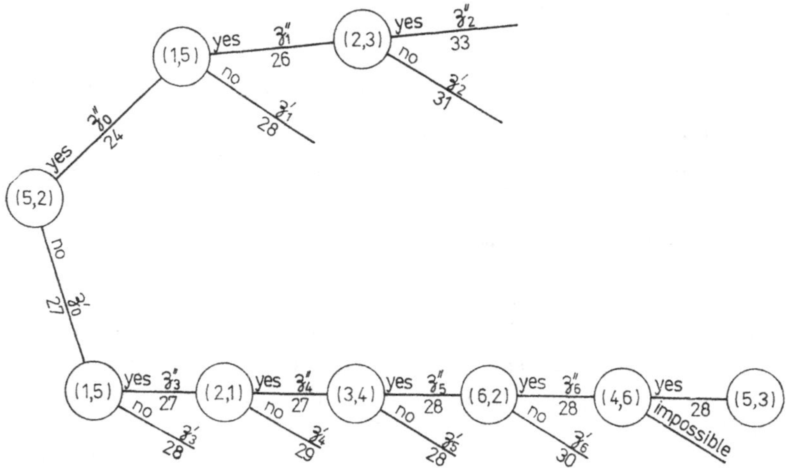

Fig. 6.2

Algorithm for solving the travelling salesman problem

(i) Set $k := 0$, $\mathfrak{Z}_k := \mathfrak{Z}$, $\boldsymbol{D}_k := \boldsymbol{D}$, $c(\mathfrak{Z}_k) := 0$, $\mathfrak{W} := \emptyset$.

(ii) Reduce $\boldsymbol{D}_k$ by means of the transformation T, i.e., if $\boldsymbol{A} = (a_{ij})$ is a matrix and $c(\mathfrak{Z}_k)$ a natural number, then form $u_i := \min_j a_{ij}$, $v_j := \min_i (a_{ij} - u_i)$, $a_{ij} := a_{ij} - u_i - v_j$,

$$c(\mathfrak{Z}_k) := c(\mathfrak{Z}_k) + \sum_{i=1}^{n} u_i + \sum_{j=1}^{n} v_j.$$

(iii) Choose an arc (i, j) (this is done by means of the reduced matrix $\boldsymbol{D}_k = (d_{rs})$ by determining the numbers w_{rs} according to $w_{rs} := \min_{t \neq s} d_{rt} + \min_{t \neq r} d_{ts}$ for all pairs r, s for which $d_{rs} = 0$ holds: $w_{ij} := \max_{r,s} w_{rs}$ with $d_{rs} = 0$).

(iv) Decompose $\mathfrak{Z}_k$ into two sets $\mathfrak{Z}_k'$ and $\mathfrak{Z}_k''$ (here, $\mathfrak{Z}_k'$ and $\mathfrak{Z}_k''$ are the sets of those hamiltonian circuits of $\mathfrak{Z}_k$ which do not contain the arc (i, j) and those which do, respectively) and form $\mathfrak{W} := \mathfrak{W} \cup \{\mathfrak{Z}_k', \mathfrak{Z}_k''\}$.

(v) From $\boldsymbol{D}_k$ form the matrices $\boldsymbol{D}_k'$, $\boldsymbol{D}_k''$ which correspond to the sets $\mathfrak{Z}_k'$ and $\mathfrak{Z}_k''$, respectively ($\boldsymbol{D}_k'$ is obtained from $\boldsymbol{D}_k$ by substituting ∞ for the element d_{ij}, with all other elements remaining unchanged; $\boldsymbol{D}_k''$ is obtained from $\boldsymbol{D}_k$ by omitting the ith row and the jth column and by substituting ∞ for that element which corresponds to an arc that would yield, together with those already chosen, a circuit of a length $< n$; all other elements, however, remain unchanged).

(vi) Reduce $\boldsymbol{D}_k'$ and $\boldsymbol{D}_k''$ according to (ii) and determine $c(\mathfrak{Z}_k')$ and $c(\mathfrak{Z}_k'')$.

(vii) Determine $\mathfrak{V} \in \mathfrak{W}$ such that $c(\mathfrak{V}) = \min_{\mathfrak{T} \in \mathfrak{W}} c(\mathfrak{T})$ holds. $\mathfrak{W} := \mathfrak{W} \setminus \{\mathfrak{V}\}$ (this ensures that the set $\mathfrak{V}$ is split up only once).

(viii) If the set $\mathfrak{V}$ contains still one hamiltonian circuit, then go to the end.

(ix) Set $k := k + 1$, $\mathfrak{Z}_k := \mathfrak{V}$; go to (iii).

End: The element from $\mathfrak{V}$ is a hamiltonian circuit of minimum length.

6.2.3. A heuristic method for solving the travelling salesman problem

In this section, let us suppose that the matrix $\boldsymbol{D} = (d_{ij})_{i,j=1,2,\ldots,n}$ underlying the travelling salesman problem is symmetrical, that means that the graph $\boldsymbol{G} = (\mathfrak{X}, \mathfrak{U})$ is undirected. This restriction is not absolutely necessary, but it makes some things simpler. Our intention is to illustrate only the principle of the approach devised by M. Held and R. M. Karp.

First, we need the term of a *1-tree*. Let $\boldsymbol{G} = (\mathfrak{X}, \mathfrak{U})$ be an undirected connected graph with n vertices and $\geqq n$ edges. We imagine any vertex (say that with the number 1) to be marked and suppose that the relation $v(1) \geqq 2$ is true for its degree. A *1-tree* $\boldsymbol{T} = (\mathfrak{X}, \mathfrak{U})$ of $\boldsymbol{G}$ is a connected subgraph with n vertices and n edges, with $v(1) = 2$ being valid in $\boldsymbol{T}$ and the only circuit contained in $\boldsymbol{T}$ (cf. chapter 1) containing the vertex 1.

One can find a 1-tree in $\boldsymbol{G}$ in the following way: Delete the vertex 1 and all edges from $\boldsymbol{G}$ which are incident with it. Then, search for a spanning tree in the residual graph and add the vertex 1 and two edges incident with it. The resulting graph is a 1-tree $\boldsymbol{T}$.

One can see in particular that a hamiltonian circuit, that is, a round trip, is a 1-tree.

For a hamiltonian circuit $\boldsymbol{H}'$ of minimum length, that is, for an optimal round trip in $\boldsymbol{G}$, it holds: The length of $\boldsymbol{H}'$ is at least that of a 1-tree of minimum length.

The purpose of the procedure to be described in the following is to have differences only in the valuations d_{ij}^k, $i, j = 1, 2, \ldots, n$. The edge valuations — preceded by a suitable valuation of the vertices — are made such that if $\boldsymbol{T}_k'$ is an optimal round trip in $\boldsymbol{G}^k$, it is also an optimal one in $\boldsymbol{G} = \boldsymbol{G}^0$.

Given the graph $\boldsymbol{G} = \boldsymbol{G}^0$ with the arc length matrix

$$\boldsymbol{D} = \boldsymbol{D}^0 = (d_{ij}^0)_{i,j=1,2,\ldots,n}.$$

Let further $\boldsymbol{p}^0 = (p_1^0, p_2^0, \ldots, p_n^0)$ be any vertex valuation of $\boldsymbol{G}$, that is, the number p_l^0, which is in general an integer, is assigned to the vertex l. From $\boldsymbol{p}^0$ we construct an edge valuation as follows:

$$D^1 = (d_{ij}^1)_{i,j=1,2,\ldots,n} \tag{1}$$
$$\text{with} \quad d_{ij}^1 = d_{ij}^0 + p_i^0 + p_j^0 \quad \text{for } i, j = 1, 2, \ldots, n.$$

Let $H = (i_1, i_2, \ldots, i_n, i_{n+1} = i_1)$ be any round trip in G with the edges (i_1, i_2), $(i_2, i_3), \ldots, (i_{n-1}, i_n), (i_n, i_1)$ and the length

$$d^0(H) = d_{i_1i_2}^0 + d_{i_2i_3}^0 + \cdots + d_{i_{n-1}i_n}^0 + d_{i_ni_1}^0.$$

Then, for the modified edge valuation and any round trip H, it follows:

$$d^1(H) = d^0(H) + 2(p_1^0 + p_2^0 + \cdots + p_n^0), \tag{2}$$

since the valuation of each vertex is counted exactly twice when making the summation over all edges of H according to (1). This, however, means: A d^0-optimal round trip is also d^1-optimal, and vice versa. Both differ from each other only in their "valuation" length by the fixed value $2\,(p_1^0 + p_2^0 + \cdots + p_n^0)$.

If we consider any 1-tree T_1' of minimum d^1-length, then the relation

$$d^1(H') \geqq d^1(T_1') \tag{3}$$

is true for an optimal round trip H' because the set of the 1-trees also contains the round trips, as already said.

Now, we calculate $d^1(T_1')$. Let $r^1 = (r_1^1, r_2^1, \ldots, r_n^1)$ be the degree vector of T_1', thus let r_i^1 be the degree of the vertex i in T_1'. Then, it holds

$$d^1(T_1') = d^0(T_1') + p_1^0r_1^1 + p_2^0r_2^1 + \cdots + p_n^0r_n^1, \tag{4}$$

since besides the original edge valuations each vertex contributes to the total length, and the vertex i delivers the value p_i^0 just as often as it is indicated by the number r_i^1 of the edges incident with it in T_1'.

From (2), (3) and (4), it follows

$$d^0(H') \geqq d^0(T_1') + p_1^0(r_1^1 - 2) + p_2^0(r_2^1 - 2) + \cdots + p_n^0(r_n^1 - 2). \tag{5}$$

Since this inequality is true in the case of any vertex valuation p^0, then, after having given p^0 and having determined a 1-tree T_1' of minimum length (relative to the edge valuation D^1 evoked by p^0), we get a lower bound for the length of an optimal round trip H'.

In the following, we will consider only those vertex valuations $p = (p_1, p_2, \ldots, p_n)$ for which $p_1 + p_2 + \cdots + p_n = 0$ is true. Then, because of (2) it even follows $d^0(T_1') = d^1(T_1')$ if T_1' is both a minimum 1-tree and a round trip.

Algorithm of Held and Karp

Given $G = (\mathfrak{X}, \mathfrak{U})$ with the arc length matrix $D^0 = (d_{ij}^0)$.

(i) Set $p := (p_1, p_2, \ldots, p_n) := (0, 0, \ldots, 0)$,
$k := 0$.

Fix l_0 (cf. (vi)).

(ii) Set $\boldsymbol{D} := (d_{ij})_{i,j=1,2,\ldots,n}$ with $d_{ij} := d_{ij}^0 + p_i + p_j$ for $i, j = 1, 2, \ldots, n$.

(iii) Determine a minimum 1-tree $\boldsymbol{T}'$ in $\boldsymbol{G}$ relative to the valuation $\boldsymbol{D}$.

(iv) Form the degree vector $\boldsymbol{r} = (r_1, r_2, \ldots, r_n)$ of $\boldsymbol{T}'$.
Form $k := \max\{k, d^0(\boldsymbol{T}') + p_1(r_1 - 2) + p_2(r_2 - 2) + \cdots + p_n(r_n - 2)\}$.
Form $\boldsymbol{p} := \boldsymbol{p} + \boldsymbol{r} - (2, 2, \ldots, 2)$.

(v) If $\boldsymbol{T}'$ is a round trip, go to the end.

(vi) If k has not improved in the course of the last l_0 iterations (that is, if it has not increased), then break it off.

(vii) Go to (ii).

End: $\boldsymbol{T}'$ is an optimal round trip of length $d^0(\boldsymbol{T}') = d(\boldsymbol{T}')$.

Breaking-off: An optimal round trip has a length $\geqq k$. An optimal round trip was not found.

To conclude, let us calculate a little example drawn from the textbook of H. Noltemeier [9]. Consider the graph represented in Fig. 6.3a. The vertices are numbered from 1 to 7, and the valuations $d_{ij} := d_{ij}^0$ are written at the edges. The unique 1-tree $\boldsymbol{T}'$ of minimum length is indicated by doubly drawn edges; it holds $d^0(\boldsymbol{T}') = d(\boldsymbol{T}') = 18$, a first lower bound for the length of a minimum round trip is thus found. We get

$$\boldsymbol{r} = (2, 3, 1, 1, 1, 4, 2),$$
$$k = 18,$$
$$\boldsymbol{p} = (0, 1, -1, -1, -1, 2, 0);$$

$\boldsymbol{T}'$ is no round trip, we go back to (ii).

The valuations p_i are inscribed in the vertices of Fig. 6.3b, the new edge valuations d_{ij} are written at the edges. Here again, a minimum 1-tree is indicated by doubly drawn edges (obviously, $\boldsymbol{T}'$ is not unique). We get

$$\boldsymbol{r} = (2, 3, 2, 3, 1, 1, 2),$$
$$k = \max(18, 22 - 1) = 21,$$
$$\boldsymbol{p} = (0, 2, -1, 0, -2, 1, 0);$$

$\boldsymbol{T}'$ is not a round trip; an optimal round trip has a length $\geqq 21$. We go back to (ii).

The new valuations p_i are inscribed in the vertices of Fig. 6.3c, the new edge valuations d_{ij} are written at the edges. A minimum 1-tree $\boldsymbol{T}'$ is again marked by doubly drawn edges ($\boldsymbol{T}'$ is not unique). Since the new degree vector $\boldsymbol{r}$ possesses only just the number 2 as components, we have finished. $\boldsymbol{T}'$ is a round trip, and thus we have found an optimal round trip of length 24.

One can easily verify that there exists still another optimal round trip (task!).

In the graph represented in Fig. 6.3c, one can also find minimum 1-trees which are not round trips. If such a tree had been found, then the calculation would have to be carried on. Of course, the vertex valuation $\boldsymbol{p}$ need not necessarily be made in the way indicated by us. The purpose of our valuation is (since in a round trip, each vertex has the degree 2) to lower the degree of the vertices having a high degree in $\boldsymbol{T}'$ by a high valuation of the edges incident with them, in the next step.

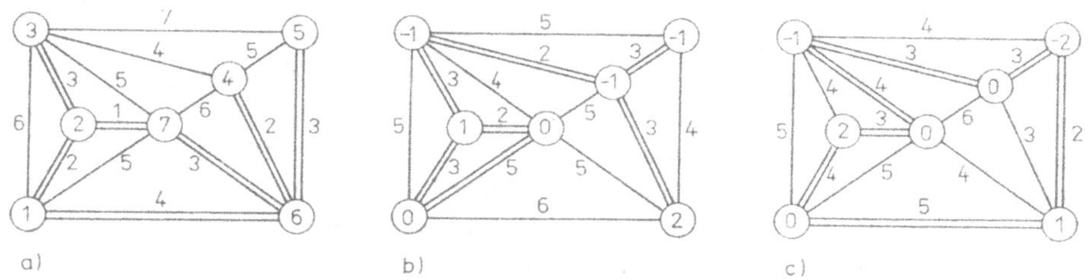

Fig. 6.3

6.3. FINAL OBSERVATIONS

From a computational point of view, too, the Hungarian method for solving the linear assignment problem can be considered satisfactory. On the other hand, the situation is quite different for the solution algorithms for the travelling salesman problem known so far.

Considering the vast amount of articles published so far on this problem, and the algorithms suggested there, one can say that no generally satisfactory algorithm has yet been set up. What ever algorithm is proposed and considered sufficient, one can always form classes of travelling salesman problems to which this algorithm can successfully be applied, but there will also exist classes in the case of which its use is hardly efficient.

As is also the case with other optimization tasks to be solved, the exact methods (e.g. the branch-and-bound algorithm for solving the travelling salesman problem) are seldom the most rapid ones.

The algorithm for solving the travelling salesman problem, which had been suggested by M. Held and R. M. Karp [5, 6] and improved later by K. H. Hansen and J. Krarup [4] and which is based on the highly efficient algorithm for determining a minimum spanning tree in a graph, leads quickly to an optimal round trip in many practical cases, but the finiteness of the algorithm is not guaranteed.

M. Schoch et al. developed for a number of combinatorial optimization problems (as also for the assignment problem and the travelling salesman problem) solution algorithms on the basis of the principle of permanence elaborated by them [12]. The algorithm based on this principle turned out to be rather efficient for their examples concerning the travelling salesman problem.

Another algorithm founded on the branch-and-bound principle is proposed by W. L. Eastman [1]. He regards the matrix corresponding to the travelling salesman problem as the matrix of an assignment problem. If the solution consists of only one cycle, then we have finished (in our example concerning the assignment problem on page 157, both optimal solutions to the assignment problem comprised three cycles each). Otherwise, any cycle $(i_1, i_2, \ldots, i_t, i_1)$ of the solution z^0 to the assignment problem is considered. By setting one after another the elements $d^0_{i_1i_2}, d^0_{i_2i_3}, \ldots, d^0_{i_{t-1},i_t}, d^0_{i_ti_1}$ of the matrix $\boldsymbol{D}^0$ of the assignment problem equal to ∞, t $(t < n)$ new assignment problems arise, with the solution z^0 being excluded from all possible solutions. The procedure is continued with that matrix for which the value of the objective function of the appertaining assignment problem becomes minimum. The task is solved if an assignment problem is found which ensures that the set of cycles corresponding to the solution consists of only one element.

An improved Eastman's method for symmetrical round trips can be found by the reader in H. Steckhan and R. Thome [15].

In [3], the reader will find another representation of the travelling salesman problem as a linear optimization task.

Long discussions about solution methods for the travelling salesman problem can be read in [2, 9, 16, 17]. General information about branch-and-bound methods can be found in [8].

In [14], complete FORTRAN programs will be found suitable for a number of problems of Operations Research as also for the assignment and the travelling salesman problems.

In recent times, it has become common practice to judge the quality of algorithms through time evaluations of the worst case. Without going into details here, we shall make a few remarks on it (the interested reader is referred to the excellent book of M. R. Garey and D. S. Johnson: Computers and Intractability: A Guide to the Theory of NP-Completeness, San Francisco, Cal., 1978):

The solution algorithm for the assignment problem (Hungarian method) is polynomial, or expressed in a somewhat lax way, it is a good algorithm.

Things are quite different in the case of all *exact* solution algorithms for the travelling salesman problem. Compared with the worst case, they are altogether non-polynomial or, in other words, *exponential*, simply bad. It must be pointed out that the finding of an exact solution algorithm (i.e., of an algorithm which yields in every exemplifying case an optimal round trip) would be equivalent to solving a fundamental mathematical hypothesis. Such an attempt is *not* recommended to the reader.

6.4. BIBLIOGRAPHY

[1] Eastman, W. L.: A solution to the Traveling Salesman Problem, American Summer Meeting of the Econometric Society, Cambridge, Mass., August 1958.

[2] Finkel'štejn, Ju. Ju.: Näherungsverfahren und Anwendungsprobleme der diskreten Optimierung (Russian), Moscow 1976.

[3] Korbut, A. A., und J. J. Finkelstein: Diskrete Optimierung, Berlin 1971. (Translated from Russian.)
[4] Hansen, K. H., and J. Krarup: Improvements of the Held-Karp-algorithm for the symmetric Traveling Salesman Problem, Math. Progr. 7 (1974), 87–96.
[5] Held, M., and R. M. Karp: The Traveling Salesman Problem and minimum spanning trees, Operations Res. 18 (1970), 1138–1162.
[6] Held, M., and R. M. Karp: The Traveling Salesman Problem and minimum spanning trees II, Math. Progr. 1 (1971), 6–25.
[7] Little, J., K. Murty, D. Sweeney and C. Karel: An algorithm for the Traveling Salesman Problem, Operations Res. 11 (1963), 972–989.
[8] Mitten, L. G.: Branch-and-bound methods: General formulation and properties, Operations Res. 18 (1970), 24–34.
[9] Noltemeier, H.: Graphentheorie mit Algorithmen und Anwendungen, Berlin/New York 1976.
[10] Piehler, J.: Ein Beitrag zum Reihenfolgeproblem, Unternehmensforschung 4 (1960), 138–142.
[11] Sachs, H.: Theorie der endlichen Graphen, Teil I, Leipzig 1970.
[12] Schoch, M.: Das Erweiterungsprinzip und seine Anwendung, Berlin and Munich/Vienna 1976.
[13] Seiffart, E.: Über Lösungsmethoden einiger Reihenfolgeprobleme, Wiss. Z. TH Magdeburg 9 (1965), 1–5.
[14] Späth, H.: Ausgewählte Operations-Research-Algorithmen in FORTRAN, Munich/Vienna 1975.
[15] Steckhan, H., und R. Thome: Vereinfachungen der Eastmanschen Branch-and-Bound-Lösung für symmetrische Traveling Salesman Probleme, in: Operations-Res. Verfahren Bd. XIV. Meisenheim 1972, p. 361–389.
[16] Thompson, G. L.: Algorithmic and computational methods for solving symmetric and asymmetric traveling salesman problems, in: Workshop in Integer Programming, Bonn 1975 (in print).
[17] Tinhofer, G.: Methoden der Angewandten Graphentheorie, Vienna/New York 1976.

Chapter 7

Coding and decision graphs

7.1. FORMULATION OF THE PROBLEM

To illustrate the problematic nature of this chapter, we shall start with a simple example.

In an urn there are 100 balls of 11 different colours. We denote them by F_1, F_2, ..., F_{11}. Let the relative frequency of the colour F_i be denoted by p_i, and let the values 30, 19, 11, 10, 9, 6, 5, 4, 3, 2 and 1 stand for the frequencies. One player takes a ball out of the urn (i.e., randomly), without the other player knowing the colour of the ball. The other player tries to find out the colour of the ball by asking a series of questions which have to be truthfully answered by the first player with "yes" or "no". (The first question could be for example: "Is the ball a member of one of the colour classes F_1, F_4 or F_9?".) The conditions of play are as follows: For each question asked by the second player, he has to pay one Mark to the first one. When he has found out the colour of the ball, he gets three Marks from the first player.

Does there exist for the second player a questioning strategy which permits him to win "at long sight" (when repeating the game sufficiently often)? In the course of this chapter, we shall see how the second player has to run the play.

To begin our study of the general problematic nature of questionnaires, we consider the following model:

Given a set

$$\mathfrak{F} = \{Y_1, \ldots, Y_N\}$$

of N *events* (in our example, it holds $N = 11$, and the event Y_i corresponds to the taking out of a ball belonging to the colour class F_i) with a given probability distribution of

$$\{p(Y_i)\}_{i=1,\ldots,N},$$

thus

$$0 < p(Y_i) < 1, \qquad \sum_{i=1}^{N} p(Y_i) = 1, \qquad N \geqq 2.$$

Further, let a set

$$\mathfrak{G} = \{X_0, X_1, \ldots, X_n\}, \qquad \mathfrak{G} \cap \mathfrak{F} = \emptyset,$$

of *questions* be given.

We consider the following directed graph $\boldsymbol{F}(\mathfrak{X}, \mathfrak{U})$ which becomes a rooted tree with the root X_0:

Let $\mathfrak{X} = \mathfrak{F} \cup \mathfrak{G}$ be the set of vertices. Let the set $\mathfrak{U}$ of the directed edges (= arcs) be a subset of $(\mathfrak{G} \times \mathfrak{G}) \cup (\mathfrak{G} \times \mathfrak{F})$, i.e., there exist only arcs of the types (X_i, X_j) and (X_k, Y_l). We denote by ΓZ the set of the immediate successors of a vertex Z and by $\Gamma^{-1}Z$ the set of the immediate predecessors of a vertex Z, and let the following relations be true:

a) $\Gamma^{-1}X_0 = \emptyset$ (that means that the first question to be asked has no predecessor);

b) $|\Gamma X| \geqq 2$ for all $X \in \mathfrak{G}$ (that means that at least two different answers can be given to each question asked, and one of the answers can result in the identification of an event, thus $\Gamma X \cap \mathfrak{F} \neq \emptyset$, or the formulation of a new question, thus $\Gamma X \cap \mathfrak{G} \neq \emptyset$);

c) $\Gamma Y = \emptyset$ for all $Y \in \mathfrak{F}$ (that means that after the identification of an event — say after the determination of the colour of the ball in our example — no other question is asked), which already follows from the definition of $\mathfrak{U}$;

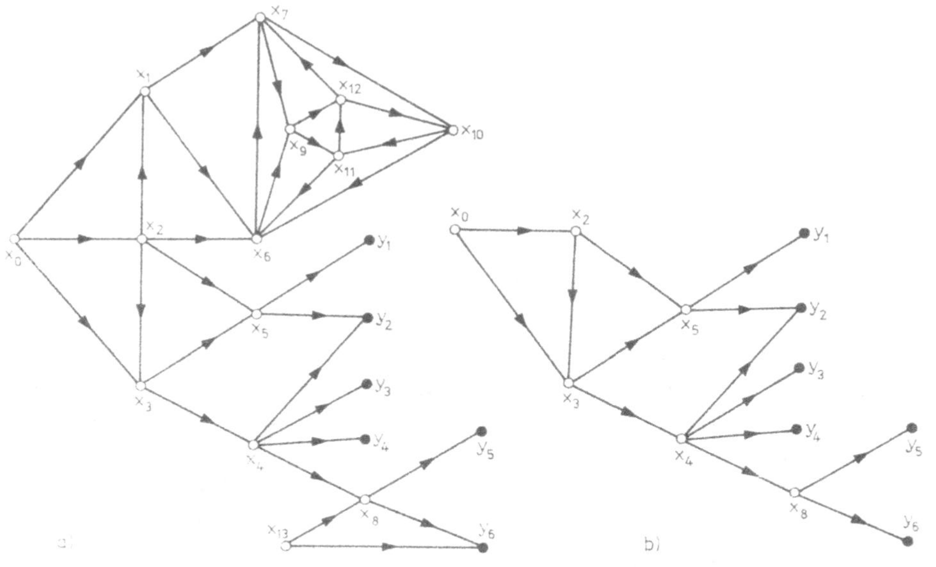

Fig. 7.1

d) to any $Y \in \mathfrak{F}$ there exists a path $(X_0 = X_{i_0}, X_{i_1}, \ldots, X_{i_k}, Y)$ with $X_{i_j} \in \mathfrak{G}$ and $(X_{i_r}, X_{i_{r+1}}) \in \mathfrak{U}$ for $r = 0, 1, \ldots, k-1$ and $(X_{i_k}, Y) \in \mathfrak{U}$ (that means that the set of the available questions must be of such a kind that each event can be identified by a finite number of questions).

Such a graph is called a *questionnaire*. The vertices $X \in \mathfrak{G}$ are called *questions*. The events $Y \in \mathfrak{F}$ are *sinks* of the graph $\boldsymbol{G}$. X_0 (the first question) is the only *source* existing in $\boldsymbol{G}$.

No special interest is attached to those questions which never have an event as immediate or mediate successor, and to those questions which can never be attained by a sequence of questions starting from X_0 (cf. Fig. 7.1). Therefore, reductions can be made according to Fig. 7.1 without any restrictions. In Fig. 7.1, we have not yet indicated any probabilities for the occurrence of the Y_i. It can be seen that, in order to save questions (and thereby in our example, to save money!), one would never ask the question X_0, since all events that can be attained via X_2 are also attainable via X_3, that is, the first question X_0 can be saved and X_3 becomes the first question (although we have not yet defined what we mean by an optimal questionnaire).

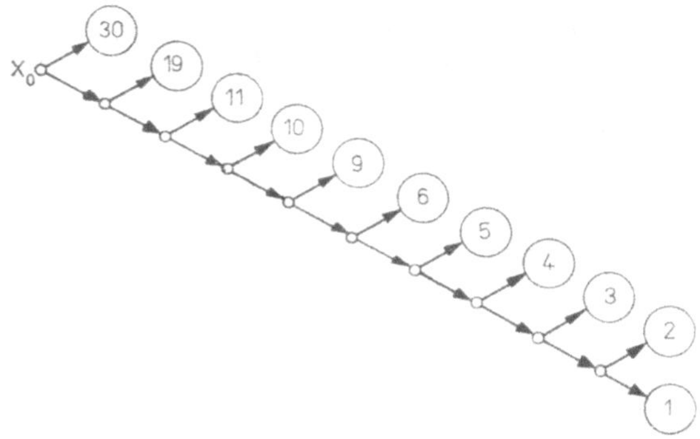

Fig. 7.2

Let us now come to the definition of an optimal questionnaire. As four our game, the point is to minimize the average number of questions needed to identify the colour of the ball drawn. It is presumably not the aim of the second player to identify a large part of the events by asking only a few questions if, on the other hand, he has to ask a large number of questions for determining the events with only small probabilities (cf. Fig. 7.2). The denotations on the graph will probably be understood without being explained in detail.

Question 1: Is the ball of colour F_1? If the answer is "Yes", then the second player has won two Marks, if the answer is "No", we go to

Question 2: Is the ball of colour F_2? If the answer is "Yes", then the second player has won one Mark, if the answer is "No", we go to

Question 3: Is the ball of colour F_3? and so on ...

The reader may find out by himself that by proceeding in this manner, 3.46 questions on average are necessary, that in the long run the second player will lose although he gets two Marks in 30% of all cases, at least one Mark in almost 50% of all cases, and although the first player wins something in only 40% of all cases.

First of all, we have to give a few concepts necessary for solving the problem of finding an optimal questioning strategy.

DEFINITION. A questionnaire is said to be *homogeneous of the order a* if for each vertex $X \in \mathfrak{G}$ the relation $|\Gamma X| = a$ holds.

As we shall see in what follows, the so-called quasi-homogeneous questionnaires are of special importance.

DEFINITION. A questionnaire is said to be *quasi-homogeneous of the order* (a, b) if for all $X \in \mathfrak{G}$, except for one vertex $X_i \in \mathfrak{G}$ at the most, the relation $|\Gamma X| = a$ holds and if for that single vertex X_i the relation $|\Gamma X_i| = b \leqq a$ is true.

In order to speak of optimal questionnaires, we first have to define that class of questionnaires which we wish to consider.

We call two quasi-homogeneous questionnaires $\boldsymbol{F}$ and $\boldsymbol{F}'$ of the order (a, b) *equivalent* if there exists an injective mapping χ of the set $\mathfrak{F}$ of sinks onto the set $\mathfrak{F}'$ of sinks in case of which the probability distributions are preserved, i.e.:

$$p(Y) = p'(\chi(Y)) \quad \text{for all} \quad Y \in \mathfrak{F}.$$

Now, we turn to the concept of the *mean length* of a questionnaire. This notion makes no difficulties at all if the graph considered is a rooted tree, thus cycle-free. Then, there exists to each event $Y \in \mathfrak{F}$ precisely one directed path going from X_0 to Y. The number of edges of such a path $\boldsymbol{W}(Y)$ is called *length* $l(\boldsymbol{W}(Y))$ *of* $\boldsymbol{W}(Y)$ or simply *length* $l(Y)$ *of* Y. By the *mean length* $L(\boldsymbol{F})$ of the questionnaire $\boldsymbol{F}(\mathfrak{X}, \mathfrak{U})$ we mean the value

$$L := \sum_{Y \in \mathfrak{F}} p(Y)\, l(\boldsymbol{W}(Y)).$$

From this, we can deduce the value 3.46 for the average number of questions of our special questionnaire depicted in Fig. 7.2.

We might now be inclined to say that a quasi-homogeneous questionnaire is optimal if it has a minimal mean path length with respect to all questionnaires

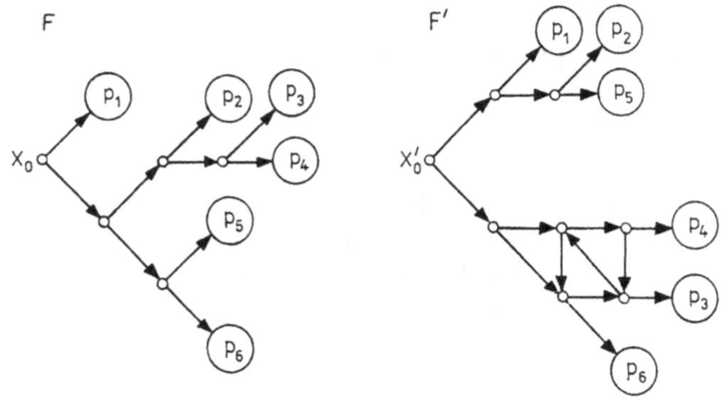

Fig. 7.3

that are equivalent to it. As is shown by Fig. 7.3, a class of equivalent questionnaires also contains questionnaires having cycles or circuits. For such graphs, however, the number of paths going from the source X_0 to an event Y may be larger than 1, and in the case of the existence of circuits, this number can even go ad infinitum.

7.2. ALGORITHM FOR THE GENERATION OF A CYCLE-FREE QUESTIONNAIRE

Looking at the two questionnaires $\boldsymbol{F}$ and $\boldsymbol{F}'$ of Fig. 7.3, we see that any path joining X_0' to Y' in $\boldsymbol{F}'$ ($Y' \in \mathfrak{F}'$ arbitrarily) is at least as long as the path joining X_0 to Y in $\boldsymbol{F}$ (we set $Y' = \chi(Y)$). We would certainly prefer the questionnaire $\boldsymbol{F}$ to the questionnaire $\boldsymbol{F}'$.

Thus, it will be our aim to generate from a given questionnaire $\boldsymbol{F}'$ having cycles or circuits a questionnaire $\boldsymbol{F}$ which has no cycles and in which the path length from X_0 to Y ($Y \in \mathfrak{F}$ arbitrarily) is not longer than any path in $\boldsymbol{F}'$ joining X_0' to Y'. Then, we restrict the equivalence classes so that only so-called *questionnaires of arborescences* (i.e. *rooted trees*) are investigated, among which we seek one that is of minimum path length (in general, we cannot expect unicity).

Algorithm for eliminating circuits and cycles

(i) Let $\boldsymbol{K}'$ be a circuit in $\boldsymbol{F}'$. Since X_0' has no predecessors, it holds $X_0' \notin \boldsymbol{K}'$. Further, let $\boldsymbol{W}'$ be a shortest path joining X_0' to a vertex, say to Z', of $\boldsymbol{K}'$ (since each vertex of $\boldsymbol{F}'$ can be reached from X_0 via at least one path, $\boldsymbol{W}'$ exists, but not necessarily uniquely). We can ascertain without difficulty that the arc of $\boldsymbol{K}'$ whose end-point is Z' can be omitted without causing a change in the distance between any vertex V' and X_0' in $\boldsymbol{F}'$ (this is the length of a shortest path going from X_0' to V').

(ii) Let all circuits of $\boldsymbol{F}'$ be eliminated by (i). Let $\boldsymbol{C}'$ be a cycle of the graph that has resulted in this way (for the sake of simplicity, we continue to denote this graph by $\boldsymbol{F}'$). Since $\boldsymbol{C}'$ is not a circuit, there exists on $\boldsymbol{C}'$ a vertex Z' with at least two immediate predecessors. Thus, there exist at least two paths from X_0' to Z' (which differ from each other by at least one arc). We omit the last arc of the longer of both paths (if the two paths are of equal length, we omit the last arc of any of them). Also in this operation, there still exists at least one path from X_0 to each vertex of the resulting graph, and the length of such a shortest path is not increased. If all cycles are eliminated in this manner, then the resulting graph is a rooted tree, but this one is not necessarily equivalent to $\boldsymbol{F}'$ (e.g., several questions might arise with a number of successors smaller than a).

(iii) Let $X' \in \mathfrak{G}'$ be a vertex of the newly generated graph (which we continue to denote by $\boldsymbol{F}'$) without successors. Then (since no path is going to any event via X') we omit X' and also the arc entering X' (only one arc can enter each vertex, since otherwise $\boldsymbol{F}'$ would still contain cycles). This operation is carried on as long as possible.

(iv) Let $X' \in \mathfrak{G}'$ be a vertex with less than a successors, and let $X_1' \in \mathfrak{G}'$ be a question which has at least one event $Y_1' \in \mathfrak{F}'$ as successor. If $l(X_1') > l(X')$, then we eliminate the arc (X_1', Y_1') and insert the arc (X', Y_1'). In this way, we "saturate" all questions with a view to their succession, except perhaps for some questions that have a maximum distance from X_0'. Obviously, after the saturation procedure no vertex is reached on a longer path than before step (iv) has been performed.

(v) All questions not yet saturated (which, as far as they exist, all have the same, and even maximum distance from X_0') are completed, except for one, perhaps so that (cf. (iv)) all remaining questions, except for one at the most, namely X_n', are saturated when (iii) is repeatedly applied (among the questions X_n' has maximum distance from X_0').

Because the questionnaire is finite, the algorithm terminates, i.e., neither (iii) nor (iv) nor (v) is then applicable.

(vi) If X_n' has exactly one successor, say Y' (of course, the successor of X_n' cannot be a question), then we remove the arcs (X_i', X_n') and (X_n', Y') in case X_i' is the immediate predecessor of X_n' as well as the question X_n', and insert the arc (X_i', Y').

If (vi) is applicable, then even a homogeneous questionnaire of order a is generated. In all other cases, however, it is a quasi-homogeneous one of order (a, b).

An example will serve to illustrate the algorithm (cf. Fig. 7.4). The events have been numbered and indicated in large circles, whereas the questions are marked by small circles.

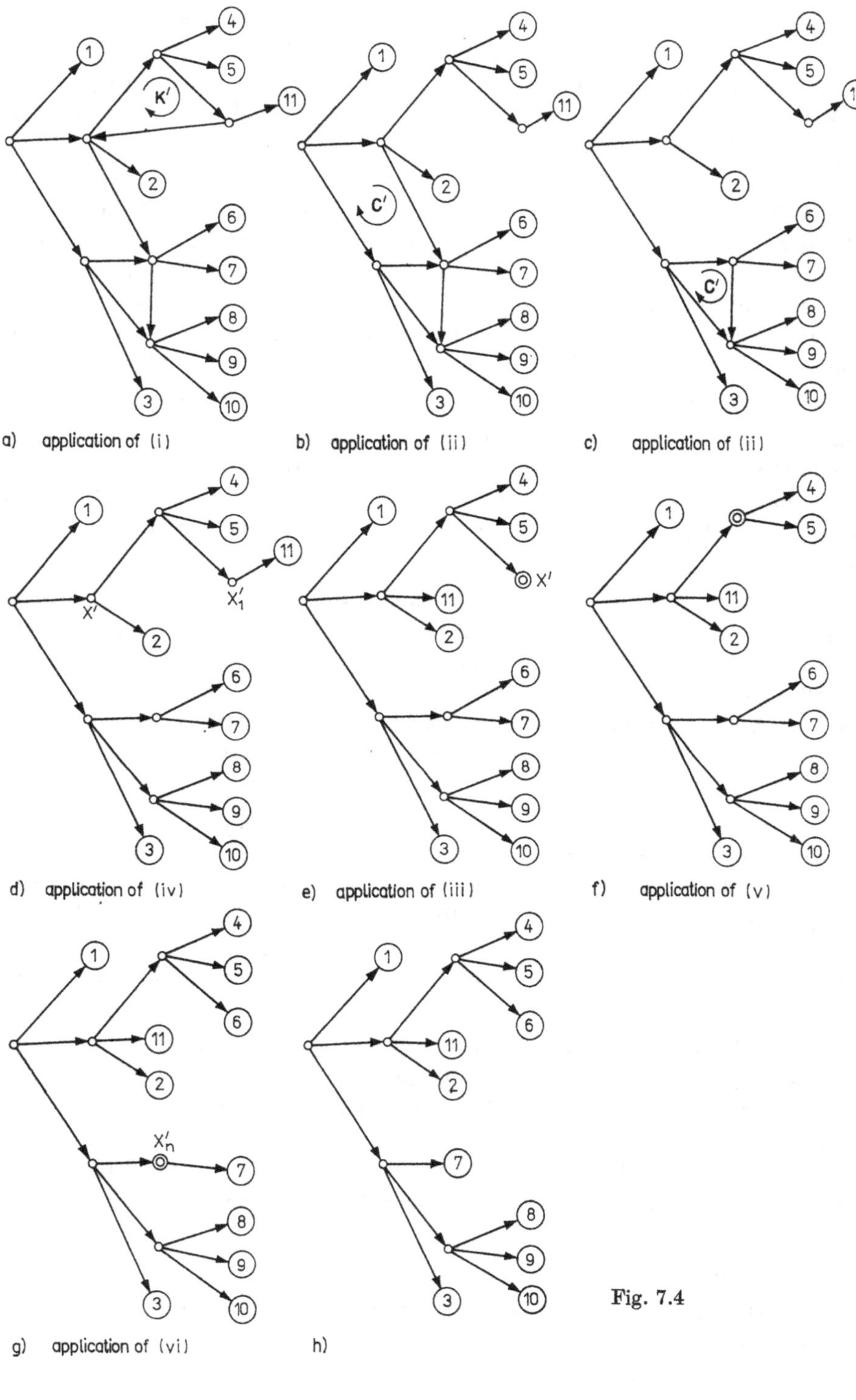
K′
C′
X′
X′₁
X′ₙ
a) application of (i)
b) application of (ii)
c) application of (ii)
d) application of (iv)
e) application of (iii)
f) application of (v)
g) application of (vi)
h)

Fig. 7.4

No definition has been made of what we understand by the mean length of a questionnaire with circuits or cycles. However, a meaningful definition has to ensure that the mean length of $\boldsymbol{F}$ (constructed from $\boldsymbol{F}'$ according to (i) to (vi)), is not longer than that of $\boldsymbol{F}'$.

All the problems relating to the definition of the mean length in questionnaires containing cycles are discussed in more detail in the book of C. F. Picard [1].

7.3. OPTIMAL QUESTIONNAIRES

Now we turn to the second problem to be solved, namely to find among all equivalent questionnaires of arborescences (i.e. rooted trees) an optimal one, that is, such a questionnaire whose mean length is minimum.

First, we make a detailed designation (coding) of the questions and events in a quasi-homogeneous questionnaire (cf. Fig. 7.5).

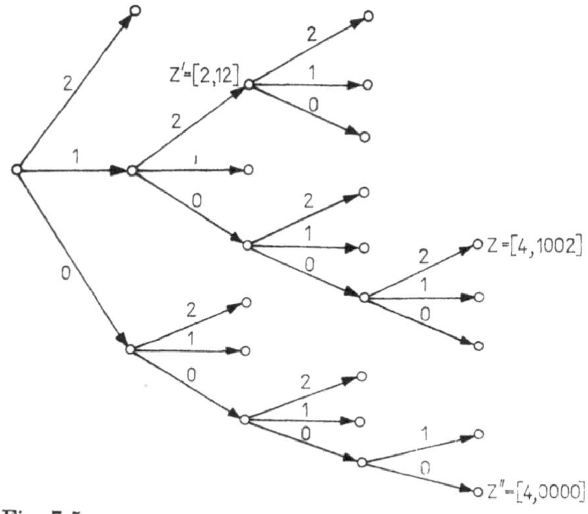

Fig. 7.5

Two natural numbers ϱ, σ are assigned to each vertex Z of the questionnaire. Thus, ϱ is the length of Z (distance from X_0) and σ a code-word consisting of ϱ figures in the number system with the base a. For fixing the figures of σ we need the path $\boldsymbol{W}$ going from X_0 to Z. If $\Gamma X_0 = \{g_i(X_0),\ i = 0, 1, \ldots, a-1\}$ is the set of the a successors of X_0 and if $\boldsymbol{W}$ contains the vertex $g_i(X_0)$, then let i be the first figure of σ. Correspondingly, the other figures of σ are then fixed according to Fig. 7.5. Note that ϱ is given in the decimal system and σ in the a-adic system (it is, of course, possible to give also σ in the decimal system). If σ is given in the a-adic system, one could easily do without the indication of the number ϱ, since the length of the vertex can also be seen from the number of figures of σ. In the case of decoding problems, however, the indication of ϱ could possibly play an essential part.

Let us recall that to each event (i.e., to each vertex of $\mathfrak{F}$) we had assigned a probability of its occurrence. If we have a questionnaire of arborescences, then we also assign recursively a probability to each question. Let X be a question of which all immediate successors already possess a probability. Then, we assign to X the sum of the probabilities of all immediate successors of it as its (own) probability.

Now, we come to the important concept of the *permutation of disjoint subarborescences*. Given a questionnaire of arborescences $\boldsymbol{F}$. A subarborescence $\boldsymbol{H}$ of $\boldsymbol{F}$ is a subgraph of $\boldsymbol{F}$ which contains not only any vertex, but also all of its successors (with respect to $\boldsymbol{F}$), as well as exactly one vertex (the root of the subarborescence) which has all other vertices of $\boldsymbol{H}$ (except for itself) as immediate or mediate successors (cf. Fig. 7.6).

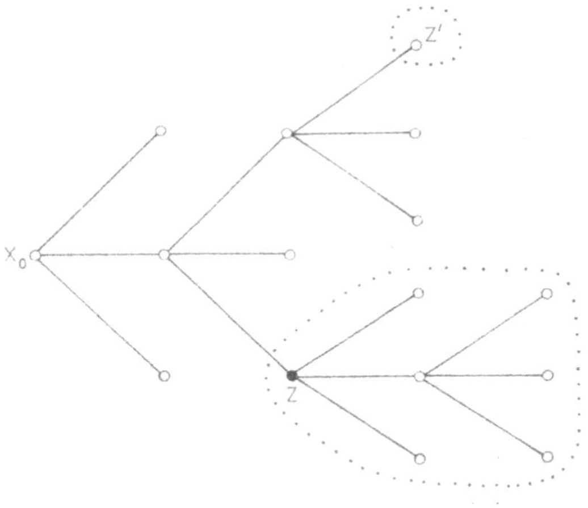

Fig. 7.6

The probabilities of the questions and of the events remain in $\boldsymbol{H}$ the same as in $\boldsymbol{F}$, but this means that a subarborescence is in general not a questionnaire.

Let Z and Z' be the roots of two disjoint subarborescences $\boldsymbol{H}$ and $\boldsymbol{H}'$, respectively, of $\boldsymbol{F}$ with the probabilities $p(Z)$ and $p(Z')$, respectively. If one exchanges $\boldsymbol{H}$ for $\boldsymbol{H}'$ including the probabilities of the vertices of $\boldsymbol{H}$ and $\boldsymbol{H}'$ (followed by the redetermination of the probabilities of the predecessor vertices of Z and Z'), then a questionnaire of arborescences equivalent to $\boldsymbol{F}$ is obviously generated (cf. Fig. 7.7).

In the following, we give a few rules by means of which one can construct from a given questionnaire of arborescences another such questionnaire with a mean length that is certainly not longer.

RULE 1 (Permutation of events): Let Z, Z' be events with $l(Z) < l(Z')$ and $p(Z) < p(Z')$. Then, we exchange the event Z for Z'. Obviously, the mean length of the questionnaire is decreased by this operation.

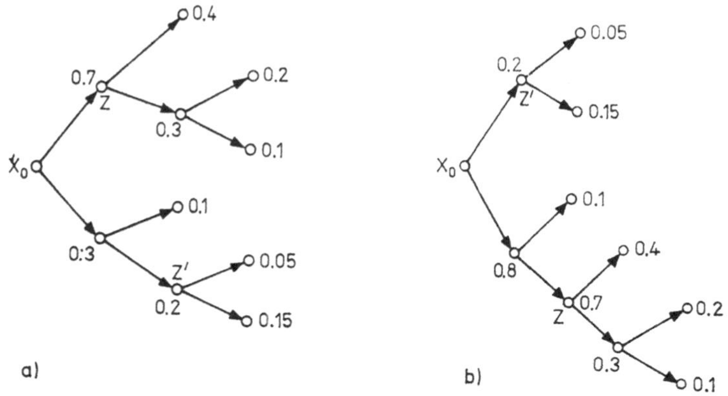

Fig. 7.7

RULE 2 (Permutation of subarborescences): Given two disjoint subarborescences $\boldsymbol{H}, \boldsymbol{H}'$ with the roots Z and Z', respectively, with at least one of the subarborescences consisting of not only one event. Let $l(Z) < l(Z')$ and $p(Z) < p(Z')$ be true. Exchanging the two subarborescences leads to an equivalent questionnaire with shorter mean length.

The truth of this assertion may be verified by the reader himself.

Likewise verify the truth of the following assertion:

LEMMA. *Let $\boldsymbol{F}$ be an optimal homogeneous questionnaire of arborescences, and let further Z and Z' be two vertices of $\boldsymbol{F}$ with $p(Z) = p(Z')$. Then, it holds $|l(Z) - l(Z')| \leqq 1$. But if $p(Z) < p(Z')$, then it holds $l(Z) \geqq l(Z')$.*

It might be thought that the repeated application of the two rules leads inevitably to an optimal questionnaire of arborescences. However, the example represented in Fig. 7.8 shows that this is not the case.

Using a third rule, which causes no change even in the mean length (namely exchange of arborescences of those subarborescences whose roots have the same length), the probabilities of some certain vertices that are no events will be altered so that it will again become possible to apply rule 1 or rule 2.

RULE 3: Rearrange the vertices so that for two vertices $Z_1 = [\varrho, \sigma_1]$ and $Z_2 = [\varrho, \sigma_2]$ with $\sigma_1 < \sigma_2$ it always holds $p(Z_1) \leqq p(Z_2)$. If possible, apply now rule 1 or rule 2, etc.

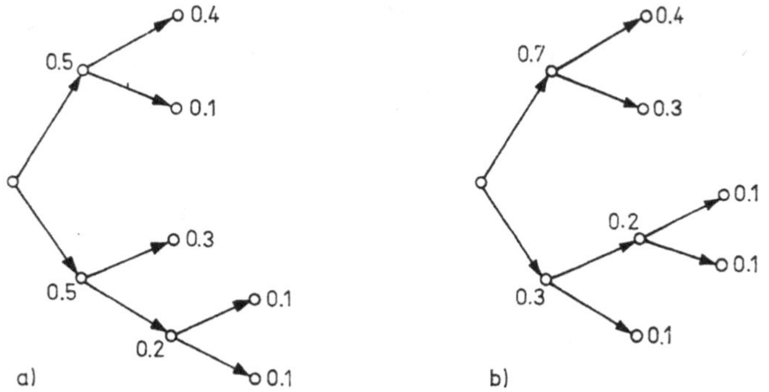

Fig. 7.8

If one applies rule **3** say to the questionnaire Fig. 7.8a, then the graph of Fig. 7.8b is generated to which rule **2** is applicable.

Now, we come to the optimality criterion.

THEOREM 7.1. *Let* $\boldsymbol{F}$ *be a quasi-homogeneous questionnaire of arborescences of order* (a, b) *to which none of the rules* 1 *to* 3 *is applicable any more. Then,* $\boldsymbol{F}$ *is optimal, that is, it has minimum mean length.*

If this theorem should be proved, we shall be able to obtain an optimal questionnaire from a quasi-homogeneous questionnaire of arborescences by using a simple method.

A new concept will now be introduced for the proof of the theorem.

DEFINITION: We call a questionnaire of arborescences (shortly QA) $\boldsymbol{F}$ *ordered* (shortly OQA) if none of the rules **1** to **3** is applicable any more.

LEMMA. *Let* $\boldsymbol{F}$ *be an optimal QA. Then, by applying the rule* 3 (the rules 1 or 2 are not applicable because of the optimality of $\boldsymbol{F}$), $\boldsymbol{F}$ *can be rearranged to form an optimal OQA* $\boldsymbol{F}_1$.

The proof of this lemma is clear.

By using the above-given acronym, our theorem takes on the following shape:

THEOREM 7.2. *Let* $\boldsymbol{F}$ *be a quasi-homogeneous OQA of order* (a, b). *Then,* $\boldsymbol{F}$ *is optimal.*

First, we mention that an optimal quasi-homogeneous QA of order (a, b) satisfies the relation $\mathbf{2} \leqq b \leqq a$, since for $b = \mathbf{1}$, one question could be saved, which would then lead to a reduction of the mean length.

Proof. Let $\boldsymbol{F}$ be an optimal questionnaire of arborescences (QA) of order (a, b), and let $\boldsymbol{F}_1$ be an OQA of same order which has emerged from $\boldsymbol{F}$ by applying the rules 1 to 3. Because of the lemma, $\boldsymbol{F}_1$ is optimal.

Let $\boldsymbol{F}_2$ be any quasi-homogeneous OQA of same order (a, b) with $2 \leqq b \leqq a$ equivalent to $\boldsymbol{F}_1$. We shall show that $\boldsymbol{F}_2$ is also optimal, which would then prove the theorem.

We consider in $\boldsymbol{F}_i$ $(i = 1, 2)$ a question X_i^1 of minimum probability. Since $\boldsymbol{F}_i$ is ordered, X_i^1 has exactly b successors Y_i^j $(j = 1, \ldots, b)$, and these b succeeding events have minimum probability (that means: For any event $Y_i \notin \{Y_i^1, \ldots, Y_i^b\}$ it holds $p(Y_i) \geqq \max\limits_{1 \leq j \leq b} p(Y_i^j)$ for $i = 1, 2$). Now, we form from $\boldsymbol{F}_i = \boldsymbol{F}_i^1$ new OQA's by omitting the b events Y_i^j and also the arcs going to them, and by making X_i^1 a new event with the probability $p(X_i^1) = \sum\limits_{j=1}^{b} p(Y_i^j)$. Since $\boldsymbol{F}_1$ and $\boldsymbol{F}_2$ are equivalent, it holds $p(X_1^1) = p(X_2^1)$. The mean lengths of the two OQA $\boldsymbol{F}_i^2$ which result in this way (the reader will prove by himself that these questionnaires of arborescences are ordered) have been decreased by the same amount, namely by $p(X_i^1)$, since it holds: the mean length of a QA equals the sum of all probabilities of the questions contained in the QA (task!).

Now, we carry on the "shrinkage" of the $\boldsymbol{F}_i$. In $\boldsymbol{F}_i^2$ $(i = 1, 2)$ there exists again a question X_i^2 of minimum probability, whose a succeeding events also have minimum probabilities (among the probabilities of the events in $\boldsymbol{F}_i^2$). We omit these a events and also the arcs leading to them and turn X_i^2 into an event of the OQA generated in this manner with a probability $p(X_i^2)$. Here again, the relation $p(X_1^2) = p(X_2^2)$ is true because the QA is ordered, and the mean length of the two OQA decreases by the same amount $p(X_i^2)$.

We carry on this shrinkage (mathematical induction!) and get two OQA $\boldsymbol{F}_i^n$ after $n - 1$ steps of shrinkage (if there exist in $\boldsymbol{F}_i$ exactly n questions). These two OQA possess only one question and a events and have the same mean lengths, namely 1. Each step of shrinkage has caused a reduction of the mean lengths of

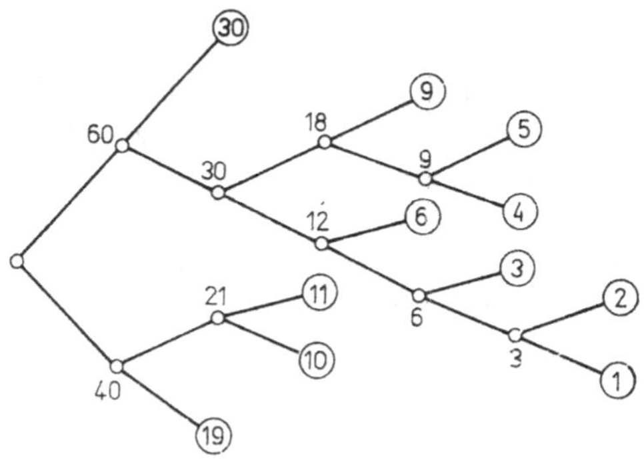

Fig. 7.9

the OQA by the same amount, which makes it clear that they already had the same mean lengths at the beginning of the shrinkage. Since $\boldsymbol{F}_1$ was optimal by premise, $\boldsymbol{F}_2$ is also optimal.

This proves the theorem, and we have also found a way to obtain an optimal QA. Either we seek any questionnaire of arborescences of a given equivalence class and repeatedly apply the rules 1 to 3 to it until the algorithm terminates, or, which will enable us to achieve our goal more quickly, we combine the b events of least probability into one question, transform this question "temporarily" into a new event, combine now the remaining events of least probabilities into a new question, transform this question again "temporarily" into a new question, transform this question again "temporarily" into a new event, and so on. This algorithm has been devised by D. A. Huffman (cf. [1]). For the game formulated in 7.1 there results an optimal questioning strategy according to Fig. 7.9 (task: Is it possible for the second player to win in the long run?).

7.4. AN EXAMPLE FROM CODING

In the case of coding, the following problem is to be solved: A communication source emits n distinct symbols S_i $(i = 1, \ldots, n)$. (In the German language for example 26 letters ignoring whether one puts $\ddot{a} = ae$, etc. and ignoring . , ! ? — etc.) A code having N symbols T_j $(j = 1, \ldots, N)$ consists of n code-words for the S_i, each consisting of a sequence of symbols T_j, with the kth code-word being composed of l_k symbols. For making a unique decoding, however, it is necessary to assure that a code-word is not the left-hand part of another code-word.

Let us take as example the Morse code (cf. Fig. 7.10).

In this case, it is $N = 3$ with the symbols *point*, *dash*, *ignore*. The numbers at the end-points indicate the probabilities (cf. [3]) for the occurrence of the single letters of the German alphabet in the texts considered by the author. In our representation, all signs such as point, comma, spacing, question-mark, etc. come under "ignore". Since the "ignore" used in the Morse code only serves to separate the single letters (from each other), a unique decoding can be guaranteed. However, the reader may realize that this way of coding is not optimal in the sense described above. An optimal Morse code containing three signs — and the reader may verify this — has the length 2.6130. Thus, the mean length of the Morse code is greater by almost 0.5 than an "optimal" alphabet containing three signs. Of course, we do not want to conceal that the "optimal alphabet" would entail some other difficulties, for example problems in the error detection in case of transmission errors. Another problem might be that certain code-words require very many symbols.

If one tried to comply more conveniently with the concrete linguistic conventions of a living language (the single letters occur by no means independently of each other, as a living language possesses a high degree of redundancy), then one would have to consider pairs, triples, quadruples, etc., of letters and their proba-

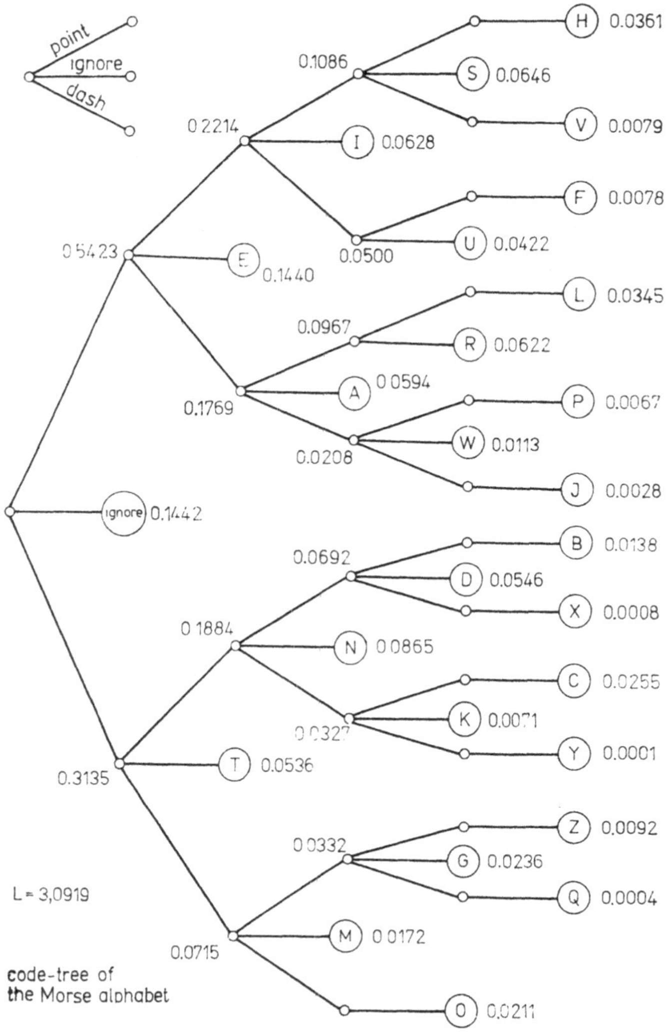

Fig. 7.10

bilities in a language. Such attempts have been made. The interested reader is referred to [2]. Naturally, the words to be coded are numerically increasing. Considering pairs of letters, one would have to code already 27^2 words (although some of these pairs are unlikely to occur, for example, in the German language, say the pair "qy"). However, with this approach it is possible to make the mean lengths of the corresponding questionnaires (with respect to a letter) considerably shorter than for instance in the above-mentioned case with only 27 "words" (in natural language, each of these "words" represents a letter).

TASK. How long is the mean length of the "optimal alphabet" in the case of two code symbols, using the probabilities according to Sacco [3] (cf. Fig. 7.10)?

An illustrative combinatorial introduction to the fundamental concepts of information theory and to codes and error correcting codes is given in [5]. Further, the reader is referred to the code developed by R. Hamming [4].

7.5. BIBLIOGRAPHY

[1] Picard, C. P.: Theorie der Fragebogen, Berlin 1973. (Translated from the French.)
[2] Fey, P.: Informationstheorie, 3rd edn., Berlin 1968.
[3] Sacco, L.: Manuel de scriptographie, Paris 1951.
[4] Hamming, R.: Error detecting and error correcting codes, Bell System Techn. J. **26** (1950), 147–160.
[5] Zemanek, H.: Elementare Informationstheorie, Munich 1959.

Chapter 8

Signal flow graphs

8.1. FORMULATION OF THE PROBLEM

Given, for example, the electrical circuit of Fig. 8.1. Let the R_i denote the ohmic resistances, the E_j electromotive forces and the I_k branch currents and their arbitrarily prescribed direction.

The branch currents are unknown quantities which can be calculated by using the Kirchhoff's laws (cf. chapter 1). The problem set leads to the solution of a linear inhomogeneous system of equations for the I_k values.

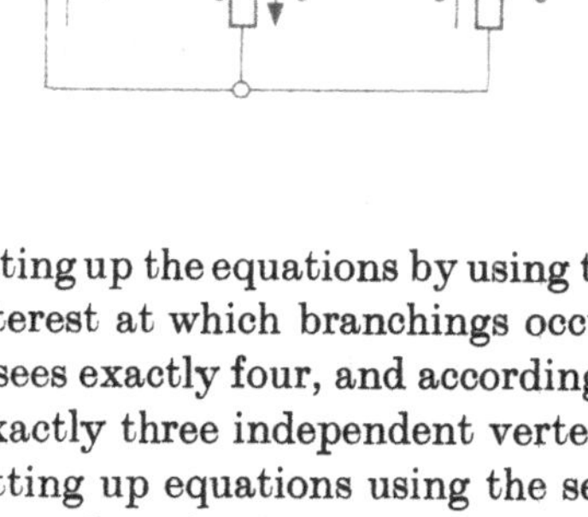

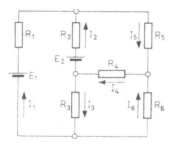

Fig. 8.1

For setting up the equations by using the Kirchhoff's node rule, only those points are of interest at which branchings occur. These points have been set off in Fig. 8.1. One sees exactly four, and according to the flow theory discussed in chapter 1, we get exactly three independent vertex equations.

For setting up equations using the second Kirchhoff's theorem (*mesh rule*), we need the number of independent circuits. In our case, there are exactly three such circuits, i.e., three mesh equations are added to the three node equations. When suitably chosen, these equations provide together a system of six equations for the six unknowns I_k ($k = 1, 2, \ldots, 6$) for which the appertaining matrix of coefficients is regular. Thus, a unique solution of the system of equations is assured.

For our example, the following system of equations would be of the required type:

$$\begin{bmatrix} R_1 & 0 & 0 & 0 & R_5 & -R_6 \\ 0 & R_2 & 0 & R_4 & R_5 & 0 \\ 0 & 0 & -R_3 & -R_4 & 0 & -R_6 \\ 1 & 1 & 0 & 0 & -1 & 0 \\ 0 & 0 & 0 & -1 & 1 & 1 \\ -1 & 0 & 1 & 0 & 0 & -1 \end{bmatrix} \cdot \begin{bmatrix} I_1 \\ I_2 \\ I_3 \\ I_4 \\ I_6 \\ I_5 \end{bmatrix} = \begin{bmatrix} E_1 \\ E_2 \\ 0 \\ 0 \\ 0 \\ 0 \end{bmatrix}.$$

The first three equations result from the mesh rule, the last three from the node rule.

In chapter 1, we saw that in a graph (network) with n vertices and exactly m arcs (branches), there exist exactly $m - n + 1$ independent cycles (meshes) and that one can always find a system of equations with a regular matrix of coefficients (which is composed of $n - 1$ node equations and $m - n + 1$ mesh equations) for the branch currents wanted.

The analysis of linear electrical networks leads either directly or indirectly to the solution of linear systems of equations (directly: see example of Fig. 8.1).

An example of the analysis of a network resulting indirectly in the solution of a system of equations will be considered at the end of paragraph 8.2.

In 1953, S. J. Mason published his work on linear flow graphs [2]. Since that time, a number of graphs have been introduced for the analysis of networks. We shall consider two types of them, namely structure and signal flow graphs. Nguyen Mong Hung [4] means by a *structure graph* a (directed or undirected) graph which has the same structure as the given network (namely an isomorphic one), and by a *signal flow graph* he means a directed arc-valuated graph reflecting the algebraic relations between the variables occurring in the network. Thus, for example, the complete graph $\boldsymbol{K}_4$ with four vertices has the same structure as the network represented in Fig. 8.1.

In most of the chapters of this book, we deal — in the sense of this definition — with structure graphs. The present section will be devoted to some remarks concerning signal flow graphs.

The main purpose of signal flow graphs is to illustrate the courses of signals, but the valuations written at the arcs must, on no account, suggest flows. In control engineering, for example, it is agreed that a fully traced point represents a branching point (cf. Fig. 8.2a) for which the relations indicated apply, as is also the case for the other figures. As shown by the Figs. 8.2c and d, not only linear algebraic operations (e.g. multiplication and division) can be represented this way. At the end of 8.2, we shall show by means of an example that it is also possible to describe by this kind of representation, for example, systems of differential equations in a clearly arranged way. It will be mentioned here that, in order to get a lucid solution of such systems, one deliberately goes over from the time domain to the frequency range using the Laplace transformation. The interested reader may find for example in [6] an introduction to these problems given in special consideration of control engineering aspects.

As for investigations conducted in the frequency range, the representation of the transfer processes in the form of block flow charts has been adopted. Figure 8.3 shows a few examples (parallel connection, series connection and feedback) with the appropriate simplification rules (to the right of them). The transfer function

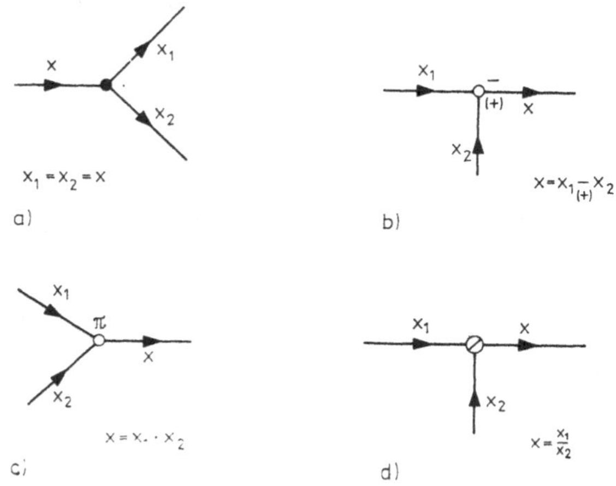

Fig. 8.2

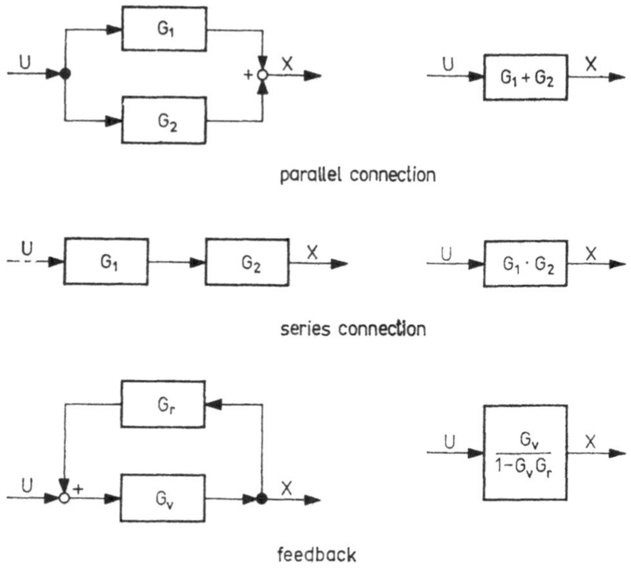

Fig. 8.3

for feedback ensues, for example, from the relation (which holds in the frequency range):

$$X(p) = G_v(p)\, U(p) + G_r(p)\, G_v(p)\, X(p),$$

from which the transfer function indicated can be derived without difficulty:

$$G(p) = \frac{G_v(p)}{1 - G_v(p)\, G_r(p)}.$$

Figure 8.4 shows the different kinds of description, namely using the block chart on the one hand and the signal flow graph on the other, and the conversion of one into the other.

Fig. 8.4

8.2. THE ALGORITHM OF MASON FOR SOLVING LINEAR SYSTEMS OF EQUATIONS

Let us now present an algorithm which makes it possible to solve a system of equations and to follow constantly the successive solving (elimination of variables), which explains why this approach is so popular.

Consider a system of equations

$$\boldsymbol{Ax} = \boldsymbol{b}$$

or, written in another way,

$$\begin{array}{l}
a_{11}x_1 + \cdots + a_{1i}x_i + \cdots + a_{1k}x_k + \cdots + a_{1n}x_n = b_1, \\
a_{21}x_1 + \cdots + a_{2i}x_i + \cdots + a_{2k}x_k + \cdots + a_{2n}x_n = b_2, \\
\cdots\cdots\cdots\cdots\cdots\cdots\cdots\cdots\cdots\cdots \\
a_{i1}x_1 + \cdots + a_{ii}x_i + \cdots + a_{ik}x_k + \cdots + a_{in}x_n = b_i, \\
\cdots\cdots\cdots\cdots\cdots\cdots\cdots\cdots\cdots\cdots \\
a_{k1}x_1 + \cdots + a_{ki}x_i + \cdots + a_{kk}x_k + \cdots + a_{kn}x_n = b_k, \\
\cdots\cdots\cdots\cdots\cdots\cdots\cdots\cdots\cdots\cdots \\
a_{n1}x_1 + \cdots + a_{ni}x_i + \cdots + a_{nk}x_k + \cdots + a_{nn}x_n = b_n.
\end{array}$$

If we choose any row, say the ith one (make sure that $a_{ii} \neq 0$), and if we solve it for x_i, then we get

$$x_i = \frac{1}{a_{ii}} (b_i - a_{i1}x_1 - \cdots - a_{i,i-1}x_{i-1} - a_{i,i+1}x_{i+1} - \cdots - a_{ik}x_k$$
$$- \cdots - a_{in}x_n).$$

If we substitute x_i in the kth row, then we get

$$a_{k1}x_1 + \cdots + a_{k,i-1}x_{i-1}$$

$$+ a_{ki}\frac{1}{a_{ii}}(b_i - a_{i1}x_1 - \cdots - a_{i,i-1}x_{i-1} - a_{i,i+1}x_{i+1} - \cdots - a_{ik}x_k - \cdots - a_{in}x_n)$$

$$+ a_{k,i+1}x_{i+1} + \cdots + a_{kk}x_k + \cdots + a_{kn}x_n = b_k$$

or in an ordered form

$$\left(a_{k1} - \frac{a_{i1}a_{ki}}{a_{ii}}\right)x_1 + \cdots + \left(a_{k,i-1} - \frac{a_{i,i-1}a_{ki}}{a_{ii}}\right)x_{i-1} + \left(a_{k,i+1} - \frac{a_{i,i+1}a_{ki}}{a_{ii}}\right)x_{i+1}$$

$$+ \cdots + \left(a_{kk} - \frac{a_{ik}a_{ki}}{a_{ii}}\right)x_k + \cdots + \left(a_{kn} - \frac{a_{in}a_{ki}}{a_{ii}}\right)x_n = b_k - \frac{a_{ki}b_i}{a_{ii}}.$$

Assign to this system of equations a graph with n vertices in the following way: each of the variables corresponds precisely to one of the vertices. In Fig. 8.5, we have drawn the vertices by large circles which contain a small circle; this one corresponds to the variable assigned to it (the index is drawn in the small circle)

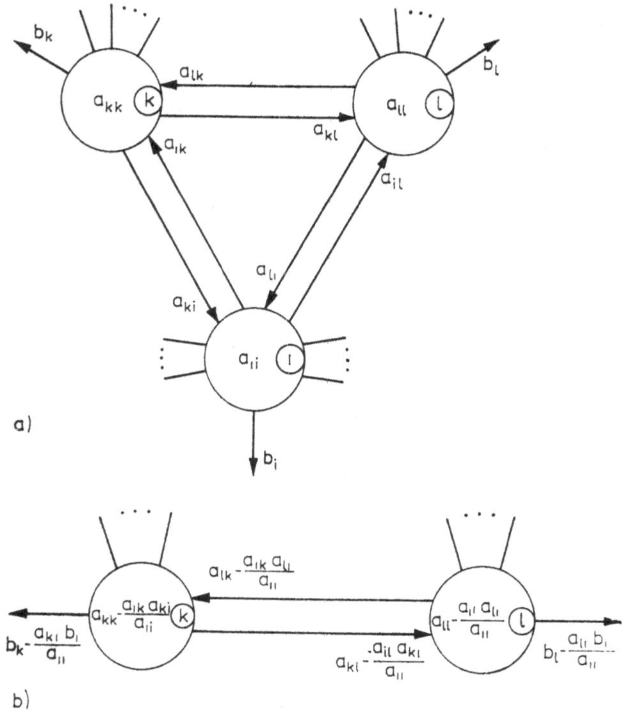

Fig. 8.5

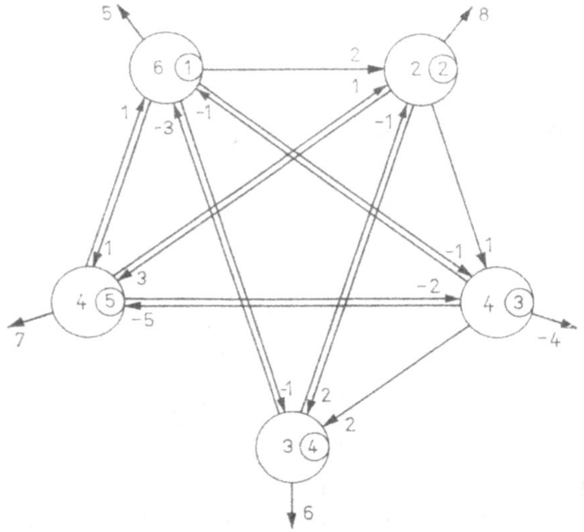

The vertex assigned to the variable x_i is provided with the coefficient a_{ii}. An arc with a valuation a_{lk} going from the vertex having the index l (in the small circle) to the vertex having the index k is then introduced. If $a_{lk} = 0$, the arc can also be omitted. The arc (whose end-point is not denoted at all) leaving the vertex with the index l is provided with the valuation b_l.

It is thus possible to read immediately from the graph the system of equations which corresponds to it (cf. Fig. 8.6), namely:

$$\begin{bmatrix} 6 & 2 & -1 & -1 & 1 \\ 0 & 2 & 1 & 2 & 3 \\ -1 & 0 & 4 & 2 & -5 \\ -3 & -1 & 0 & 3 & 0 \\ 1 & 1 & -2 & 0 & 4 \end{bmatrix} \cdot \begin{bmatrix} x_1 \\ x_2 \\ x_3 \\ x_4 \\ x_5 \end{bmatrix} = \begin{bmatrix} 5 \\ 8 \\ -4 \\ 6 \\ 7 \end{bmatrix}.$$

Now, we consider the substitution of a variable of the graph, say x_i (Fig. 8.5a). When x has been eliminated, the coefficients change as follows: for the modified right-hand sides of the originally kth row we get

$$b_k' = b_k - \frac{a_{ki} b_i}{a_{ii}},$$

for the modified element a'_{kl}

$$a'_{kl} = a_{kl} - \frac{a_{il} a_{ki}}{a_{ii}} \qquad (l \neq k)$$

and for the new diagonal element a'_{kk}

$$a'_{kk} = a_{kk} - \frac{a_{ik}a_{ki}}{a_{ii}},$$

which represents the same expression as for $l = k$.

These new values are indicated in the reduced graph in Fig. 8.5 b (that is, in the graph reduced by the vertex with the index i). Figure 8.7 shows the graph which belongs to the system of equations after the elimination of x_1 in our example. It

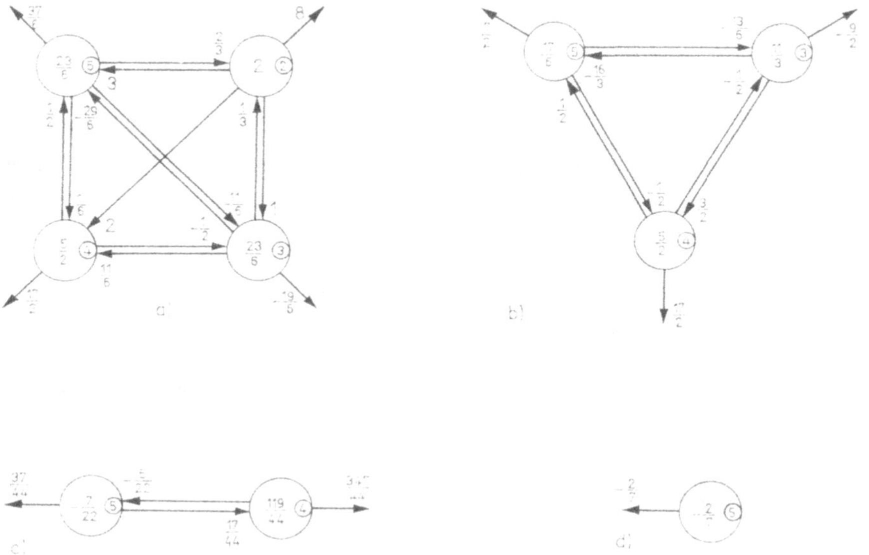

Fig. 8.7

will not be difficult for the reader to follow up the single elimination steps. According to the Gaussian algorithm, one can now go the other way round and derive the solution from the reduced graphs. Then, one gets

$$x_5 = 1, \quad x_4 = 3, \quad x_3 = -1, \quad x_2 = 0, \quad x_1 = 1.$$

Kai Wai Chen [7] suggests a "generalized" Mason's algorithm making it possible to eliminate not only one single vertex (i.e., only one variable). Describing this algorithm seemed to us too sophisticated (an example with only seven variables requires, according to this method, very many drawings and calculation filling several pages). As we have already stated at the beginning of this chapter, Mason's algorithm does not serve exclusively for solving systems of equations, but rather for signal tracing and monitoring. Therefore, it will be necessary for the interested reader to familiarize himself with some special literature. Our concern is to give only some ideas and an introduction to certain methods.

Let us consider an example (which we have taken from the textbook [6] of K. Reinisch) that leads to a linear system of equations containing one parameter and that can be solved by means of the algorithm given by Mason (but, of course, also by other means).

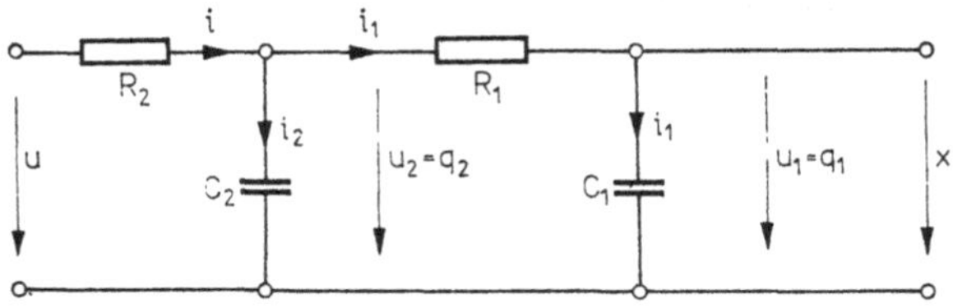

Fig. 8.8

We consider the electrical transfer element of Fig. 8.8. Using Kirchhoff's rules and the relations

$$i_k = C_k \dot{u}_k = C_k \dot{q}_k \qquad (k = 1, 2)$$

and introducing the time constants

$$T_1 = R_1 C_1, \qquad T_2 = R_2 C_2, \qquad T_{21} = R_2 C_1,$$

we get the system of linear differential equations of the first order for the state variables q_1 and q_2:

$$\begin{aligned} \dot{q}_1 &= -\frac{1}{T_1} q_1 + \frac{1}{T_1} q_2, \\ \dot{q}_2 &= \frac{T_{21}}{T_1 T_2} q_1 - \frac{T_1 + T_{21}}{T_1 T_2} q_2 + \frac{1}{T_2} u \end{aligned} \tag{0}$$

with the output equation $x = q_1$. A system of linear ordinary differential equations of the first order of type (0) can be transformed into a system of linear equations by applying the Laplace transformation. In our example, the following system of equations results:

$$\begin{bmatrix} p + \dfrac{1}{T_1} & -\dfrac{1}{T_1} \\ -\dfrac{T_{21}}{T_1 T_2} & p + \dfrac{T_1 + T_{21}}{T_1 T_2} \end{bmatrix} \cdot \begin{bmatrix} Q_1 \\ Q_2 \end{bmatrix} = \begin{bmatrix} 0 \\ \dfrac{1}{T_2} \end{bmatrix} U.$$

Here, we denote by $Q_k(p)$ the Laplace transform of $q_k(t)$ $(k = 1, 2)$ and by $U(p)$ the Laplace transform of $u(t)$ (unilateral transform in either case if we imagine a tension $u(t)$ to be applied to the input of the transfer element at the time $t = 0$).

If we assign, as agreed above, a graph to the system of équations, then we obtain the graph represented in Fig. 8.9. Solving this system according to the specifications given in Fig. 8.5 or also according to the formulae calculated results for the variable $Q_1(p)$ wanted in

$$Q_1(p) = \frac{2U(p)}{3T}\left[\frac{1}{p+\dfrac{1}{2T}} - \frac{1}{p+\dfrac{2}{T}}\right]$$

(with given time constants say $T_1 = T_2 = 2T_{21} = T$), as can be verified by the reader.

Since there are products of Laplace transforms in the transformed domain, the output function $x(t) = q_1(t)$ results in the form of a convolution integral in the

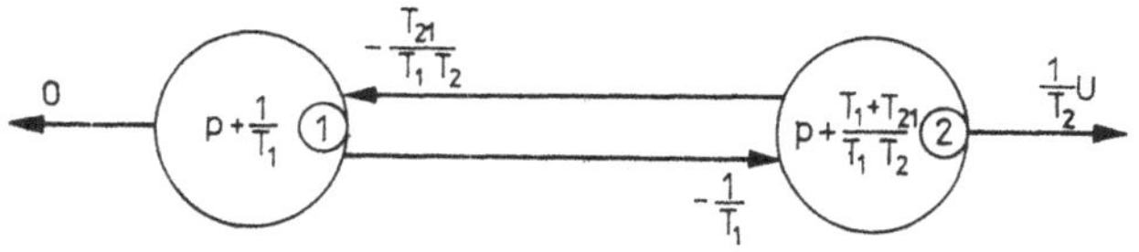

Fig. 8.9

case of any $u(t)$ as input tension. In order to get an insight into the solution, one can choose for instance a step function

$$u(t) = \sigma(t) = u_0 \quad (\text{for } t \geqq 0)$$

as input. Hence it follows for the solution of our specialization (in consideration of the fact that the Laplace transform of $\sigma(t)$ is equal to $\frac{1}{p}$):

$$x(t) = q_1(t) = u_0\left[\sigma(t) - \frac{4}{3}\exp\left(-\frac{t}{2T}\right) + \frac{1}{3}\exp\left(-\frac{t}{\frac{1}{2}T}\right)\right].$$

Our concern in this chapter was not to deal with problems relating to the application of the Laplace transformations. We only wanted to present a possible way (which is often applied by control engineers) of not only solving, by using the idea of Mason (which ensued from the treatment of the transmission of signals), systems of equations (for this, the well-known Gaussian algorithm is certainly more suitable), but furthermore, of following up constantly the signal courses and reciprocal dependences when solving systems of equations.

Mason's graphs and generalized Mason's graphs, as, for example, the normalized Mason's graph introduced by M. Roth, are widely applied in the investigation of electrical and electronic networks (for example in the analysis of RC operational amplifier networks). The normalized Mason's graph is a vertex-valuated graph, with the vertex valuations causing a certain relativization of the valuations of arcs which are incident with these vertices.

The reader who is interested in this subject is referred to [4] and [5].

In addition, the textbook [7] of Wai Kai Chen gives a useful survey of the applications of Coates' graphs and Mason's graphs.

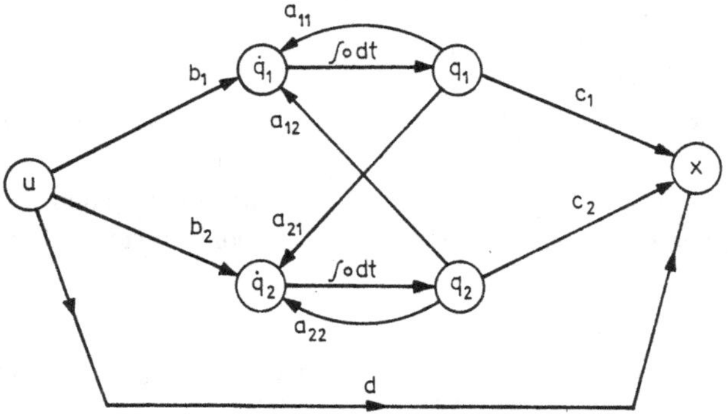

Fig. 8.10

In conclusion, we indicate a possible variant of representing systems of, for example, linear differential equations by using signal flow graphs. Figure 8.10 shows the signal flow graph of the system

$$\dot{q}_1 = a_{11}q_1 + a_{12}q_2 + b_1u,$$
$$\dot{q}_2 = a_{21}q_1 + a_{22}q_2 + b_2u$$

with the output equation

$$x = c_1q_1 + c_2q_2 + du.$$

8.3. BIBLIOGRAPHY

[1] Busacker, R. G., and T. L. Saaty: Finite Graphs and Networks, An Introduction with Applications. New York 1965 (German: Munich/Vienna 1968; Russian: Moscow 1974).

[2] Mason, S. J.: Feedback theory: Some properties of linear flow graphs, Proc. IRE **41** (1953), 1144–1156.

[3] Mason, S. J.: Feedback theory: Further properties of linear flow graphs, Proc. IRE **44** (1956), 920–926.

[4] Nguyen Mong Hung: Entwicklung und Anwendung von Graphenmethoden zur Analyse von *RC*-Operationsverstärker-Netzwerken und zur Untersuchung des Realentwurfes derartiger Netzwerke, Diss. TH Ilmenau 1973.

[5] Roth, M., und Nguyen Mong Hung: Zur topologischen Analyse linearer Netzwerke, Wiss. Z. TH Ilmenau **17** (1971), 115–127.

[6] Reinisch, K.: Kybernetische Grundlagen und Beschreibung kontinuierlicher Systeme, Berlin 1974.

[7] Wai Kai Chen: Applied Graph Theory, Amsterdam/London 1971.

Chapter 9

Minimum sets of feedback arcs

9.1. FORMULATION OF THE PROBLEM

During the running of algorithms, for instance for determining an optimal order of calculation, loops often play a decisive role. How many feedback arcs have to be dissected in order to make the running loop-free or cycle-free? How can minimum sets of feedback arcs be found?

Using the terminology of graph theory we can formulate the problem dealt with in this chapter in the following way:

Given a directed graph $\boldsymbol{G}(\mathfrak{X}, \mathfrak{U})$.

1. *Find a set of arcs* $\mathfrak{V} \subseteqq \mathfrak{U}$ *such that the graph resulting after the removal of the arcs from* $\mathfrak{V}$ *does not contain a circuit* (i.e., a uniformly directed cycle). *Here,* $\mathfrak{V}$ *shall be minimal in the sense that no proper subset of* $\mathfrak{U}$ *represents* (contains) *all circuits.* $\mathfrak{V}$ *is called minimal set of feedback arcs.*

2. *Among all minimal sets representing all circuits, find a set with minimum number of arcs. Such a set is called minimum set of feedback arcs.*

3. *Find now algorithms which solve the problems* 1 *and* 2.[1])

The reader who is interested in more theoretic questions relating to this field is referred to monograph [3] (part II). There, a number of statements are made on arc and vertex numbers representing all circuits in an undirected graph, but mainly with a given number of independent circuits, that is, of circuit sets of which two circuits are pairwise disjoint at a time.

An interesting case of application of the subject introduced above is known to us from the field of chemical engineering. Certain flows (of material) correspond to the arcs, and to the vertices there correspond process units between which there are flows. In addition, an arc valuation is given to which correspond parameters for determining the flows. First, the total graph is to be decomposed into a sequence

[1]) The formulated problem belongs to the class of the NP-hard-problems, that means that an exact algorithm always leads to a complete enumeration [5].

of complexes, with a complex being regarded as calculable if all complexes lying before it in the sequence are calculated (it is, however, not our intention to discuss this decomposition and determination of the order of calculation). In our terminology, a complex is a strongly connected graph. In order to calculate by way of iteration the resulting system of equations in a complex, arcs have to be cut off, the flows at the cut points have to be evaluated and calculation has to be started with these estimates. It is desirable to execute the dissection so that the resulting graph is not only circuit-free, but, in addition, the sum of the arc valuations of the dissected arcs is minimized.

In what follows, we will deal with two algorithms for solving the problems set. Section 9.2 is then devoted to an algorithm found by Lempel and Cederbaum, and in 9.3 we shall be concerned with an algorithm devised by D. H. Younger. The literature used for this latter paragraph especially is not so easy to read.

9.2. THE ALGORITHM OF LEMPEL AND CEDERBAUM

The method described here can be called a direct one. Each particular step can be seen directly from the task to be solved. To what extent this algorithm seems to us advantageous will be discussed at the end of this chapter.

In the first part of this section, we shall deal with the determination of all circuits of a simple directed graph.

Let the graph G be given in the form of its adjacency matrix A. Let the vertices of G be numbered from 1 to n. The element a_{ii} is equated to zero, for $i \neq j$ we set $a_{ij} = 1$ if the vertex i is connected with the vertex j by an arc directed towards j, otherwise we set $a_{ij} = 0$. From the adjacency matrix A we set up a new matrix B in the following way: Let the arcs of G be numbered from 1 to m. We set $b_{ii} = 1$, $b_{ij} = 0$ if $a_{ij} = 0$ $(i \neq j)$ holds, and $b_{ij} = x_r$ if the arc with the number r is going from the vertex i to the vertex j. Consider the example of the graph represented in Fig. 9.1. The following matrix B is assigned to this graph:

$$B = \begin{bmatrix} 1 & 0 & x_{11} & x_{13} & 0 & 0 \\ 0 & 1 & x_1 & 0 & x_{12} & 0 \\ 0 & 0 & 1 & x_2 & 0 & x_8 \\ 0 & x_5 & 0 & 1 & x_3 & x_4 \\ x_7 & 0 & 0 & 0 & 1 & x_6 \\ x_{10} & x_9 & 0 & 0 & 0 & 1 \end{bmatrix}.$$

Let $\{i_1, i_2, \ldots, i_n\}$ be a non-trivial permutation of the first n natural numbers (n indicates the number of the vertices of G). Consider a product $b_{1i_1} b_{2i_2} \ldots b_{ni_n}$ (as everybody knows, such products occur when forming the determinant of B). A product of this kind either represents a set of edge-disjoint circuits (if it is unequal to 0), or it is equal to 0 and would then be of no interest for us.

In our example, say $b_{14}b_{23}b_{36}b_{45}b_{51}b_{62}$ represents two disjoint circuits, namely (x_3, x_7, x_{13}) and (x_8, x_9, x_1).

How can we now determine all circuits in $\boldsymbol{G}$?

Calculate the determinant of the matrix $\boldsymbol{B}$. Examine each summand which is different from zero and find out whether it represents a circuit or a set of disjoint circuits. A summand of the value zero does not yield a circuit (nor does the circuit of the value $1 = b_{11} \ldots b_{nn}$).

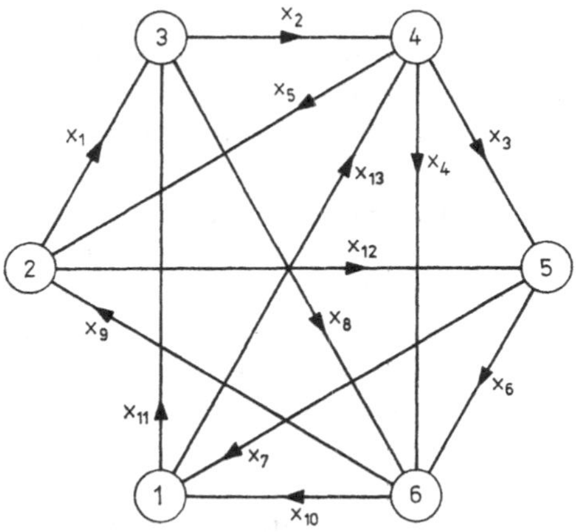

Fig. 9.1

As *determinant* of $\boldsymbol{B}$ we get for our example (here, we write for the products the sign "+" and for the sums the sign "(...)" for reasons which become evident only when determining minimum sets of arcs which represent all circuits):

$$\overset{+}{|}\boldsymbol{B}\overset{+}{|} = (1 + 2 + 5)\,(9 + 1 + 8)\,(9 + 1 + 2 + 3 + 6)\,(9 + 1 + 2 + 4)$$
$$(12 + 9 + 6)\,(7 + 9 + 11 + 8 + 12)\,(7 + 9 + 11 + 2 + 4 + 12)$$
$$(7 + 9 + 13 + 1 + 3 + 8)\,(7 + 9 + 13 + 4 + 12)$$
$$(7 + 5 + 12 + 13)\,(7 + 5 + 12 + 2 + 11)\,(10 + 13 + 4)$$
$$(10 + 11 + 8)\,(10 + 11 + 2 + 4)\,(10 + 1 + 5 + 13 + 8)$$
$$(10 + 6 + 3 + 13)\,(10 + 6 + 3 + 2 + 11)\,(10 + 6 + 5 + 12 + 13)$$
$$(10 + 6 + 5 + 12 + 11 + 2)\,(3 + 7 + 13)\,(2 + 3 + 7 + 11).$$

It can easily be seen by trying out that all circuits are in fact contained. It may also happen that some certain circuits occur several times, namely those circuits whose determinant summand does not only represent a circuit. Such circuits occurring several times may be treated just as if they are encountered only once.

It can be seen directly that really no circuit is omitted when using this method.

Now, we deal with the question of how to determine minimal sets of arcs representing all circuits, that is, those sets which do not contain a proper subset already representing all circuits. The method described in the following can also be defined to be a direct one.

Assume there exist in $\boldsymbol{G}$ precisely k circuits $\boldsymbol{K}_1, \ldots, \boldsymbol{K}_k$. Further, we assume that we have ascertained all minimal sets of arcs representing those $k-1$ circuits $\boldsymbol{K}_1, \ldots, \boldsymbol{K}_{k-1}$ (that is, not only those sets of arcs which have a minimum number of circuits). Let for instance $\mathfrak{M} = \{x_{i_1}, x_{i_2}, \ldots, x_{i_r}\}$ be such a minimal set of arcs. If $\boldsymbol{K}_k$ contains at least one of these arcs x_{i_j}, then $\mathfrak{M}$ is also representative of all k circuits. But if $\mathfrak{M}$ and $\boldsymbol{K}_k$ do not contain an arc in common, then each set $\mathfrak{M}'$ of arcs is representative of all k circuits if it contains not only all arcs of $\mathfrak{M}$, but also one arc of $\boldsymbol{K}_k$. This, however, does not mean that $\mathfrak{M}'$ must also be minimal.

The following can be stated: Let $\mathfrak{M}' = \{x_{i_1}, x_{i_2}, \ldots, x_{i_r}, x\}$ be representative of all circuits, but not minimal (with $x \in \boldsymbol{K}_k$, and no arc of $\mathfrak{M}$ is contained in $\boldsymbol{K}_k$). Then, without restriction of generality, $\mathfrak{M}'' = \{x_{i_2}, \ldots, x_{i_r}, x\}$ is already representative of all circuits. If $\mathfrak{M}''$ is minimal, then it represents also all circuits $\boldsymbol{K}_1, \ldots, \boldsymbol{K}_{k-1}$ and is minimal for this set of circuits or it contains a proper minimal subset representing the circuits $\boldsymbol{K}_1, \ldots, \boldsymbol{K}_{k-1}$. In the first case, we can do without $\mathfrak{M}'$ for representation, and we find among the minimal sets of arcs representing the first $k-1$ circuits a set which has $\mathfrak{M}'$ as superset. Of course, this applies also if $\mathfrak{M}''$ is not minimal.

In order to make this fact profitable for our algorithm, we only need to include successively all circuits in the examination and to introduce new minimal sets if there exists no single proper subset among the hitherto existing ones.

For this, we take our example. We have taken down all circuits. We imagine the circuits to be numbered in the order indicated. For representing the circuit $(1+2+5)$ each of its arcs can be used. Moreover, it is a minimal-representing one, thus

$$R_1\colon \{1\}, \{2\}, \{5\}.$$

Obviously,

$$R_2\colon \{1\}, \{2, 8\}, \{2, 9\}, \{5, 8\}, \{5, 9\}$$

are minimal-representative of $\boldsymbol{K}_1$ and $\boldsymbol{K}_2$. Furthermore,

$$R_3\colon \{1\}, \{2, 8\}, \{2, 9\}, \{3, 5, 8\}, \{5, 6, 8\}, \{5, 9\}$$

are minimal-representative of $\boldsymbol{K}_1$, $\boldsymbol{K}_2$ and $\boldsymbol{K}_3$. We continue in this way.

The reader will verify that for representing all 20 circuits ($(7+9+13+1+3+8)$ was the sum of two circuits) the following sets turn out to be minimal (this, however, already involves much calculation):

{1, 3, 10, 12}, {1, 6, 7, 10}, {1, 6, 11, 13}, {1, 7, 9, 10}, {1, 7, 10, 12}, {1, 9, 11, 13}, {1, 11, 12, 13}, {2, 6, 8, 13}, {2, 7, 9, 10}, {2, 8, 9, 13}, {2, 8, 12, 13}, {2, 9, 10, 13}, {2, 9, 11, 13}, {3, 5, 9, 10}, {5, 7, 9, 10}, {5, 9, 11, 13}, {1, 2, 10, 12, 13}, {1, 3, 4, 8, 12}, {1, 3, 4, 11, 12}, {1, 4, 6, 7, 8}, {1, 4, 6, 7, 11}, {1, 5, 6, 11, 13}, {1, 6, 7, 8, 13}, {2, 3, 4, 8, 12}, {2, 3, 8, 10, 12}, {2, 3, 9, 10, 12}, {2, 4, 6, 7, 8}, {2, 6, 7, 8, 10}, {2, 7, 8, 10, 12}, {3, 4, 5, 6, 8}, {3, 4, 5, 8, 9}, {3, 4, 5, 8, 12}, {3, 4, 5, 9, 11}, {4, 5, 6, 7, 8}, {1, 3, 4, 5, 6, 11}, {4, 5, 6, 7, 9, 11}, {4, 5, 6, 8, 11, 13}.

If we want to find minimum sets, we examine the sets consisting of four elements each.

It can immediately be seen that for representing all circuits at least four arcs are in fact required, because the four circuits (1, 2, 5), (6, 9, 12), (8, 10, 11), (3, 7, 13) are pairwise disjoint, and for representing them one arc is necessary for each circuit.

If the arcs are weighted (that is, if it is desired to take certain arcs in a minimal set wanted into consideration), all minimal sets have to be calculated, which makes it possible to determine among them the minimum with respect to the weighting.

9.3. THE IDEA OF YOUNGER

In this section, we are concerned with a method for determining minimum sets of feedback arcs published by D. H. Younger in 1963. To run the algorithm devised by him up to the end, one would certainly need more computing operations than for the approach suggested by Cederbaum and Lempel which itself requires a lot of computation. Nevertheless, some of his ideas are very interesting, so we want to present them in what follows.

On the whole, the algorithm can be rather sophisticated (and it will be so), since the well-known *branch-and-bound* idea is used here in a modified form.

First of all we compile a few lemmata which are more or less directly clear (task!).

LEMMA 1. *Let G be a circuit-free directed graph. Then, the vertices can be numbered so that for any arc x it holds: the initial point of x has a smaller number than the terminal point.*

The reader is referred to 1.5 where a numeration is given.

LEMMA 2. *Let* (according to lemma 1) *the vertices $X_1, X_2, \ldots, X_n$ of a circuit-free graph G be numbered so that for any arc (X_i, X_j) it holds: $i < j$. Then, the adjacency* *matrix appertaining to this numeration has only zeros below the main diagonal* (and also in the main diagonal because the graph possesses no loops).

Let $\mathfrak{N}$ be any numeration of the vertices of a graph G with n vertices, that is, an injective mapping $\mathfrak{N}$ of the vertices of G into the first n natural numbers.

We set

$$\mathfrak{F}_{\mathfrak{N}} = \{(P, Q) : \mathfrak{N}(P) \geqq \mathfrak{N}(Q); P, Q \in \mathfrak{X}\}.$$

Thus, $\mathfrak{F}_{\mathfrak{N}}$ is the set of the arcs which are oppositely directed to the numeration $\mathfrak{N}$. We choose any vertex numeration $\mathfrak{N}$ in G and consider the adjacency matrix $A_{\mathfrak{N}}(G)$ of the graph G relative to the numeration $\mathfrak{N}$. If we remove from G all those arcs for which non-zero (natural) numbers can be found on the main diagonal (if there exist loops) or below it, then the resulting graph is obviously circuit-free. This set of arcs is thus a set of feedback arcs, namely $\mathfrak{F}_{\mathfrak{N}}$.

LEMMA 3. *Let $\mathfrak{F}_{\mathfrak{N}}$ be a set of feedback arcs (of a connected graph) with minimum number of arcs. Then, the graph resulting after the removal of the arcs from $\mathfrak{F}_{\mathfrak{N}}$ is connected.*

Otherwise, we could remove from $\mathfrak{F}_{\mathfrak{N}}$ an arc (joining two components of the graph which is assumed to be unconnected) without violating the property that there exists a set of feedback arcs. This, however, contradicts the minimality.

DEFINITION. A vertex numeration $\mathfrak{N}$ is called *optimal* if $\mathfrak{F}_{\mathfrak{N}}$ is a set of feedback arcs with minimum number of arcs.

LEMMA 4. *The set of all optimal vertex numerations is invariant relative to the deletion of loops and circuits of length* 2.

In other words: A numeration which is optimal for the total graph yields an optimal numeration also for the graph if we remove the loops and circuits of length 2.

The proof of this statement is immediately clear, since loops are contained in each set of feedback arcs, and as for a circuit of length 2 consisting for instance of the arcs (X, Y) and (Y, X), at least one of the two arcs always lies in a set of feedback arcs.

When examining sets of feedback arcs, however, we can restrict ourselves always to the *reduced graph* from which the loops and circuits of length 2 have been removed. Therefore, we shall consider in the following only reduced graphs.

For further discussion we need the notion of an $\mathfrak{N}$-*sequential subgraph*. We imagine a vertex numeration $\mathfrak{N}$ to be given, say as a permutation $\{i_1, \ldots, i_n\}$. The form of the adjacency matrix $A_{\mathfrak{N}}(G)$ induced by $\mathfrak{N}$ will then be such that in the first row all arcs leaving X_{i_1} are taken into consideration, in the second row all arcs leaving X_{i_2}, etc.

DEFINITION. Given a graph with the vertices $X_1, \ldots, X_n$ and a vertex numeration $\mathfrak{N} = (i_1, i_2, \ldots, i_n)$. A subgraph of $\boldsymbol{G}$ spanned by the vertices

$$X_{i_j}, X_{i_{j+1}}, \ldots, X_{i_{j+k}} \qquad (1 \leqq j,\ k \geqq 0,\ j + k \leqq n)$$

is called $\mathfrak{N}$-*sequential.*

LEMMA 5. *Let $\mathfrak{N}$ be an optimal vertex numeration and $\boldsymbol{G}'$ an $\mathfrak{N}$-sequential subgraph of a graph $\boldsymbol{G}$.*

a) *Then, the set of arcs*

$$\mathfrak{F}_{\mathfrak{N}}' = \{(P, Q): \mathfrak{N}(P) \geqq \mathfrak{N}(Q);\ P, Q \in \boldsymbol{G}'\}$$

is a set of feedback arcs with minimum number of arcs in $\boldsymbol{G}'$.

b) *If we delete in $\boldsymbol{G}$ all arcs of $\boldsymbol{G}'$ and contract all vertices of $\boldsymbol{G}'$ to one vertex, which is then labelled with the number of one of the vertices of $\boldsymbol{G}'$, the following applies to the resulting graph $\boldsymbol{H}$: the set of arcs*

$$\mathfrak{F}_{\mathfrak{N}}'' = \{(P, Q): \mathfrak{N}(P) \geqq \mathfrak{N}(Q);\ P, Q \in \boldsymbol{H}\}$$

is a set of feedback arcs of $\boldsymbol{H}$ with minimum number of arcs.

Before carrying on, let us explain by means of an example the notions introduced so that it can be left up to the reader to prove the lemmata given above.

Consider the graph of Fig. 9.1. The adjacency matrix

$$\boldsymbol{A}_{\mathfrak{N}}(\boldsymbol{G}) = \begin{array}{c} \\ 2 \\ 5 \\ 1 \\ 3 \\ 4 \\ 6 \end{array} \begin{array}{c} \begin{array}{cccccc} 2 & 5 & 1 & 3 & 4 & 6 \end{array} \\ \begin{bmatrix} 0 & 1 & 0 & 1 & 0 & 0 \\ 0 & 0 & 1 & 0 & 0 & 1 \\ 0 & 0 & 0 & 1 & 1 & 0 \\ 0 & 0 & 0 & 0 & 1 & 1 \\ 1 & 1 & 0 & 0 & 0 & 1 \\ 1 & 0 & 1 & 0 & 0 & 0 \end{bmatrix} \end{array}$$

corresponds to a vertex numeration

$$\mathfrak{N} = (i_1, i_2, i_3, i_4, i_5, i_6) = (2, 5, 1, 3, 4, 6).$$

The set of feedback arcs relative to $\mathfrak{N}$ is

$$\mathfrak{F}_{\mathfrak{N}} = \{x_3, x_5, x_9, x_{10}\}.$$

As calculated in 9.2, this set of feedback arcs is minimal (i.e., no subset is also a set of feedback arcs). In addition, it is a minimum set of feedback arcs (i.e., among all sets of feedback arcs it has a minimum number of arcs). Thus, $\mathfrak{N}$ is an optimal vertex numeration.

If we remove from G the arcs of $\mathfrak{F}_{\mathfrak{R}}$, then the resulting graph is connected, which is in conformity with lemma 3 (the reader may easily verify this).

For our given $\mathfrak{R}$, for instance the graph generated by the vertices X_5, X_1, X_3 is $\mathfrak{R}$-sequential, whereas the graph spanned by the vertices X_1, X_2, X_3 is not $\mathfrak{R}$-sequential.

Lemma 5a says, for example, that for the subgraph G' spanned by the vertices X_1, X_3, X_4, X_6 a minimum set of feedback arcs can be obtained by considering those rows and columns (with the numbers 3, 4, 5, 6) of the adjacency matrix $A_{\mathfrak{R}}(G)$ which correspond to these four vertices, by determining all non-zero elements below the main diagonal (in our case, there is only one, namely that in the sixth row and third column) and by going to the corresponding arcs in the original graph. We get the arc x_{10} as minimum set in G'.

If we contract the four vertices X_1, X_3, X_4, X_6 to one vertex X and consider the graph H (which is formed according to lemma 5b), then the minimum set of arcs relative to $\mathfrak{R}$ contains precisely three elements, namely the original arcs x_5, x_9, x_3.

From lemma 5a) it follows that in the case of an optimal vertex numeration $\mathfrak{R}$ in a reduced graph (i.e., in a graph without loops and without circuits of length 2), only zeros will turn up directly below the main diagonal of $A_{\mathfrak{R}}(G)$.

The reader may make sure of the truth of the latter statement by making a better vertex numeration if in case of a reduced graph G, for instance the relation $a_{k+1,k} > 0$ holds in $A_{\mathfrak{R}}(G)$. For the case of a graph which is not yet reduced, it follows from lemma 5a that in the case of an optimal numeration $\mathfrak{R}$, the adjacency matrix $A_{\mathfrak{R}}(G)$ has the property that $a_{k,k+1} \geqq a_{k+1,k}$ is true for $a_{k+1,k} > 0$.

Consider now in a graph G two $\mathfrak{R}$-sequential subgraphs G_1, G_2 in the case of an optimal vertex numeration $\mathfrak{R}$. Let the vertices belonging to $G_1 \cup G_2$ span an $\mathfrak{R}$-sequential subgraph, too, but G_1 and G_2 must have no vertex in common, i.e., the vertices of G_1 and G_2 lie one after another with respect to $\mathfrak{R}$. Using the vertex numeration $\mathfrak{R} = (2, 5, 1, 3, 4, 6)$ in our example, we could imagine G_1 to be spanned for instance by the vertices X_5, X_1, X_3 and G_2 by the vertices X_4, X_6 (or also only X_4).

LEMMA 6. *Let G_1 and G_2 be two $\mathfrak{R}$-sequential subgraphs of G with n_1 and n_2 vertices, respectively, with $\mathfrak{R}$ being optimal. Let the graph spanned by the vertices of G_1 and G_2 be $\mathfrak{R}$-sequential, too, with $n_1 + n_2$ vertices. Then, it holds for the numbers c_{12}, c_{21} of the arcs going from G_1 to G_2* (and from G_2 to G_1, respectively):

a) $c_{12} \geqq c_{21}$.

b) *If $c_{12} = c_{21}$, then also the vertex numeration $\mathfrak{R}'$ resulting from $\mathfrak{R}$ is optimal, with this numeration being generated when „exchanging" G_1 for G_2 and maintaining the order in G_1 and G_2.*

To illustrate this, we consider in the optimal numeration $\mathfrak{R} = (2, 5, 1, 3, 4, 6)$ the $\mathfrak{R}$-sequential graph G_1 spanned by the vertex X_5 and the $\mathfrak{R}$-sequential subgraph

G_2 spanned by the vertices X_1, X_3, X_4. If we exchange G_1 for G_2, the vertex numeration $\mathfrak{N}' = (2, 1, 3, 4, 5, 6)$ results with the appertaining adjacency matrix

$$A_{\mathfrak{N}'}(G) = \begin{array}{c} \\ 2 \\ 1 \\ 3 \\ 4 \\ 5 \\ 6 \end{array} \begin{array}{c} \begin{array}{cccccc} 2 & 1 & 3 & 4 & 5 & 6 \end{array} \\ \begin{bmatrix} 0 & 0 & 1 & 0 & 1 & 0 \\ 0 & 0 & 1 & 1 & 0 & 0 \\ 0 & 0 & 0 & 1 & 0 & 1 \\ 1 & 0 & 0 & 0 & 1 & 1 \\ 0 & 1 & 0 & 0 & 0 & 1 \\ 1 & 1 & 0 & 0 & 0 & 0 \end{bmatrix} \end{array}.$$

We get again a minimum set of feedback arcs (represented below the main diagonal) consisting of the arcs x_5, x_7, x_9, x_{10}, which has already been found in section 9.2.

We are now capable of constructing a set of feedback arcs which is not necessarily minimum, but quite often minimal.

(i) We start out from a minimal set of feedback arcs (i.e., a set which contains no proper subset that is a set of feedback arcs as well). For finding such a set, we proceed in the following way:

We start out from any numeration $\mathfrak{N}$ and consider the set $\mathfrak{F}_{\mathfrak{N}}$ of the feedback arcs (these arcs can be made out in the adjacency matrix below the main diagonal). If we remove from G the arcs of $\mathfrak{F}_{\mathfrak{N}}$, then a circuit-free graph G' results. If possible, we add now to G' any arc of $\mathfrak{F}_{\mathfrak{N}}$, as far as the graph generated in this way is circuit-free. We carry on adding further arcs as long as the resulting graph remains circuit-free. The remaining set of arcs of $\mathfrak{F}_{\mathfrak{N}}$ is minimal.

(ii) Now, applying lemma 6, we try to reduce the set of feedback arcs by exchanging $\mathfrak{N}$-sequential successive subgraphs G_1, G_2. This can easily be done by hand by covering the adjacency matrix.

Let us take an example. We start out from the numeration $\mathfrak{N} = (2, 3, 6, 1, 4, 5)$ of the graph represented in Fig. 9.1. As adjacency matrix $A_{\mathfrak{N}}(G)$ we get

$$\begin{array}{c} \\ 2 \\ 3 \\ 6 \\ 1 \\ 4 \\ 5 \end{array} \begin{array}{c} \begin{array}{cccc|cc} 2 & 3 & 6 & 1 & 4 & 5 \end{array} \\ \left[\begin{array}{cccc|cc} 0 & 1 & 0 & 0 & 0 & 1 \\ 0 & 0 & 1 & 0 & 1 & 0 \\ 1 & 0 & 0 & 1 & 0 & 0 \\ 0 & 1 & 0 & 0 & 1 & 0 \\ \hline 1 & 0 & 1 & 0 & 0 & 1 \\ 0 & 0 & 1 & 1 & 0 & 0 \end{array}\right] \end{array}.$$

Below the main diagonal, there is represented the minimal set $\{x_9, x_{11}, x_5, x_4, x_6, x_7\}$ ascertained on page 201. When we cover the matrix following the lines drawn, we see that the exchange of the $\mathfrak{R}$-sequential graph $\boldsymbol{G}_1$ (formed by the vertices X_2, X_3, X_6, X_1) for $\boldsymbol{G}_2$ (spanned by X_4, X_5) reduces the number of arcs of a set of feedback arcs by 1.

We obtain the modified numeration $\mathfrak{R}' = (4, 5, 2, 3, 6, 1)$, and the appertaining adjacency matrix $\boldsymbol{A}_{\mathfrak{R}'}(\boldsymbol{G})$ reads:

$$\begin{array}{c} \\ 4 \\ 5 \\ 2 \\ 3 \\ 6 \\ 1 \end{array} \begin{array}{c} \begin{array}{cc|cc|cc} 4 & 5 & 2 & 3 & 6 & 1 \end{array} \\ \left[\begin{array}{cc|cc|cc} 0 & 1 & 1 & 0 & 1 & 0 \\ 0 & 0 & 0 & 0 & 1 & 1 \\ \hline 0 & 1 & 0 & 1 & 0 & 0 \\ 1 & 0 & 0 & 0 & 1 & 0 \\ \hline 0 & 0 & 1 & 0 & 0 & 1 \\ 1 & 0 & 0 & 1 & 0 & 0 \end{array}\right] \end{array}.$$

The set of the feedback arcs still consists of five arcs, namely $x_2, x_9, x_{11}, x_{12}, x_{13}$.

It is interesting to note that this set of feedback arcs is not minimal, since the arc x_{12} can be removed from it, which leads to a minimum set $\{x_2, x_9, x_{11}, x_{13}\}$. However, this minimum set may also be obtained by exchanging the second and third rows and columns. But if we make a covering as in the last matrix (according to lemma 6), thus obtaining a numeration $\mathfrak{R}'' = (2, 3, 4, 5, 6, 1)$, we get the adjacency matrix

$$\boldsymbol{A}_{\mathfrak{R}''}(\boldsymbol{G}) = \begin{array}{c} \\ 2 \\ 3 \\ 4 \\ 5 \\ 6 \\ 1 \end{array} \begin{array}{c} \begin{array}{cccccc} 2 & 3 & 4 & 5 & 6 & 1 \end{array} \\ \begin{bmatrix} 0 & 1 & 0 & 1 & 0 & 0 \\ 0 & 0 & 1 & 0 & 1 & 0 \\ 1 & 0 & 0 & 1 & 1 & 0 \\ 0 & 0 & 0 & 0 & 1 & 1 \\ 1 & 0 & 0 & 0 & 0 & 1 \\ 0 & 1 & 1 & 0 & 0 & 0 \end{bmatrix} \end{array}$$

with the minimum set of feedback arcs $\{x_5, x_9, x_{11}, x_{13}\}$.

Note that if a set of feedback arcs is given which is minimal or even a minimum set, the vertex numeration inducing this set is not unique. The reader may verify without difficulty that both $\mathfrak{R} = (4, 2, 3, 5, 6, 1)$ and $\mathfrak{R}' = (4, 2, 5, 3, 6, 1)$ induce the minimum set $\{x_2, x_9, x_{11}, x_{13}\}$.

It would now appear that the repeated application of the following two operations a) and b), inevitably entails that a minimum set of feedback arcs is found:

Given a set $\mathfrak{F}_{\mathfrak{R}}$ of feedback arcs (i.e., $\mathfrak{F}_{\mathfrak{R}}$ is a set of arcs after the removal of which from $\boldsymbol{G}$ there exists no circuit in the residual graph $\boldsymbol{G}'$).

a) Choose from $\mathfrak{F}_{\mathfrak{R}}$ any arc u and add it to G'. If the resulting graph G'' contains no circuit, remove u from $\mathfrak{F}_{\mathfrak{R}}$.

b) If the number of arcs of $\mathfrak{F}_{\mathfrak{R}}$ can be reduced by exchanging two successive $\mathfrak{R}$-sequential subgraphs, carry out this exchange.

As the example of Fig. 9.2 shows, the operations a) and b) are not sufficient to guarantee that the minimum set will be found.

For the vertex numeration $\mathfrak{R}$ of a graph G let the following adjacency matrix

$$\begin{array}{c|cccccc} & 1 & 2 & 3 & 4 & 5 & 6 \\ \hline 1 & 0 & 1 & 1 & 0 & 0 & 0 \\ 2 & 0 & 0 & 0 & 1 & 0 & 0 \\ 3 & 0 & 0 & 0 & 0 & 1 & 1 \\ 4 & 1 & 0 & 0 & 0 & 1 & 0 \\ 5 & 0 & 1 & 0 & 0 & 0 & 1 \\ 6 & 0 & 0 & 0 & 1 & 0 & 0 \end{array}$$

with a set of feedback arcs $\mathfrak{F}_{\mathfrak{R}} = \{(4, 1), (5, 2), (6, 4)\}$ be given.

The reader may ascertain without difficulty that applying a) or b) does not lead to a smaller set of feedback arcs although for instance the two arcs (4, 1) and (4, 5) contain all circuits. Nevertheless, by performing the operations a) and b), one will obtain in general a set of feedback arcs which is not too far from being a minimum set.

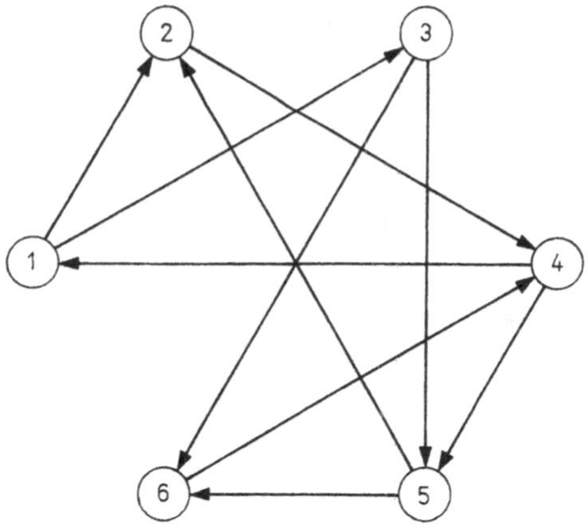

Fig. 9.2

D. H. Younger suggests an algorithm for further reducing sets of feedback arcs. The interested reader is referred to the original work [4] which is by no means easy to read. As no example is worked out there, we can only quote the author's opinion that the algorithm is also suitable for calculation by hand.

We call a numeration *feasible* if neither a) nor b) is applicable.

Algorithm of Younger

(i) Choose a feasible numeration $\mathfrak{R}^0 = \mathfrak{R} = (i_1, i_2, \ldots, i_n)$ of the n vertices. Determine the set $\mathfrak{F}_\mathfrak{R}$ of the feedback arcs (represented by the non-zero elements below the main diagonal of the adjacency matrix $\boldsymbol{A}_\mathfrak{R} = \boldsymbol{A}_\mathfrak{R}(\boldsymbol{G})$ of the graph $\boldsymbol{G}$).

Put the arcs of $\mathfrak{F}_\mathfrak{R}$ on the branching level 1.

Introduce a counting variable z and set $z = 0$.

Denote by $b(\mathfrak{R})$ the number of the arcs of the set of feedback arcs with respect to $\mathfrak{R}$, thus $b(\mathfrak{R}) = |\mathfrak{F}_\mathfrak{R}|$.

(ii) Choose from $\mathfrak{F}_\mathfrak{R}$ a feedback arc $r = (i_{j+k}, i_j)$ with $k > 0$ and make the new numeration $\mathfrak{R}' = (i_1, i_2, \ldots, i_{j-1}, i_{j+k}, i_j, i_{j+1}, \ldots, i_{j+k-1}, i_{j+k+1}, i_{j+k+2}, \ldots, i_n)$ (i.e., the initial point i_{j+k} of the arc r becomes the immediate predecessor of i_j in the numeretion $\mathfrak{R}'$, and all other positions of $\mathfrak{R}$ remain unchanged). Thus, r is no longer a feedback arc, but it is possible that new feedback arcs arise.

Delete the arc r and identify the two vertices i_{j+k} and i_j.

Delete any circuits of length 2 which may arise and add the number of these circuits to the counting variable z. Let $\mathfrak{R}''$ result from $\mathfrak{R}'$ by setting $i_{j+k} = i_j$ (thus, $\mathfrak{R}''$ becomes a numeration of $l - 1$ vertices as far as $\mathfrak{R}'$ is a numeration of l vertices).

(iii) Form $b(\mathfrak{R}^0) - z$. If $b(\mathfrak{R}^0) - z < 0$, go to (iv), otherwise to (v).

(iv) Dismiss the vertex numeration $\mathfrak{R}'$. Go to (ii), with the next feedback arc being chosen. Make z assume the value that it had before the last possible change (if z had been augmented during the last running of (ii), then we undo this augmentation; if z had not been increased, we do not change it). If on the current branching level all feedback arcs have been taken into consideration, choose the next feedback arc on the next lower level.

If there are no more arcs to be considered on the first branching level, the algorithm terminates.

(v) Change the numeration $\mathfrak{R}''$ until it is feasible. We continue to denote this feasible numeration by $\mathfrak{R}''$.

(vi) a) If $b(\mathfrak{R}'') < b(\mathfrak{R}^0) - z$, choose the numeration $\mathfrak{R}_0''$ as new numeration $\mathfrak{R}^0 = \mathfrak{R}$ and go to (i). Here, $\mathfrak{R}_0''$ arises from $\mathfrak{R}''$ by undoing the vertex contractions which have been effected each time in step (ii). However, make sure that i_{j+k} is the immediate predecessor of i_j. It turns out that the $\mathfrak{R}_0''$ determined during this procedure possesses less feedback arcs than the original $\mathfrak{R}^0 = \mathfrak{R}$ considered in step (i).

b) If $b(\mathfrak{R}'') = b(\mathfrak{R}^0) - z$, put $\mathfrak{R}_0''$ on the list of the potentially optimal numerations (as the number of the feedback arcs of $\mathfrak{R}_0''$ is equal to that of the original numeration $\mathfrak{R}^0$).

c) If $b(\mathfrak{R}'') \geqq b(\mathfrak{R}^0) - z$, determine the set $\mathfrak{F}_{\mathfrak{R}''}$ of the feedback arcs with respect to $\mathfrak{R}''$ (here, note that $\mathfrak{R}''$ is not a numeration of all n vertices), put this set of feedback arcs on a branching level that is higher by 1 than the level which resulted from $\mathfrak{R}''$, and go with $\mathfrak{R} = \mathfrak{R}''$ back to (ii) (by taking the current z).

When all feedback arcs on the branching level 1 are taken into account, the numerations put on the list of the potentially optimal numerations represent minimum sets of feedback arcs, thus optimal numerations.

To illustrate the algorithm, which is by no means a simple one, an example is given.

Consider the following vertex numeration $\mathfrak{R}^0 = \mathfrak{R} = (1, 2, 3, 4, 5, 6, 7, 8)$ of the graph G with the following adjacency matrix $A_{\mathfrak{R}}(G)$ (zeros have to be inserted in the free cells!):

$$
\begin{array}{c|cccccccc|}
 & 1 & 2 & 3 & 4 & 5 & 6 & 7 & 8 \\
\hline
1 & 0 & 1 & & & & 1 & & 1 \\
2 & & 0 & 1 & & 1 & & & \\
3 & & & 0 & 1 & & & 1 & \\
4 & & 1 & & 0 & 1 & & & 1 \\
5 & & & & & 0 & 1 & & \\
6 & & & 1 & & & 0 & 1 & \\
7 & 1 & & & & 1 & & 0 & 1 \\
8 & & 1 & & & & & & 0 \\
\hline
\end{array}
$$

Make sure that $\mathfrak{R}$ is feasible with the set of feedback arcs

$$\mathfrak{F}_{\mathfrak{R}} = \{(4, 2), (6, 3), (7, 1), (7, 5), (8, 2)\}.$$

These five arcs are to be found on the branching level 1, we set $z = 0$, it holds $b(\mathfrak{R}^0) = 5$. Now, we go to (ii).

0. We choose (4, 2). Thus, it results $\mathfrak{R}' = (1, 4, 2, 3, 5, 6, 7, 8)$. For the sake of clarity, let us write down the adjacency matrix of the reduced graph to be used for further calculation:

$$
A_{\mathfrak{R}''}(G) = \begin{array}{r|ccccccc|}
 & 1 & \begin{matrix}4\\2\end{matrix} & 3 & 5 & 6 & 7 & 8 \\
\hline
1 & 0 & 1 & & & 1 & & 1 \\
4 = 2 & & 0 & 1^* & 2 & & & 1^* \\
3 & & 1^* & 0 & & & 1 & \\
5 & & & & 0 & 1 & & \\
6 & & & 1 & & 0 & 1 & \\
7 & 1 & & & 1 & & 0 & 1 \\
8 & & 1^* & & & & & 0 \\
\hline
\end{array}.
$$

The unit elements provided with an asterisk indicate the resulting circuits of length 2. Thus, z becomes 2. We get $\mathfrak{R}'' = (1, 4 = 2, 3, 5, 6, 7, 8)$. Now, we get to (iii). It holds $b(\mathfrak{R}^0) - z = 5 - 2 = 3 > 0$.

We carry on at (v) and make the numeration $\mathfrak{R}''$ feasible (this procedure, however, can result in various feasible numerations). For example, we get the following feasible $\mathfrak{R}'' = (1, 4 = 2, 5, 6, 3, 7, 8)$ with the adjacency matrix

$$A_{\mathfrak{R}''}(G) = \begin{array}{c} \\ \\ 1 \\ 4=2 \\ 5 \\ 6 \\ 3 \\ 7 \\ 8 \end{array} \begin{array}{c} \begin{array}{ccccccc} & 4 & & & & & \\ 1 & 2 & 5 & 6 & 3 & 7 & 8 \end{array} \\ \begin{bmatrix} 0 & 1 & & 1 & & & 1 \\ & 0 & 2 & & & & \\ & & 0 & 1 & & & \\ & & & 0 & 1 & 1 & \\ & & & & 0 & 1 & \\ 1 & & 1 & & & 0 & 1 \\ & & & & & & 0 \end{bmatrix} \end{array}.$$

Since $b(\mathfrak{R}'') = 2 < b(\mathfrak{R}^0) - z = 3$, the case (vi), a), occurs. Thus, we get as improved vertex numeration $\mathfrak{R}_0'' = (1, 4, 2, 5, 6, 3, 7, 8)$, and we go on at (i), or which is even better, we begin anew at (i).

We put down the adjacency matrix $A_{\mathfrak{R}}(G)$ for $\mathfrak{R}^0 = \mathfrak{R}_0''$ and set $\mathfrak{R}^0 = \mathfrak{R}$:

$$\begin{array}{c} \\ 1 \\ 4 \\ 2 \\ 5 \\ 6 \\ 3 \\ 7 \\ 8 \end{array} \begin{array}{c} \begin{array}{cccccccc} 1 & 4 & 2 & 5 & 6 & 3 & 7 & 8 \end{array} \\ \begin{bmatrix} 0 & & 1 & & 1 & & & 1 \\ & 0 & 1 & 1 & & & & 1 \\ & & 0 & 1 & & 1 & & \\ & & & 0 & 1 & & & \\ & & & & 0 & 1 & 1 & \\ & 1 & & & & 0 & 1 & \\ 1 & & & 1 & & & 0 & 1 \\ & & 1 & & & & & 0 \end{bmatrix} \end{array}.$$

This arrangement is feasible with the set of feedback arcs

$$\mathfrak{F}_{\mathfrak{R}} = \{(3, 4), (7, 1), (7, 5), (8, 2)\}.$$

The arcs of $\mathfrak{F}_{\mathfrak{R}}$ are put on the branching level 1, we set $z = 0$, and $b(\mathfrak{R}) = 4$ holds.

1. We choose (3, 4) according to (ii) as feedback arc and get $\mathfrak{R}' = (1, 3, 4, 2, 5, 6, 7, 8)$. Hence it follows $\mathfrak{R}'' = (1, 3 = 4, 2, 5, 6, 7, 8)$ with the appertaining

adjacency matrix

	1	3 4	2	5	6	7	8
1	0		1		1		1
3 = 4		0	1*	1		1	1
2		1*	0	1			
5				0	1		
6		1			0	1	
7	1			1		0	1
8			1				0

A circuit of length 2 has resulted, thus z becomes equal to 1.

According to (iii), we form $b(\mathfrak{R}^0) - z = 4 - 1 = 3$. We make $\mathfrak{R}''$ feasible and get as adjacency matrix of the $\mathfrak{R}''$ which is now feasible:

	1	6	3 4	7	8	2	5
1	0	1			1	1	
6		0	1	1			
3 = 4			0	1	1		1
7	1			0	1		1
8					0	1	
2						0	1
5		1					0

With $b(\mathfrak{R}'') = 2$ it results $b(\mathfrak{R}'') = 2 < b(\mathfrak{R}^0) - z = 4 - 1 = 3$, thus case (vi), a) occurs.

We choose $\mathfrak{R}^0 = \mathfrak{R} = (1, 6, 3, 4, 7, 8, 2, 5)$ as new feasible starting numeration and go back to (i) (we forget all other open cases). As adjacency matrix with respect to $\mathfrak{R} = \mathfrak{R}^0$ we get

	1	6	3	4	7	8	2	5
1	0	1				1	1	
6		0	1		1			
3			0	1	1			
4				0		1	1	1
7	1				0	1		1
8						0	1	
2			1				0	1
5		1						0

It holds $\mathfrak{F}_{\mathfrak{R}} = \{(7, 1), (2, 3), (5, 6)\}$. These three arcs are put on the branching level 1, we set $z = 0$ and get $b(\mathfrak{R}^0) = 3$.

2. We choose (7, 1) and find $\mathfrak{R}' = (7, 1, 6, 3, 4, 8, 2, 5)$, $\mathfrak{R}'' = (7 = 1, 6, 3, 4, 8, 2, 5)$. As adjacency matrix $\boldsymbol{A}_{\mathfrak{R}''}$ we get

$$
\begin{array}{r|ccccccc|}
 & \overset{7}{1} & 6 & 3 & 4 & 8 & 2 & 5 \\
\hline
7=1 & 0 & 1^* & & & 2 & 1 & 1 \\
6 & 1^* & 0 & 1 & & & & \\
3 & 1 & & 0 & 1 & & & \\
4 & & & & 0 & 1 & 1 & 1 \\
8 & & & & & 0 & 1 & \\
2 & & & 1 & & & 0 & 1 \\
5 & & 1 & & & & & 0 \\
\hline
\end{array}.
$$

There results a circuit of length 2, thus it holds $z = 1$, and $b(\mathfrak{R}^0) - z = 3 - 1 = 2 > 0$ is true. We carry on at (v).

A feasible $\mathfrak{R}''$ follows from the adjacency matrix

$$
\begin{array}{r|ccccccc|}
 & 6 & 3 & \overset{7}{1} & 4 & 8 & 2 & 5 \\
\hline
6 & 0 & 1 & & & & & \\
3 & & 0 & 1 & 1 & & & \\
7=1 & & & 0 & & 2 & 1 & 1 \\
4 & & & & 0 & 1 & 1 & 1 \\
8 & & & & & 0 & 1 & \\
2 & & 1 & & & & 0 & 1 \\
5 & 1 & & & & & & 0 \\
\hline
\end{array}
$$

with $\mathfrak{R}'' = (6, 3, 7 = 1, 4, 8, 2, 5)$. We get $b(\mathfrak{R}'') = 2 = b(\mathfrak{R}^0) - z = 3 - 1$, thus with $\mathfrak{R}_0'' = (6, 3, 7, 1, 4, 8, 2, 5)$ and the appertaining adjacency matrix

$$
\boldsymbol{A}_{\mathfrak{R}_0''} =
\begin{array}{r|cccccccc|}
 & 6 & 3 & 7 & 1 & 4 & 8 & 2 & 5 \\
\hline
6 & 0 & 1 & 1 & & & & & \\
3 & & 0 & 1 & & 1 & & & \\
7 & & & 0 & 1 & & 1 & & 1 \\
1 & 1 & & & 0 & & 1 & 1 & \\
4 & & & & & 0 & 1 & 1 & 1 \\
8 & & & & & & 0 & 1 & \\
2 & & 1 & & & & & 0 & 1 \\
5 & 1 & & & & & & & 0 \\
\hline
\end{array}
$$

it holds $b(\mathfrak{R}_0'') = 3 = b(\mathfrak{R}^0)$, i.e., the case (vi), b), occurs. Thus, this $\mathfrak{R}_0''$ can be put on the list of the potentially optimal numerations. Now, we carry on at (vi), c), and determine $\mathfrak{F}_{\mathfrak{R}''} = \{(2, 3), (5, 6)\}$. These arcs are put on the branching level 2, and we carry on at (ii) with $z = 1$.

2.1. We choose (2, 3), $\mathfrak{R}' = (6, 2, 3, 7 = 1, 4, 8, 5)$, $\mathfrak{R}'' = (6, 2 = 3, 7 = 1, 4, 8, 5)$, the adjacency matrix $\boldsymbol{A}_{\mathfrak{R}''}$ follows as

$$\begin{array}{r} \\ \\ 6 \\ 2=3 \\ 7=1 \\ 4 \\ 8 \\ 5 \end{array}
\begin{array}{c}
\begin{array}{cccccc} & 2 & 7 & & & \\ 6 & 3 & 1 & 4 & 8 & 5 \end{array} \\
\left[\begin{array}{cccccc}
0 & 1 & & & & \\
 & 0 & 1^* & 1^* & & 1 \\
 & 1^* & 0 & & 2 & 1 \\
 & 1^* & & 0 & 1 & 1 \\
 & 1 & & & 0 & \\
1 & & & & & 0
\end{array}\right]
\end{array}.$$

Two new circuits of length 2 develop, thus z becomes equal to 3, further it follows $b(\mathfrak{R}^0) - z = 3 - 3 = 0$, i.e., we continue at (v) and make $\mathfrak{R}''$ feasible. The adjacency matrix of a feasible $\mathfrak{R}''$ results for instance as

$$\begin{array}{r} \\ \\ 7=1 \\ 4 \\ 8 \\ 5 \\ 6 \\ 2=3 \end{array}
\begin{array}{c}
\begin{array}{cccccc} 7 & & & & & 2 \\ 1 & 4 & 8 & 5 & 6 & 3 \end{array} \\
\left[\begin{array}{cccccc}
0 & & 2 & 1 & & \\
 & 0 & 1 & 1 & & \\
 & & 0 & & & 1 \\
 & & & 0 & 1 & \\
 & & & & 0 & 1 \\
 & & & 1 & & 0
\end{array}\right]
\end{array}.$$

From this, we get $b(\mathfrak{R}'') = 1 > b(\mathfrak{R}^0) - z = 0$ and carry on at (vi), c). The $\mathfrak{R}_0''$ appertaining to $\mathfrak{R}''$ has the form $\mathfrak{R}_0'' = (7, 1, 4, 8, 5, 6, 2, 3)$, and $b(\mathfrak{R}'') = 4$ holds. The set of the feedback arcs with respect to the current $\mathfrak{R}''$ has the form $\mathfrak{F}_{\mathfrak{R}''} = \{(2 = 3, 5)\}$. We go on at (ii) on the branching level 3. The variable z has now taken on the value 3.

2.1.1. We choose $(2 = 3, 5)$ (of course, there is only one feedback arc), $\mathfrak{R}' = (7 = 1, 4, 8, 2 = 3, 5, 6)$, $\mathfrak{R}'' = (7 = 1, 4, 8, 2 = 3 = 5, 6)$. The adjacency matrix appertaining to $\mathfrak{R}''$ results as

$$\begin{array}{r|cccccl}
 & \begin{smallmatrix}7\\1\end{smallmatrix} & 4 & 8 & \begin{smallmatrix}2\\3\\5\end{smallmatrix} & 6 & \\
\hline
7=1 & 0 & & 2 & 1 & & \\
4 & & 0 & 1 & 1 & & \\
8 & & & 0 & 1 & & \\
2=3=5 & & & & 0 & 1^* & \\
6 & & & & 1^* & 0 &
\end{array}.$$

A new circuit of length 2 develops, thus $z = 4$ holds. Now, it follows $b(\mathfrak{R}^0) - z = 3 - 4 < 0$. We go on at (iv), dismiss $\mathfrak{R}'$ and reduce z to 3.

As no more feedback arc is to be considered on the branching level 3, we go back to the level 2, choosing the next feedback arc.

2.2. We choose (5, 6,) $\mathfrak{R}' = (5, 6, 3, 7 = 1, 4, 8, 2)$, $\mathfrak{R}'' = (5 = 6, 3, 7 = 1, 4, 8, 2)$. The appertaining adjacency matrix is

$$\begin{array}{r|cccccc}
 & \begin{smallmatrix}5\\6\end{smallmatrix} & 3 & \begin{smallmatrix}7\\1\end{smallmatrix} & 4 & 8 & 2 \\
\hline
5=6 & 0 & 1 & & & & \\
3 & & 0 & 1 & 1 & & \\
7=1 & 1 & & 0 & & 2 & 1 \\
4 & 1 & & & 0 & 1 & 1 \\
8 & & & & & 0 & 1 \\
2 & 1 & 1 & & & & 0
\end{array}.$$

No circuits of length 2 develop, z remains equal to 1 and $b(\mathfrak{R}^0) - z = 3 - 1 = 2 > 0$ holds. A feasible numeration $\mathfrak{R}'' = (3, 7 = 1, 4, 8, 2, 5 = 6)$ results with

$$b(\mathfrak{R}'') = 2 = b(\mathfrak{R}) - z = 3 - 1, \qquad b(\mathfrak{R}_0'') = 3 = b(\mathfrak{R}^0)$$

(as can easily be checked by the reader).

$\mathfrak{R}_0''$ is put on the list of the potentially optimal numerations, and we carry on branching $\mathfrak{R}''$ with

$$\mathfrak{F}_{\mathfrak{R}''} = \{(2, 3), (5 = 6, 3)\}.$$

We are now on the branching level 3. We set $\mathfrak{R} = \mathfrak{R}''$ and go on at (ii).

2.2.1. We choose (2, 3). The reader will verify that three new circuits of length 2 occur, and therefore, the numeration $\mathfrak{R}'$ is dismissed according to (iv).

2.2.2. We choose (5 = 6, 3). Make sure of the two new circuits of length 2 which arise. Thus, z becomes 3, and the adjacency matrix of a feasible $\mathfrak{R}'' = (7 = 1, 4, 8, 2, 5 = 6 = 3)$ results as

$$\begin{array}{r|ccccc|} & \begin{matrix}7\\1\end{matrix} & 4 & 8 & 2 & \begin{matrix}5\\6\\3\end{matrix} \\ \hline 7=1 & 0 & & 2 & 1 & \\ 4 & & 0 & 1 & 1 & \\ 8 & & & 0 & 1 & \\ 2 & & & & 0 & 2 \\ 5=6=3 & & & & & 0 \end{array}$$

with $b(\mathfrak{R}'') = 0 = b(\mathfrak{R}) - z$ and $\mathfrak{R}_0'' = (7, 1, 4, 8, 2, 5, 6, 3)$ and $b(\mathfrak{R}_0'') = 3$, causing $\mathfrak{R}_0''$ to be put on the list of the potentially optimal numerations.

Thus, case 2 is completed, and we carry on at $z = 0$, $b(\mathfrak{R}^0) = 3$.

3. We choose (2, 3). This case and also case 4, where (5, 6) is chosen, are treated analogously.

This example is relatively simple, but it shows that a lot of calculation is required.

The interested reader may verify that using the algorithm given, no set of feedback arcs can be found containing less than three arcs. However, a considerable number of sets with precisely three arcs can be ascertained, even though not all of them are generally found.

D. H. Younger says that this algorithm yields a minimum set in every case, but does not completely prove his assertion. Establishing this proof, however, does not seem imperative. As is already indicated by our example, the above algorithm goes beyond the determination of feasible numerations in most cases.

In particular, we doubt the following: If we have found, say in case of our last example, a feasible numeration with the set of feedback arcs

$$\mathfrak{F}_{\mathfrak{R}} = \{(i_1, j_1), (i_2, j_2), \ldots, (i_r, j_r)\}$$

and if we do not succed in finding a better numeration in the course of the algorithm, then all numerations that are optimal in the sense of the algorithm (and that can be found) have the property that there exists an index t such that i_t is the immediate predecessor of j_t in this optimal numeration (although it could be possible that i_t is the predecessor of j_t, in fact, but not necessarily the immediate one; this means for our example that in each optimal numeration found, 7 is the immediate predecessor of 1, or 2 the immediate predecessor of 3, or 5 the immediate predecessor of 6). But what will happen if the minimum sets of feedback arcs and the set of feedback arcs found so far are disjoint and if we are in a "local minimum"?

Of course, by using lemma 6b, it is possible to find further numerations with the same number of feedback arcs as those with minimal number of feedback arcs found in the course of the algorithm (which is also said by D. H. Younger), but those other numerations are not considered in the algorithm.

Despite our doubts, we have discussed this latter algorithm in great detail, since it is really quite promising. In comparison with the algorithm of Lempel and Cederbaum, where it was not simply a determinant that had to be calculated (in the case of n vertices of circumference n), but a determinant such that one variable per arc had to be taken along in the graph, the algorithm given by Younger seems us to be more efficient and also more suitable especially for machine computation.

9.4. BIBLIOGRAPHY

[1] Lempel, A., and I. Cederbaum: On directed circuits and cutsets of a linear graph, Faculty of Electrical Engineering, Technion — Israel Inst. Technology, Haifa (Israel), Publ. 31. June 1965.

[2] Lempel, A., and I. Cederbaum: Minimum feedback arc and vertex sets of a directed graph, IEEE Trans. Circuit Theory **CT-13** (1966), 399—403.

[3] Walther, H., und H.-J. Voss: Über Kreise in Graphen, Berlin 1974.

[4] Younger, D. H.: Minimum feedback arc sets for a directed graph, IEEE Trans. Circuit Theory **CT-10** (1963), 238—245.

[5] Garey, M. R., and D. S. Johnson: Computers and Intractability, A guide to the Theory of NP-Completeness, San Francisco 1979.

Chapter 10

Embedding of planar graphs in the plane

10.1. FORMULATION OF THE PROBLEM

In practice, there often arises the problem of drawing in the plane graphs which exist, say, in the form of electric networks or time schedules, without any crossing. Such an embedding without crossing is important for the design of printed circuits. If all of the components of an electric network are left unconsidered, the graph G of the network remains and the embedding of this yields a solution to the problem.

By trial and error, one could attempt to obtain a representation without crossings, which would possibly lead to a quick solution in the case of graphs containing only few vertices. However, for many vertices and edges such an approach would be troublesome, and a decision on whether or not the given graph is embeddable could hardly be made.

In this chapter, we shall deal with some methods which make it possible to embed a given graph in the plane provided that such an embedding is possible at all.

10.2. THEOREMS OF KURATOWSKI, MACLANE AND WHITNEY

DEFINITION. A graph G is said to be *planar* if it can be embedded in the plane. The *embedding* of G is called a *plane graph* G'.

Figure 10.1b shows a plane graph G' to the planar graph G of Fig. 10.1a. We shall restrict ourselves to undirected simple graphs, i.e., to graphs without loops and multiple edges. If a given graph is not simple, then remove loops from it, replace the multiple edges by a single edge and examine whether the graph which results in this way is planar. If this graph is planar, then the original graph is also planar. But if not, the original graph is not planar either (task!).

We may also restrict ourselves to connected graphs, since in case of an undirected graph, each component has to be considered separately. Obviously, the graph is planar if and only if each of its components is planar.

Before addressing ourselves to the first planarity criterion, we still wish to show that it is sufficient to consider only 3-connected graphs. For this purpose, we give without proof a theorem that can be read in H. Sachs [21]:

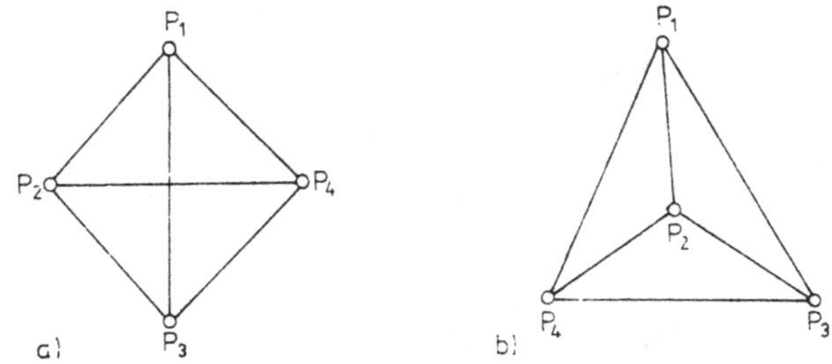

Fig. 10.1

THEOREM 10.1.

a) *Let* G *be exactly* 1*-connected. Then,* G *possesses a cut-vertex* P*, and* G *is union of two connected subgraphs* $\tilde{G}_1$ *and* $\tilde{G}_2$ *which have the vertex* P *in common, and only this one* (cf. Fig. 10.2). *It holds:* G *is planar if and only if* $\tilde{G}_1$ *and* $\tilde{G}_2$ *are planar.*

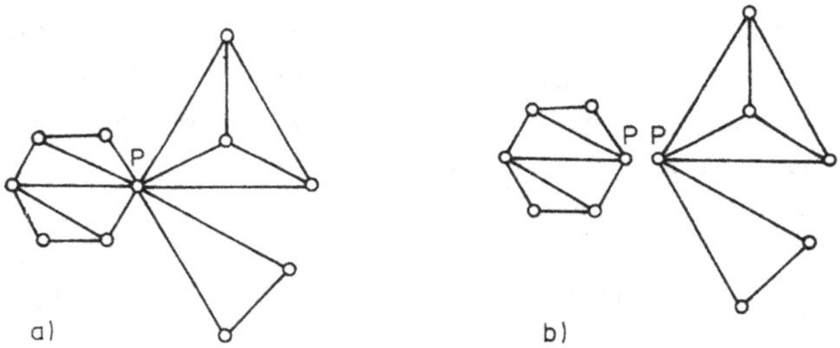

Fig. 10.2

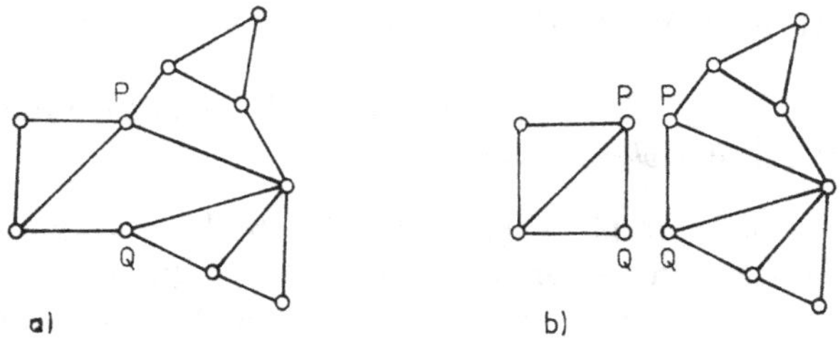

Fig. 10.3

b) *Let* G *be exactly* 2*-connected. Then, it holds:* G *is union of two connected subgraphs* G_1 *and* G_2 *which have precisely two vertices* P *and* Q *in common and in addition to this, also the edge* (P, Q) *if* P *and* Q *are adjacent. If, however,* P *and* Q *are not adjacent, then we add the edge* (P, Q) *to both* G_1 *and* G_2*. Thus, let the graphs* G_1' *and* G_2' *develop. If* (P, Q) *is contained in* G *and thus also in* G_1 *and*

G_2, then let

$$G_1' = G_1, \qquad G_2' = G_2$$

be true.

In this case, it holds: G is planar if and only if G_1' and G_2' are planar (cf. Fig. 10.3).

We have thus shown that we can restrict ourselves to *3-connected* graphs when making *planarity examinations* (it is no secret that a great deal of work might be required for verifying that a graph is 3-connected). Let us call such graphs *primitive*.

DEFINITION. Two manifolds (of topological spaces) are called *homeomorphic* if there exists an injective mapping of the two manifolds one onto the other which is together with its inverse mapping continuous (neighbourhood-preserving).

A theorem which is of importance for embedding is given without proof:

THEOREM 10.2. *A primitive planar graph is uniquely embeddable in the plane except for homeomorphism.*

DEFINITION. A graph H is called *subdivision* of a graph G if H results from G by inserting vertices in the interior of edges of G.

It will be agreed that for the case $G = H$, the graph H is also a subdivision of G (cf. Fig. 10.4).

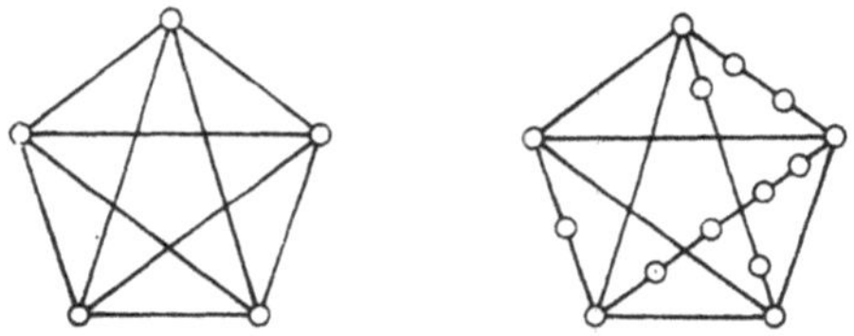

Fig. 10.4

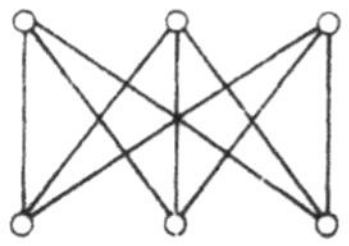

Fig. 10.5

THEOREM 10.3 (Kuratowski [14]). *A graph is not planar if and only if it contains a subdivision of the complete graph K_5 with five vertices or of the complete bipartite graph $K_{3,3}$ with three vertices in each of the two classes* (cf. Figs. 10.4 and 10.5).

Of course, the application of Kuratowski's theorem in practical planarity studies involves the difficulty that it will not always be easy to recognize in a graph subdivisions of the $K_{3,3}$ or K_5. Therefore, this planarity criterion is hardly recommendable for practical purposes. On the other hand, Kuratowski's criterion is often used in theoretical reflections.

Let $\boldsymbol{G}$ be an at least 2-connected graph. A graph $\boldsymbol{DG}$ is called *dual* to $\boldsymbol{G}$ if there exists an injective mapping of the edges of $\boldsymbol{G}$ onto $\boldsymbol{DG}$ satisfying the following condition: A set of edges of $\boldsymbol{G}$ forms a *cut* (cf. section 1.4.; in the case of any orientation of the edges of $\boldsymbol{G}$, a *cut* becomes an *elementary cocycle*) if and only if the set of edges corresponding to it in $\boldsymbol{DG}$ forms a circuit.

THEOREM 10.4 (Whitney [27, 28]). *A graph* $\boldsymbol{G}$ *is planar if and only if there exists a graph* $\boldsymbol{DG}$ *which is dual to it.*

It turns out that the graph which is dual to a planar graph $\boldsymbol{G}$ is also planar.

A graph which is dual to a planar graph can be generated in the following way (cf. Fig. 10.6):

Let $\boldsymbol{G}$ be at least 2-connected and embedded in the plane. By this embedding, the plane is subdivided in several *domains*, with each of them being bounded by a *circuit*. The set of vertices of $\boldsymbol{DG}$ is obtained by inserting a *vertex* (that belongs to $\boldsymbol{DG}$) in the *interior* of each of the domains of $\boldsymbol{G}$. Two vertices of $\boldsymbol{DG}$ are connected by an edge (of $\boldsymbol{DG}$) if and only if the domains corresponding to them in $\boldsymbol{G}$ are adjacent (if two domains have precisely k edges in common, the vertices corresponding to them in $\boldsymbol{DG}$ are connected by k edges).

The interested reader is referred to H. Whitney [27].

We still want to give a third planarity criterion.

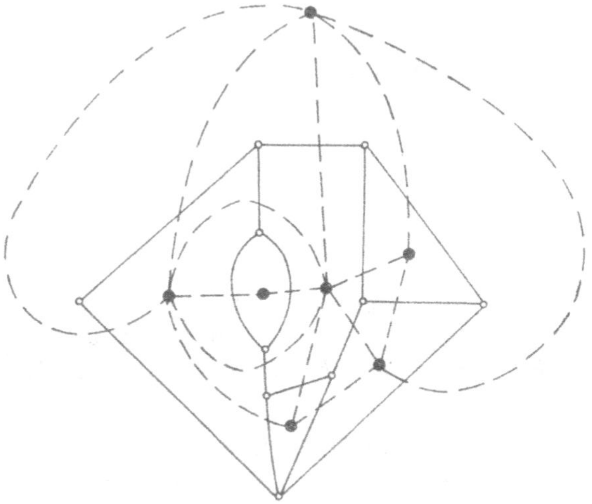

Fig. 10.6

Let C_i be a circuit of a graph G. We assign to C_i a *vector* with m components if G possesses precisely m edges (cf. the definition of a vector assigned to a *cycle*, chapter 1). The jth component of the vector $\mathbf{c}_i = (c_i^1, c_i^2, \ldots, c_i^m)$ assigned to C_i is 1 if the jth edge of G lies in C_i (in case of any fixed numeration of the edges of G). If, however, the jth edge is not contained in C_i, then we set $c_i^j = 0$.

DEFINITION. We say that *k circuits* $C_1, C_2, \ldots, C_k$ form a *complete system of circuits* in G if any circuit C of G can uniquely be represented in the form

$$\mathbf{c} = (c_1, \ldots, c_m) = \sum \mathbf{c}_{i_j} \quad (\text{mod. } 2)$$

with the i_j being taken from the set $\{1, \ldots, k\}$.

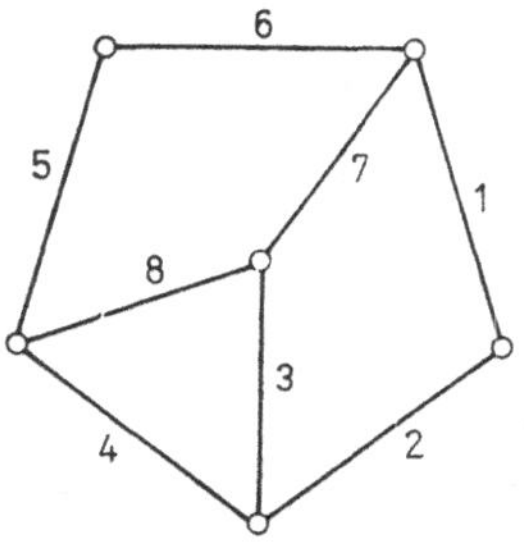

Fig. 10.7

Each graph contains at least one complete system of k circuits with

$$k = \textit{number of edges} - \textit{number of vertices} + \textit{number of components}.$$

For this, consider the graph of Fig. 10.7. The three circuits

$$C_1 = (1, 2, 3, 7), \qquad \mathbf{c}_1 = (1, 1, 1, 0, 0, 0, 1, 0),$$
$$C_2 = (1, 2, 4, 8, 7), \qquad \mathbf{c}_2 = (1, 1, 0, 1, 0, 0, 1, 1),$$
$$C_3 = (5, 6, 7, 8), \qquad \mathbf{c}_3 = (0, 0, 0, 0, 1, 1, 1, 1)$$

obviously form a complete system, since for the remaining four circuits, it holds:

$$C_4 = (1, 2, 3, 8, 5, 6), \qquad \mathbf{c}_4 \equiv \mathbf{c}_1 + \mathbf{c}_3 \quad (\text{mod. } 2),$$
$$C_5 = (1, 2, 4, 5, 6), \qquad \mathbf{c}_5 \equiv \mathbf{c}_2 + \mathbf{c}_3,$$
$$C_6 = (3, 4, 8), \qquad \mathbf{c}_6 \equiv \mathbf{c}_1 + \mathbf{c}_2,$$
$$C_7 = (3, 4, 5, 6, 7), \qquad \mathbf{c}_7 \equiv \mathbf{c}_1 + \mathbf{c}_2 + \mathbf{c}_3.$$

THEOREM 10.5 (MacLane [17]). *A graph is planar if and only if it possesses a complete system of circuits such that no edge is contained in more than two of these circuits.*

10.3. THE PLANARITY ALGORITHM OF DAMBITIS

Let $\boldsymbol{G}'$ be a *primitive* graph with n vertices and m edges, and let $\boldsymbol{L}'$ be a *spanning tree* of $\boldsymbol{G}'$. We number the edges *not* lying in $\boldsymbol{L}'$ (called *chords*) from 1 to $m - n + 1$, and the edges contained in $\boldsymbol{L}'$ from $m - n + 2$ to m. Let $\boldsymbol{R}'$ be the *incidence matrix* of $\boldsymbol{G}'$. We arrange $\boldsymbol{R}'$ such that the last $n - 1$ columns correspond to the spanning tree edges of $\boldsymbol{L}'$. We partition $\boldsymbol{R}'$ in the following way:

$$
\begin{array}{c}
\\
\boldsymbol{R}' =
\end{array}
\begin{array}{c}
\begin{array}{cc} \text{chords} & \text{spanning tree edges} \end{array} \\
\left[\begin{array}{c|c}
\boldsymbol{A}_{11} & \boldsymbol{A}_{12} \\
\hline
\boldsymbol{A}_{21} & \boldsymbol{A}_{22}
\end{array}\right]
\begin{array}{l} 1 \\ 2 \\ \vdots \\ n-2 \\ n-1 \\ n \end{array}
\quad \text{vertices} \\
\begin{array}{cc} 1\ 2\ \ldots\ m-n+1 & m-n+2\ldots m \end{array}
\end{array}
$$

Starting from $\boldsymbol{R}'$ we calculate a matrix $\boldsymbol{B}'$ representing $m - n + 1$ *linearly independent circuits* (linearly independent in the sense of the definition of a complete system of circuits):

$$\boldsymbol{B}' = [\boldsymbol{E}_{m-n+1}, \boldsymbol{A}_{11}^T(\boldsymbol{A}_{12}^{-1})^T].$$

Here, $\boldsymbol{E}_i$ is an $(i \times i)$-*identity matrix*. When forming the products and when adding up, we always calculate modulo 2. The edges of the graph are assigned to the columns of $\boldsymbol{B}'$ as in case of $\boldsymbol{R}'$.

Let us consider an example for illustrating this. The construction represented in Fig. 10.8 ensures that each circuit occurring in $\boldsymbol{B}'$ consists of *exactly one chord* and *one path as part of* $\boldsymbol{L}'$, with this path connecting the end-points of the chord with each other. This chord can be denoted by *representing the edge* of the circuit

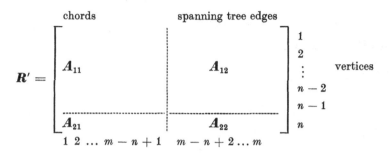

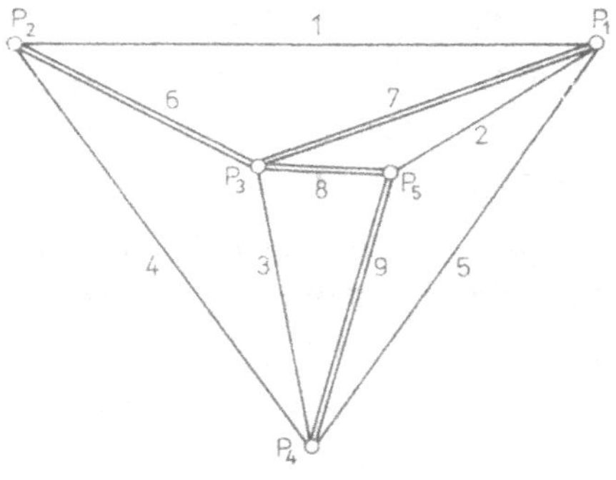

Fig. 10.8

(with a given invariant spanning tree $\boldsymbol{L}'$). The reader is also referred to chapter 1 where the notion of the *cycle basis* is explained. The representing edges can immediately be read from the identity matrix contained in $\boldsymbol{B}'$. In the case of our example given by Fig. 10.8, we get

$$\boldsymbol{R}' = \left[\begin{array}{ccccc|cccc} 1 & 1 & 0 & 0 & 1 & 0 & 1 & 0 & 0 \\ 1 & 0 & 0 & 1 & 0 & 1 & 0 & 0 & 0 \\ 0 & 0 & 1 & 0 & 0 & 1 & 1 & 1 & 0 \\ 0 & 0 & 1 & 1 & 1 & 0 & 0 & 0 & 1 \\ \hline 0 & 1 & 0 & 0 & 0 & 0 & 0 & 1 & 1 \end{array}\right],$$

$$\boldsymbol{B}' = \left[\begin{array}{ccccc|cccc} 1 & 0 & 0 & 0 & 0 & 1 & 1 & 0 & 0 \\ 0 & 1 & 0 & 0 & 0 & 0 & 1 & 1 & 0 \\ 0 & 0 & 1 & 0 & 0 & 0 & 0 & 1 & 1 \\ 0 & 0 & 0 & 1 & 0 & 1 & 0 & 1 & 1 \\ \hline 0 & 0 & 0 & 0 & 1 & 0 & 1 & 1 & 1 \end{array}\right].$$

Assume a planar graph $\boldsymbol{G}'$ to be given. Let $\boldsymbol{G}'$ be *primitive*. Then, $\boldsymbol{G}'$ possesses a dual graph $\boldsymbol{G}$.

A cut of $\boldsymbol{G}$ corresponds to a circuit of $\boldsymbol{G}'$, and vice versa.

We set $\boldsymbol{Q} = \boldsymbol{B}'$ and call $\boldsymbol{Q}$ the *cut-set matrix* of $\boldsymbol{G}$.

J. J. Dambitis gives an algorithm which makes it possible to calculate from $\boldsymbol{Q}$ the *incidence matrix* $\boldsymbol{R}$ of $\boldsymbol{G}$.

It turns out that $\boldsymbol{G}'$ is planar if and only if the following algorithm can be run up to the end, which then leads to the matrix $\boldsymbol{R}$. This algorithm utilizes the *planarity criterion of Withney.*

However, let us clarify a few concepts and notions before turning to the algorithm.

Representing edge: In the course of dualizing, a *spanning tree edge* of $\boldsymbol{G}'$ is transformed into a *chord* of $\boldsymbol{G}$, and vice versa. The edges with the numbers $1, 2, \ldots, m - n + 1$ thus form in $\boldsymbol{G}$ a *spanning tree* $\boldsymbol{L}$. In the course of dualizing, an *edge representing* in $\boldsymbol{G}'$ a *circuit* (i.e., which is a chord there) is transformed into a *cut edge* that lies in the *spanning tree* $\boldsymbol{L}$ induced by $\boldsymbol{L}'$ in $\boldsymbol{G}$. This edge is thus an *edge representing the cut* (these notions have already been discussed in detail in chapter 1; thus, there corresponds to the set of *chords a cotree*, with the edges being oriented in any way). If the edge i represents the cut $\boldsymbol{Q}_i$ in $\boldsymbol{G}$, then the spanning tree $\boldsymbol{L}$ divides into two trees $\boldsymbol{L}_1$ and $\boldsymbol{L}_2$ after the removal of i. The cut $\boldsymbol{Q}_i$ contains not only the edge i, but also all chords in $\boldsymbol{G}$, with one of their end-points lying in $\boldsymbol{L}_1$ and the other in $\boldsymbol{L}_2$.

Pendant row: A row q_i of the cut-set matrix $\boldsymbol{Q}$ is called *pendant* if the representing edge i is a *terminal edge* of the spanning tree $\boldsymbol{L}$, that is, if it is incident with an end-point (i.e., with a vertex of the degree 1) in $\boldsymbol{L}$. Let P be the end-point of a terminal edge of $\boldsymbol{L}$. Then, the edges of $\boldsymbol{G}$ incident with P just form the cut $\boldsymbol{Q}_i$.

The pendant rows of the matrix $\boldsymbol{Q}$ are determined in the following way:

Let us find out for example whether the ith row q_i of $\boldsymbol{Q}$ is pendant. For this, we delete all those columns of $\boldsymbol{Q}$ in which q_i has a 1, subsequently the row q_i is also deleted. We denote the matrix which results in this way by $\boldsymbol{Q}^i$. Neglecting the identity matrix contained in $\boldsymbol{Q}^i$, this latter represents a *system of chords.*

We call two *rows* of $\boldsymbol{Q}^i$ *joined* if there exists a column in which both rows have a 1.

Now, we try to combine the rows of $\boldsymbol{Q}^i$ according to a certain rule. Let v and w be two joined rows of $\boldsymbol{Q}^i$. Add these two rows by using the following rules:

$$1 + 1 = 1, \qquad 1 + 0 = 1, \qquad 0 + 1 = 1, \qquad 0 + 0 = 0.$$

The resulting row is then substituted for either of the two rows v and w, with the other one being deleted. This addition of rows is performed as long as possible. If it turns out that all rows of $\boldsymbol{Q}^i$ can be combined to one row, then the *row* q_i is *pendant.* If this is not possible, however, we get several connected systems $\boldsymbol{H}_{i1}$, ..., $\boldsymbol{H}_{ik}$ of chords with $1 < k < m - n + 1$ into which the graph $\boldsymbol{G}$ is decomposed after the removal of the cut $\boldsymbol{Q}_i$. Such a *connected system of chords* is characterized by a *maximal* set of rows which can be combined.

We consider the example of Fig. 10.8. We need not write down the identity matrix $\boldsymbol{E}_{m-n+1}$, since this part of the matrix is of no importance for the addition of the rows:

Is the row q_1 in $\boldsymbol{Q} = \boldsymbol{B}'$ pendant?

$$\boldsymbol{Q}^1 = \begin{bmatrix} 1 & 0 \\ 1 & 1 \\ 1 & 1 \\ 1 & 1 \end{bmatrix} \begin{matrix} \text{2nd} \\ \text{3rd} \\ \text{4th} \\ \text{5th} \end{matrix} \left. \vphantom{\begin{matrix} 1\\1\\1\\1 \end{matrix}} \right\} \text{row with respect to } \boldsymbol{Q}$$

8. 9.
column

Obviously, q_1 is pendant.

Is q_4 pendant?

$$\boldsymbol{Q}^4 = \begin{bmatrix} 1 \\ 1 \\ 0 \\ 1 \end{bmatrix} \begin{matrix} \text{1st} \\ \text{2nd} \\ \text{3rd} \\ \text{5th} \end{matrix} \left. \vphantom{\begin{matrix} 1\\1\\1\\1 \end{matrix}} \right\} \text{row with respect to } \boldsymbol{Q}$$

7.
column

It can be seen that the rows 1, 2, 5 are joined. The row 3 is not joined to any row of $\boldsymbol{Q}^4$.

Since not all rows can be combined, the row q_4 is not pendant. The following connected systems of chords arise:

$$\boldsymbol{H}_{41} = (1, 2, 5), \qquad \boldsymbol{H}_{42} = (3).$$

Checking up all rows yields:

$$q_1, q_2, q_3 \text{ pendant},$$
$$q_4 \text{ not pendant with } \boldsymbol{H}_{41} = (1, 2, 5),\ \boldsymbol{H}_{42} = (3),$$
$$q_5 \text{ not pendant with } \boldsymbol{H}_{51} = (1, 4),\ \boldsymbol{H}_{52} = (2),\ \boldsymbol{H}_{53} = (3).$$

Thus, each row of $\boldsymbol{Q}$ can be characterized as "pendant" or "not pendant plus corresponding systems of chords".

Termination condition: Since each spanning tree containing not less than three vertices possesses at least two terminal edges, there must exist in $\boldsymbol{Q}$ at least two pendant rows. Otherwise, there would be a contradiction revealing that the assumption of planarity of $\boldsymbol{G}'$ is not justified. That is, $\boldsymbol{G}'$ is not planar in this case.

If there is in $\boldsymbol{Q}$ a column in which three or more pendant rows have a 1, then this would be in contradiction to the fact that an edge is incident with exactly two distinct vertices. From this, we also conclude that $\boldsymbol{G}'$ is not planar.

These two breakoff conditions bring the algorithm to a stop and indicate the nonplanarity of $\boldsymbol{G}'$.

Now, we come to the construction of the incidence matrix $\boldsymbol{R}$. To describe the planarity algorithm of J. J. Dambitis, we shall use a special method introduced by Ljapunov for the description of an algorithm. For the sake of simplicity, however, we shall write the single steps one beneath the other. In doing so, the letter A means an instruction and the letter P a test. We shall not discuss in detail why the algorithm indicated always makes it possible to decide whether or not a given graph is planar. The interested reader is referred to [4].

Algorithm of Dambitis

(A1) Calculate the matrix $\boldsymbol{Q} = \boldsymbol{B}'$ (the circuit matrix of $\boldsymbol{G}'$).

(A2) Determine all pendant rows of $\boldsymbol{Q}$ and use them to form the matrix $\boldsymbol{R}^0$ (which already constitutes a part of the incidence matrix $\boldsymbol{R}$ of $\boldsymbol{G}$ to be determined).

(P3) Check whether $\boldsymbol{R}^0$ consists of less than two rows. If so, then the graph $\boldsymbol{G}'$ is not planar,
go to the End.
If not,

(P4) Check whether there exists in $\boldsymbol{R}^0$ a column containing more than two 1's. If so, then the graph $\boldsymbol{G}'$ is not planar,
got to the End,
If not,

(P5) Check whether there is no column of $\boldsymbol{Q}$ with more than two 1's.
If so, then make up the matrix $\boldsymbol{R} = \begin{bmatrix} \boldsymbol{Q} \\ \boldsymbol{T} \end{bmatrix}$, with $\boldsymbol{T}$ resulting as one row, namely as the sum mod. 2 of all rows of $\boldsymbol{Q}$. $\boldsymbol{R}$ is then the required incidence matrix of $\boldsymbol{G}$.
Go to the End.
If not,

(A6) Form for each non-pendant row q_i of $\boldsymbol{Q}$ the connected systems of chords $\mathfrak{S}_i = \{\boldsymbol{H}_{i1}, \boldsymbol{H}_{i2}, \ldots, \boldsymbol{H}_{ip_i}\}$ as well as the set $\mathfrak{N} = \{q_i: q_i \text{ not pendant}\}$.

(A7) Put the matrix $\boldsymbol{Q}$ on a waiting list.

(P8) Check whether the waiting list is empty.
If so, then $\boldsymbol{R} = \boldsymbol{R}^0$ is the required incidence matrix of $\boldsymbol{G}$.
If not,

(A9) Choose a matrix from the waiting list (e.g. in the order of their inscription) and denote it by $\boldsymbol{Q}$.

(P10) Check whether there exists in $\boldsymbol{Q}$ a non-pendant row q_i with $p_i = 2$ (thus $|\mathfrak{S}_i| = 2$).
If so, then go to (A16).
If not,

(A11) Make up for each $q_k \in \mathfrak{N}$ the number z_k of those connected systems of chords from $\mathfrak{S}_k$ which contain denotations for non-pendant rows of $\boldsymbol{Q}$.

(A12) Form $z := \min\limits_{q_k \in \mathfrak{N}} z_k$.

(A13) Form $\mathfrak{N}' = \{q_r : q_r \in \mathfrak{N} \text{ and } z_r = z\}$.

(A14) Find a $q_i \in \mathfrak{N}'$ with $|\mathfrak{S}_i| \leqq |\mathfrak{S}_r|$ for all $q_r \in \mathfrak{N}'$ (that is, among all non-pendant rows belonging to $\mathfrak{N}'$ find one (namely q_i) for which the number of the connected systems of chords is minimum).

(P15) Check whether $\mathfrak{S}_i$ can be decomposed into two non-empty disjoint classes $\mathfrak{S}_i^1$, $\mathfrak{S}_i^2$ in the following manner: Form from q_i and the rows with denotations from $\mathfrak{S}_i^1$ (and $\mathfrak{S}_i^2$) a matrix $\boldsymbol{Q}_1$ ($\boldsymbol{Q}_2$, respectively) such that in none of the submatrices $\boldsymbol{Q}_1' \subseteqq \boldsymbol{Q}_1$ and $\boldsymbol{Q}_2' \subseteqq \boldsymbol{Q}_2$ (made up of q_i and the pendant rows of $\boldsymbol{Q}_1$ and $\boldsymbol{Q}_2$, respectively) more than two 1's occur in any column.
If so, then go to (A17).
If not, then $\boldsymbol{G}'$ is not planar,
go to the End.

(A16) Form from q_i and the rows with denotations from $\boldsymbol{H}_{i1}$ (and $\boldsymbol{H}_{i2}$) the matrix $\boldsymbol{Q}_1$ ($\boldsymbol{Q}_2$, respectively).

(A17) Mark the row q_i as pendant in both $\boldsymbol{Q}_1$ and $\boldsymbol{Q}_2$.

(A18) If q_k is a row of $\boldsymbol{Q}_1$ ($\boldsymbol{Q}_2$, respectively) and if q_k is pendant in $\boldsymbol{Q}$, then mark also q_k as pendant in $\boldsymbol{Q}_1$ ($\boldsymbol{Q}_2$, respectively).

(P19) Check whether there is in the submatrix made up of the pendant rows of $\boldsymbol{Q}_1$ a column containing more than two 1's.
If so, then G' is not planar,
go to the End.
If not,

(P20) Check whether there is in the submatrix made up of the pendant rows of $\boldsymbol{Q}_2$ a column containing more than two 1's.
If so, then G' is not planar,
go to the End.
If not,

(A21) If q_k is a row of $\boldsymbol{Q}_1$ ($\boldsymbol{Q}_2$, respectively) and not pendant in $\boldsymbol{Q}$, then mark q_k also in $\boldsymbol{Q}_1$ ($\boldsymbol{Q}_2$, respectively) as not pendant (for $q_k \neq q_i$). Make up for each of the matrices $\boldsymbol{Q}_1$, $\boldsymbol{Q}_2$ the set $\mathfrak{N} = \{q_k : q_k \text{ not pendant}\}$.

(A22) Take from $\boldsymbol{Q}$ the connected systems of chords for non-pendant rows (thus not for q_i any more) going to $\boldsymbol{Q}_1$ and to $\boldsymbol{Q}_2$ (according to which matrix such a row is contained in after the decomposition of $\boldsymbol{Q}$ into $\boldsymbol{Q}_1$ and $\boldsymbol{Q}_2$) and delete in the connected systems of chords of non-pendant rows of $\boldsymbol{Q}_1$ and $\boldsymbol{Q}_2$ all denotations k if the row q_k does not lie in $\boldsymbol{Q}_1$ and $\boldsymbol{Q}_2$, respectively.

(P23) Check whether in $\boldsymbol{Q}_1$ all rows are pendant.
If so, form a row

$$r = \sum_{q_k \in \boldsymbol{Q}_1} q_k \quad (\text{mod. } 2)$$

and enlarge $\boldsymbol{R}^0$ by the row r. Then, go to (**P24**).
If not, put $\boldsymbol{Q}_1$ on the waiting list and go on at

(P24) Check whether in $\boldsymbol{Q}_2$ all rows are pendant.
If so, form a row

$$r = \sum_{q_k \in \boldsymbol{Q}_2} q_k \quad (\text{mod. } 2)$$

and enlarge $\boldsymbol{R}^0$ by the row r. Then, go to (**P8**).
If not, put $\boldsymbol{Q}_2$ on the waiting list and go on at (**P8**).

End: Termination of the algorithm.

To help the reader understand the algorithm, let us calculate an example. Consider the graph in Fig. 10.9. For the spanning tree drawn here with double lines, the matrix $\boldsymbol{Q} = \boldsymbol{B}'$ (circuit matrix) results as (fill in the blank cells by a 0):

$\boldsymbol{B}' =$

	1	2	3	4	5	6	7	8	9	10	11	12	
1	1							1	1	1	1	1	(2) (3) (4) (5) (6) (7)
2		1							1	1	1	1	(1, 3, 7) (4) (5) (6)
3			1					1	1				pendant
4				1						1	1		pendant
5					1				1	1			pendant
6						1					1	1	pendant
7							1	1	1	1			(1, 2, 4, 6) (3) (5)

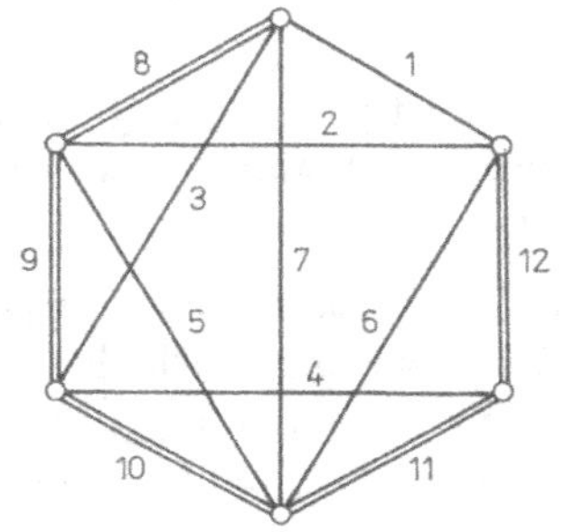

Fig. 10.9

The instructions are executed in the order given below:

1. (A1) $\boldsymbol{B}'$ is given above.
2. (A2) The pendant rows are indicated at $\boldsymbol{B}'$, and $\boldsymbol{R}^0$ consists of the four pendant rows.
3. (P3) No.
4. (P4) No.
5. (P5) No.
6. (A6) $\mathfrak{S}_i$ are indicated at $\boldsymbol{B}'$ (confusions in the denotation are presumably not possible), the set $\mathfrak{N}$ can also be read.
7. (A7) $\boldsymbol{Q} = \boldsymbol{B}'$ is put on the waiting list.
8. (P8) No, since $\boldsymbol{B}'$ is on the waiting list.
9. (A9) $\boldsymbol{B}'$ becomes the matrix $\boldsymbol{Q}$.
10. (P10) No.
11. (A11) $z_1 = 2$, $z_2 = 1$, $z_7 = 1$.
12. (A12) $z = 1$.
13. (A13) $\mathfrak{N}' = \{q_2, q_7\}$.
14. (A14) $q_i = q_7$.

15. (P15) Yes, since we can write in $\boldsymbol{Q}_1$ the rows $q_7, q_1, q_2, q_4, q_6, q_3$ and in $\boldsymbol{Q}_2$ the rows q_7, q_5 (the following representation is probably clear):

$$\boldsymbol{Q}_1 = \begin{array}{c} \\ 7 \\ 1 \\ 2 \\ 4 \\ 6 \\ 3 \end{array} \begin{array}{c} \begin{array}{ccccc} 8 & 9 & 10 & 11 & 12 \end{array} \\ \left[\begin{array}{ccccc} 1 & 1 & 1 & & \\ 1 & 1 & 1 & 1 & 1 \\ & 1 & 1 & 1 & 1 \\ & & 1 & 1 & \\ & & & 1 & 1 \\ 1 & 1 & & & \end{array}\right] \end{array} \begin{array}{l} \\ \text{pendant} \\ (2)\ (3)\ (4)\ (6)\ (7) \\ (1, 3, 7)\ (4)\ (6) \\ \text{pendant} \\ \text{pendant} \\ \text{pendant} \end{array}$$

$$\boldsymbol{Q}_2 = \begin{array}{c} \\ 7 \\ 5 \end{array} \begin{array}{c} \begin{array}{ccccc} 8 & 9 & 10 & 11 & 12 \end{array} \\ \left[\begin{array}{ccccc} 1 & 1 & 1 & & \\ & 1 & 1 & & \end{array}\right] \end{array} \begin{array}{l} \\ \text{pendant} \\ \text{pendant} \end{array}$$

16. (A17) Has been executed in the above matrices $\boldsymbol{Q}_1$, $\boldsymbol{Q}_2$.
17. (A18) Has been executed in the above matrices.
18. (P19) No.
19. (P20) No.
20. (A21) Has been executed in the above matrices.
21. (A22) Has been executed in the above matrices.
22. (P23) No, $\boldsymbol{Q}_1$ is put ont he waiting list.
23. (P24) Yes, we add to $\boldsymbol{R}^0$ the new row $r = q_7 + q_5$; this row has thus a 1 in the columns with the numbers 5, 7, 8, in all other columns a 0.
24. (P8) No.
25. (A9) The matrix $\boldsymbol{Q}_1$ madeup by the rows $q_7, q_1, q_2, q_4, q_6, q_3$ becomes the matrix $\boldsymbol{Q}$.
26. (P10) No.
27. (A11) $z_1 = z_2 = 1$.
28. (A12) $z = 1$.
29. (A13) $\mathfrak{N}' = \{q_1, q_2\}$.
30. (A14) $q_i = q_2$.
31. (P15) No, since however we decompose the connected systems of chords of q_i, one of the two matrices, say $\boldsymbol{Q}_1$, would still have to contain at least q_1, q_3, q_7 in addition to q_2, thus

$$\begin{array}{c} \\ 2 \\ 1 \\ 3 \\ 7 \end{array} \begin{array}{c} \begin{array}{ccccc} 8 & 9 & 10 & 11 & 12 \end{array} \\ \left[\begin{array}{ccccc} & 1 & 1 & 1 & 1 \\ 1 & 1 & 1 & 1 & 1 \\ 1 & 1 & & & \\ 1 & 1 & 1 & & \end{array}\right] \end{array} \begin{array}{l} \\ \\ \text{non-pendant} \\ \text{pendant} \\ \text{pendant} \end{array}$$

In the column belonging to the edge 9, there is a 1 in the row 2 and also in the two pendant rows, that is, $\boldsymbol{G}'$ is not planar.

Thus, the algorithm terminates.

According to the theorem of Kuratowski, G' would have a subdivision of the $\boldsymbol{K}_5$ or $\boldsymbol{K}_{3,3}$. This is the case, in fact. Deleting say, the edge 11, yields a subdivision of the $\boldsymbol{K}_5$. By deleting the edges 6, 8, 10, one also gets the $\boldsymbol{K}_{3,3}$.

Let us give, in an abridged form, another example of a graph which will turn out to be planar and which will also be embedded by us.

Given the following circuit matrix:

	1	2	3	4	5	6	7	8	9	10	11	
1	1						1	1	1	1	1	(2) (3) (4) (5) (6)
2		1					1	1	1			pendant
3			1					1	1	1	1	(1, 2, 4) (5) (6)
4				1			1	1	1	1		(1, 3) (2) (5) (6)
5					1			1	1	1		(1, 2, 3, 4) (6)
6						1			1	1		pendant

The instructions and tests up to (A9) run well, we state "Yes" at (P10), go on at (A16) and get the matrices $\boldsymbol{Q}_1$, $\boldsymbol{Q}_2$ with $q_i = q_5$:

$\boldsymbol{Q}_1 =$

	1	2	3	4	5	6	7	8	9	10	11	
1	1						1	1	1	1	1	(2) (3) (4) (5)
2		1					1	1	1			pendant
3			1					1	1	1	1	(1, 2, 4) (5)
4				1			1	1	1	1		(1, 3) (2) (5)
5					1			1	1	1		pendant

$\boldsymbol{Q}_2 =$

	1	2	3	4	5	6	7	8	9	10	11	
5					1			1	1	1		pendant
6						1			1	1		pendant

Since all rows from $\boldsymbol{Q}_2$ are pendant, we determine from $\boldsymbol{Q}_2$ a row of the matrix $\boldsymbol{R}^0$:

$$r = q_5 + q_6.$$

Next we determine the row q_3 with the splitting of $\boldsymbol{Q}_1$ into

$\boldsymbol{Q}_1 =$

	1	2	3	4	5	6	7	8	9	10	11	
1	1						1	1	1	1	1	(2) (3) (4)
2		1					1	1	1			pendant
3			1					1	1	1	1	pendant
4				1			1	1	1	1		(1, 3) (2)

$\boldsymbol{Q}_2 =$

	1	2	3	4	5	6	7	8	9	10	11	
3			1					1	1	1	1	pendant
5					1			1	1	1		pendant

Q_2 cannot be decomposed further and supplies a row of $\boldsymbol{R}^0$, namely

$$r = q_3 + q_5 .$$

The reader can easily verify that a further decomposition of the matrices successively put on the waiting list is possible and that the following rows still missing can be obtained for $\boldsymbol{R}^0$:

$$\begin{aligned} r &= q_2 + q_4 , \\ &= q_1 + q_3 , \\ &= q_1 + q_4 . \end{aligned}$$

Thus, we find as incidence matrix $\boldsymbol{R}$ of the graph $\boldsymbol{G}$ which is dual to $\boldsymbol{G}'$, the following matrix (however, let us point out that because of the theorem of Whitney, the mere existence of the incidence matrix $\boldsymbol{R}$ of $\boldsymbol{G}$ assures that $\boldsymbol{G}'$ is planar):

$$\boldsymbol{R} = \begin{array}{c|ccccccccccc|} & 1 & 2 & 3 & 4 & 5 & 6 & 7 & 8 & 9 & 10 & 11 \\ \hline 1 & & 1 & & & & & 1 & 1 & 1 & & \\ 2 & & & & & & 1 & & & 1 & 1 & \\ 3 & & & & & 1 & 1 & & 1 & & & \\ 4 & & & 1 & & 1 & & & & & & 1 \\ 5 & & 1 & & 1 & & & & & & 1 & \\ 6 & 1 & & 1 & & & & 1 & & & & \\ 7 & 1 & & & 1 & & & & & & & 1 \end{array}$$

Now, we turn to an algorithm which, knowing the matrix $\boldsymbol{R}$, makes it possible to embed $\boldsymbol{G}'$.

First, let us take into account that exactly one edge in $\boldsymbol{G}$ corresponds to one edge in $\boldsymbol{G}'$, and vice versa, and that to each vertex in $\boldsymbol{G}'$ there corresponds exactly one face, and vice versa (here, that face is also to be taken into consideration which is not finite, which is why the number of rows in $\boldsymbol{R}$ is greater by 1 than in $\boldsymbol{B}'$). If thus $\boldsymbol{R}'$ is the incidence matrix of $\boldsymbol{G}'$, then this matrix can also be taken as a matrix of circuits of $\boldsymbol{G}$ (that is, of all those circuits which bound faces).

Consider any vertex P_i' in $\boldsymbol{G}'$ of the degree $v(P_i') = v_i'$. Let $\mathfrak{K}_i' = \{k'_{i_j}: j = 1, \ldots, v_i'\}$ be the set of the edges of $\boldsymbol{G}'$ which are incident with P_i'. The embedding method discussed in the following utilizes consistently the duality between $\boldsymbol{G}$ and $\boldsymbol{G}'$ described in chapter 1.

Embedding algorithm of Dambitis

(i) Choose any row q_i' in $\boldsymbol{R}'$ (the unities of this row correspond to the edges k'_{i_j} which are incident with the vertex P_i' in $\boldsymbol{G}'$). The set $\mathfrak{K}_i$ of edges, which corresponds to the set $\mathfrak{K}_i'$ of edges of $\boldsymbol{G}'$ in case of dualizing, forms a circuit $\boldsymbol{C}_i$ in $\boldsymbol{G}$. From the incidence matrix $\boldsymbol{R}$ of $\boldsymbol{G}$ one can determine the set of those vertices which are incident with the edges of $\boldsymbol{C}_i$. Embed $\boldsymbol{C}_i$ in the plane and orient all edges of C_i so that C_i is traversed in anti-clockwise direction (that is, a circuit is generated in the directed graph). Now, embed in the plane, together with the edges from $\mathfrak{R}_i'$ (in the order of the traversal of the corresponding edges of C_i), the vertex P_i' which corresponds to the circuit C_i in case of dualizing (this embedding, however, is carried out independently of that of G).

(ii) Choose in R' any row q_j' other than q_i', with q_j' and q_i' having a common unity in at least one column. In the embedding of G realized so far, reorient all those (common) edges which correspond to the columns of R' in which q_i' and q_j' have 1 each. Now, embed the elementary circuit C_j corresponding to the row q_j' in G in such a way that the circuit that corresponds to the row $q_i' + q_j'$ in G is traversed in anti-clockwise direction.

Delete q_i' and q_j' in R', add to R' the new row $q_i' + q_j'$ (mod. 2) and denote it by q_i'.

From the order of the edges of C_j we get an embedding of the vertex P_j' of G' together with all edges which are incident with P_j', again in consideration of the order of the edges. Now, go back to (ii).

If R' still contains one single row, we break off, thus having embedded both G and G'.

Let us illustrate the embedding algorithm by giving a further example.

In the preceding one, we have determined the incidence matrix R of a graph G dual to a planar graph G'. If we apply once again the algorithm of Dambitis, that is, if we denote in the following R by R', form the matrix B' and then determine R (task!), we obtain the incidence matrix of the original graph G':

$$R' = \begin{array}{c} \\ 1 \\ 2 \\ 3 \\ 4 \\ 5 \\ 6 \end{array} \begin{array}{c} \begin{array}{ccccccccccc} 1 & 2 & 3 & 4 & 5 & 6 & 7 & 8 & 9 & 10 & 11 \end{array} \\ \left[\begin{array}{ccccccccccc} 1 & 1 & & 1 & & & 1 & & & & \\ & & 1 & & 1 & & 1 & 1 & & & \\ & & & & & 1 & & 1 & 1 & & \\ & 1 & & & & & & & 1 & 1 & \\ & & & 1 & 1 & 1 & & & & 1 & 1 \\ 1 & & 1 & & & & & & & & 1 \end{array}\right] \end{array}$$

To embed immediately as many vertices as possible, choose as q_i' a row which contains as many unities as possible, say $q_i' = q_5'$. From the incidence matrix R we find that the order of edges in C_5 is 4, 11, 5, 6, 10.

Figure 10.10a shows the embedding of C_5, Fig. 10.10b that of the vertex P_5'. As next row choose for instance q_1'. The result of the embedding is represented in Fig. 10.10c and in Fig. 10.10d. Now, make up the sum $q_5' + q_1'$ (mod. 2), which leads to

$$(1, 1, 0, 0, 1, 1, 1, 0, 0, 1, 1).$$

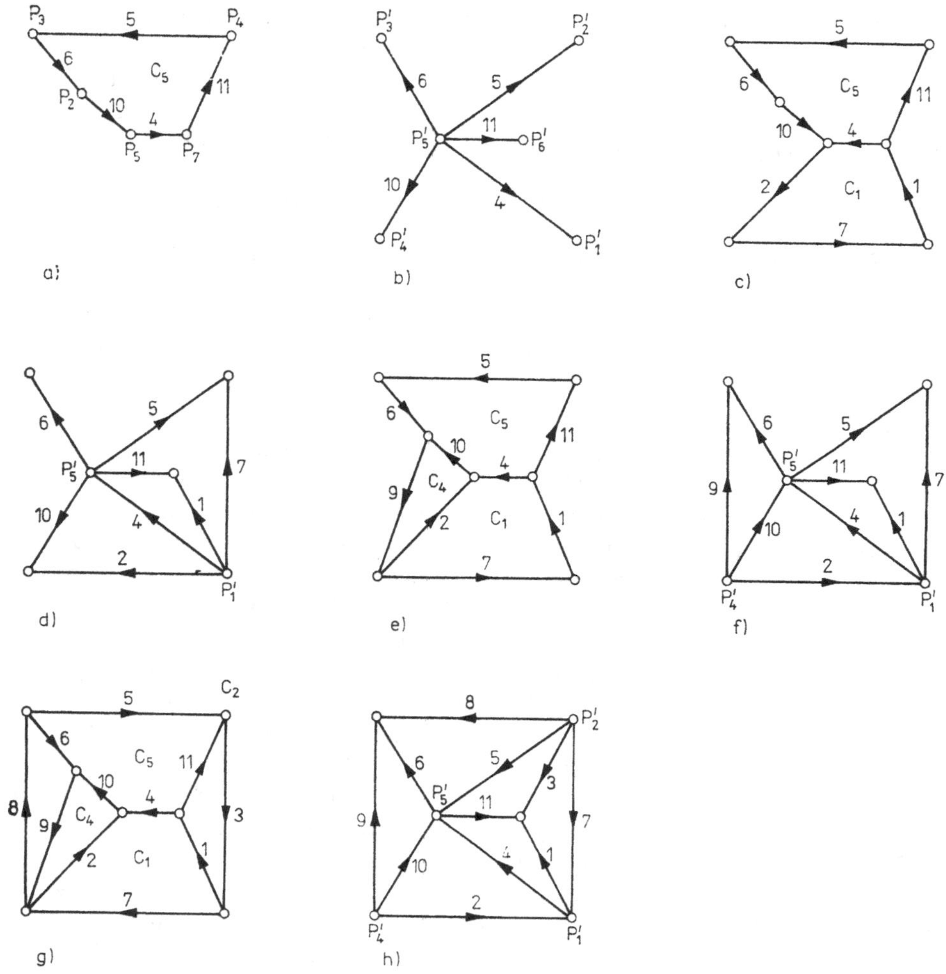

Fig. 10.10

Obviously, to this corresponds in G the *exterior circuit* of Fig. 10.10c. Now, choose say the row q_4' as q_i'. Figures 10.10e and 10.10f illustrate the result of the embedding. If we choose next the row q_2', then there results the complete embedding of G and G' (represented in Figs. 10.10g and h) which we sought at the beginning. Although there are still a few rows of R' left, all edges of G' are already embedded; thus we can stop.

Obviously, the algorithm found by Dambitis requires only matrix operations to be performed for its realization. Thus, no topological investigations at all are made. That means that this procedure offers some advantages for its implementation from the point of view of computer engineering. Of course, larger matrices entail some difficulties in so far as the memory capacity might be exceeded.

The examples given so far have shown that the non-planarity of the graphs under examination can be verified within a relatively short time, whereas a lot of time is necessary to determine the incidence matrix of $\boldsymbol{G}$ in case of planarity.

10.4. PLANARITY STUDIES MADE BY DECOMPOSING GRAPHS

Let $\boldsymbol{G}(\mathfrak{X}, \mathfrak{U})$ be a connected simple graph and $\boldsymbol{C}$ any subgraph of $\boldsymbol{G}$. Let l be any edge of $\boldsymbol{G}$ which does not belong to $\boldsymbol{C}$. We consider the set of all paths of $\boldsymbol{G}$ containing l, with the condition that no interior vertex of a path belongs to $\boldsymbol{C}$. Let $\boldsymbol{H}(\boldsymbol{C}, l)$ denote the subgraph spanned by all these paths (cf. Fig. 10.11a). We call $\boldsymbol{H}(\boldsymbol{C}, l)$ a *bridge* of $\boldsymbol{G}$ *with respect to* $\boldsymbol{H}$. Figure 10.11a contains exactly four bridges (if we choose for $\boldsymbol{C}$ that subgraph whose edges are drawn with double lines). Bridges can be generated in the following manner: Consider the set of all edges which do not belong to $\boldsymbol{C}$ and which are incident with at least one vertex of $\boldsymbol{C}$. Let k be such an edge with the end-points P and Q, of which P belongs to $\boldsymbol{C}$ and Q does not (if also Q belongs to $\boldsymbol{C}$, k forms by itself a bridge). Detach k from P and provide it with a terminal vertex P' (which has thus the degree 1). If Q also lies on $\boldsymbol{C}$, perform this operation also with Q (cf. Fig. 10.11b). Each component which results in this way and which does not belong to $\boldsymbol{C}$ represents a bridge.

It can immediately be seen that an edge k which lies in a bridge $\boldsymbol{H}(\boldsymbol{C}, l)$ generated by the edge l also generates the bridge $\boldsymbol{H}(\boldsymbol{C}, l) = \boldsymbol{H}(\boldsymbol{C}, k)$ with respect to $\boldsymbol{C}$.

Therefore, we say in the following that $\boldsymbol{C}$ *generates the bridges* $\boldsymbol{H}_1, \ldots, \boldsymbol{H}_r$ in $\boldsymbol{G}$, and do not denominante a generating edge (cf. [20]).

Vertices of a bridge which also belong to $\boldsymbol{C}$ are called *boundary points of the bridge*, and all other vertices (if any) of a bridge are called *interior points* of it.

Let us now define a few graphs. Let $\boldsymbol{G}$ be a graph, $\boldsymbol{C}$ a subgraph and $\{\boldsymbol{H}_1, \ldots, \boldsymbol{H}_r\}$ the set of bridges of $\boldsymbol{G}$ with respect to $\boldsymbol{C}$ (that is, the bridges generated by $\boldsymbol{C}$ in $\boldsymbol{G}$). Let the graph $\boldsymbol{G}_c$ result from taking $\boldsymbol{C}$, assigning to each of the $\boldsymbol{H}_i$ a vertex P_i of $\boldsymbol{G}_c$ and joining P_i in $\boldsymbol{G}_c$ to each of the boundary points of $\boldsymbol{H}_i$ in $\boldsymbol{G}_c$ by an edge (cf. Fig. 10.11c for the graph of Fig. 10.11a).

Let the graph $\boldsymbol{G}(\boldsymbol{H}_i)$ be generated by combining all edges and vertices belonging to $\boldsymbol{C}$ and $\boldsymbol{H}_i$ (cf. Fig. 10.11d).

The following theorem holds, it is proved by W. Bader, G. J. Fischer, O. Wing, L. Auslander and S. V. Parter:

THEOREM 10.6 ([1], [2], [6]). *A graph* $\boldsymbol{G}$ *is planar if and only if the following conditions are satisfied:*

a) *The graph* $\boldsymbol{G}_c$ *is planar.*

b) *Each of the graphs* $\boldsymbol{G}(\boldsymbol{H}_i)$ *is planar.*

Here, $\boldsymbol{C}$ *is any circuit of* $\boldsymbol{G}$, *and* $\boldsymbol{H}_i$ $(i = 1, \ldots, r)$ *are the bridges generated by* $\boldsymbol{C}$ *in* $\boldsymbol{G}$.

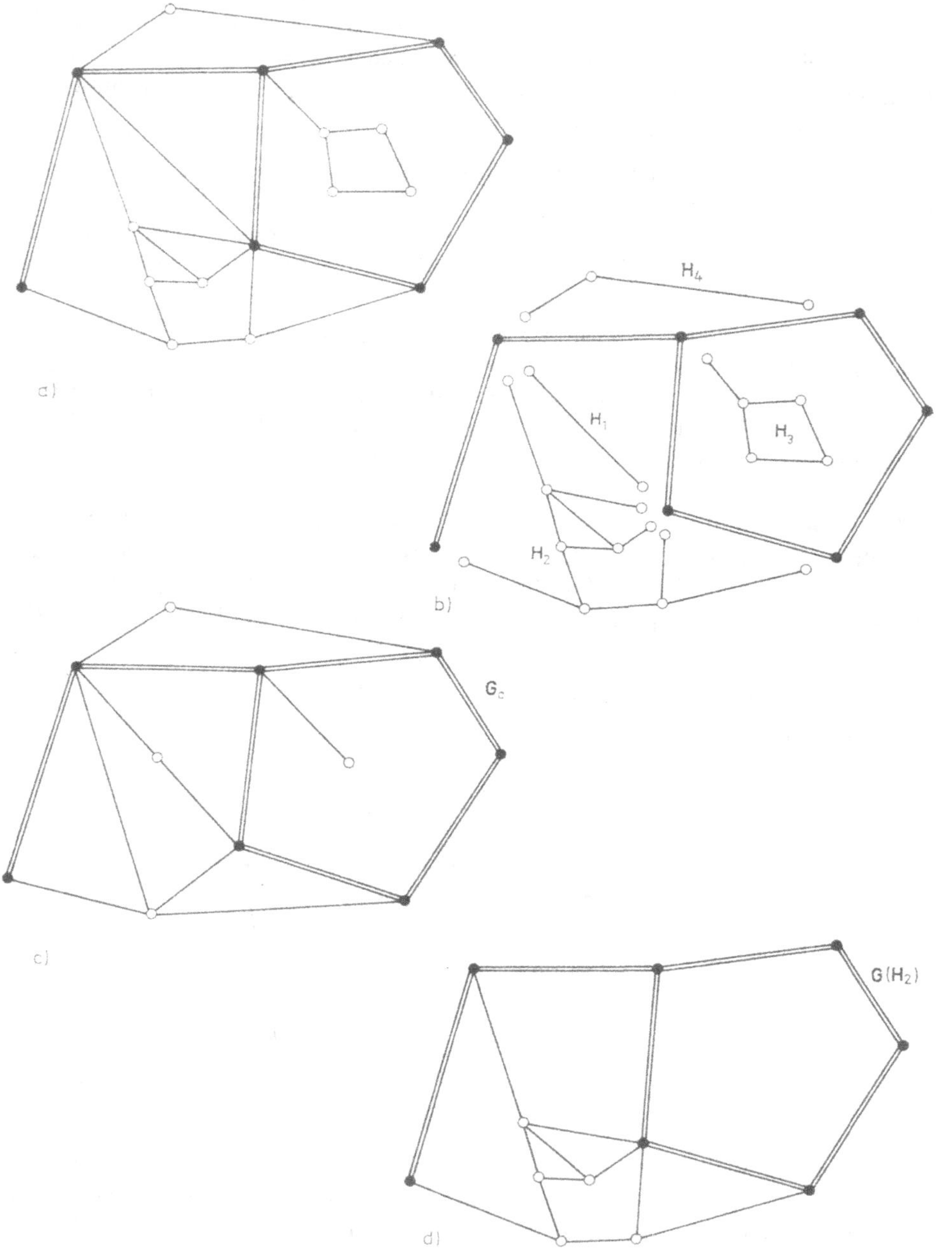

Fig. 10.11

Proof. The necessity of the condition is immediately clear.

Conversely, let the conditions a) and b) be satisfied. First we embed G_c in the plane, which is possible by premise. After this, each of the vertices P_i is "blown up" to form that bridge H_i from which it originally emerged. Thus, we obtain an embedding of the whole graph G in the plane.

The basic idea of the procedure discussed in the following is that on the basis of the preceding theorem the graph which is to be examined for planarity is decomposed into parts, and each of these parts undergoes a separate planarity test. The planarity of the large graph can then be deduced from the possible planarity of the smaller ones.

First, let us suppose that the given graph G possesses a *hamiltonian circuit*, that is, a circuit which contains all vertices of G. (We also know some graphs, even with restricting conditions, which have no hamiltonian circuit. Therefore, such a premise is very drastic. If it is known that the graph in question possesses a hamiltonian circuit, then it is by no means easy to identify such a hamiltonian circuit).

Let C be a hamiltonian circuit of G. Then, the bridges generated by C in G are all edges. We consider two edges $l_1 = (P_1, Q_1)$ and $l_2 = (P_2, Q_2)$ which do not belong to C (which are therefore bridges). We say that l_1 and l_2 are *incompatible* if the vertices P_1, P_2, Q_1, Q_2 are pairwise distinct from each other (that is, if the edges are not adjacent) and if they lie on C in the order

$$P_1 \dots P_2 \dots Q_1 \dots Q_2 \dots \quad \text{or} \quad P_1 \dots Q_2 \dots Q_1 \dots P_2 \dots$$

(cf. Fig. 10.12).

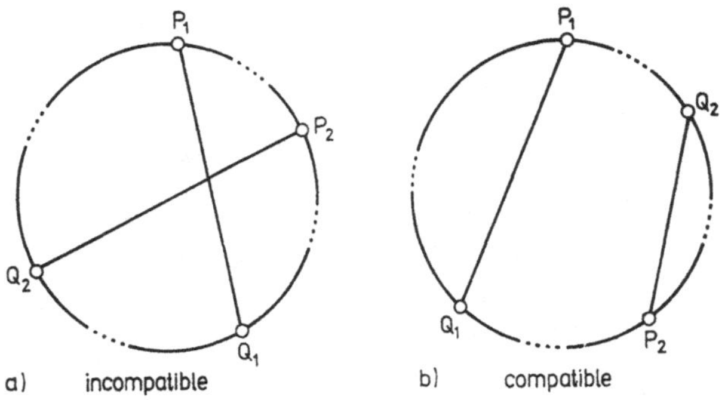

Fig. 10.12

The hamiltonian circuit C being embedded in the plane, it decomposes this plane into an interior (finite) and an exterior (infinite) domain. To embed two incompatible edges l_1, l_2 together with C, one of the edges has to be embedded in the interior domain and the other in the exterior one. Otherwise, planarity is not assured.

Now, we construct an auxiliary graph $\boldsymbol{F}$. We assign to each bridge (that is, to each edge that does not belong to $\boldsymbol{C}$) a vertex from $\boldsymbol{F}$ and join two vertices in $\boldsymbol{F}$ by an edge if the edges corresponding to them in $\boldsymbol{G}$ are incompatible. If the graph $\boldsymbol{F}$ which results in this way is *bipartite*, i.e., if the vertices of $\boldsymbol{F}$ can be subdivided into two classes such that vertices of the same class are not joined, then the initial graph $\boldsymbol{G}$ is planar. We only have to embed the edges corresponding to the vertices of the one class in the interior domain and those corresponding to the vertices of the other class in the exterior domain of $\boldsymbol{C}$.

If the auxiliary graph $\boldsymbol{F}$ is not bipartite, then, in each attempt to embed $\boldsymbol{G}$, there will be two incompatible edges in at least one of the domains. But then, $\boldsymbol{G}$ is not planar.

To check whether $\boldsymbol{F}$ is bipartite, we can proceed as follows: First, we look for a spanning tree in $\boldsymbol{F}$ (if $\boldsymbol{F}$ is not connected, we look for a spanning tree in each component of $\boldsymbol{F}$). This spanning tree is bipartite. We can assign to each of the vertices of $\boldsymbol{F}$ a label along the spanning tree, for example the label "+" or "−" such that vertices with the same label are never joined in the spanning tree. Now, we try to add one after the other all edges of $\boldsymbol{F}$ which are still missing without joining vertices with the same label. If this is possible, then $\boldsymbol{F}$ is bipartite and thus, $\boldsymbol{G}$ is planar. If the edges of $\boldsymbol{F}$ cannot be added in this way, then $\boldsymbol{G}$ is not planar.

Now, we discuss the case in which $\boldsymbol{G}$ possesses no hamiltonian circuit or in which it is difficult to obtain such a circuit.

Choose in $\boldsymbol{G}$ any circuit $\boldsymbol{C}_1$ (if there exists no circuit in $\boldsymbol{G}$, then $\boldsymbol{G}$ is a forest, i.e., a union of trees, and is thus planar). Let Q_1 and Q_2 be two successive vertices lying on $\boldsymbol{C}_1$. If there exists a path $\boldsymbol{W}$ joining Q_1 to Q_2 which has no vertex, besides Q_1 and Q_2, in common with $\boldsymbol{C}_1$, we delete the edge (Q_1, Q_2) and add the path $\boldsymbol{W}$. Obviously, a circuit is generated that is longer than $\boldsymbol{C}_1$ (such a path $\boldsymbol{W}$ exists if and only if there is a bridge of $\boldsymbol{G}$ with respect to $\boldsymbol{C}_1$ which contains Q_1, Q_2 and at least still another vertex). Denote the circuit resulting in this way by $\boldsymbol{C}_2$. We carry on enlarging the circuits until they cannot be enlarged in this way any more. We denote the circuit which has eventually been generated by $\boldsymbol{C}$. If $\boldsymbol{C}$ is a hamiltonian circuit, we continue the procedure described above. We may thus assume that $\boldsymbol{C}$ is not a hamiltonian circuit.

We consider the set of bridges $\boldsymbol{H}_i$ of $\boldsymbol{G}$ with respect to $\boldsymbol{C}$ as well as the graph $\boldsymbol{G}_c$ which is formed by assigning to each bridge $\boldsymbol{H}_i$ a vertex P_i and by joining P_i in $\boldsymbol{G}_c$ to all those vertices of $\boldsymbol{C}$ which were adjacent to vertices from $\boldsymbol{H}_i$, that is, by joining P_i to all boundary vertices of $\boldsymbol{H}_i$ (cf. Fig. 10.11). Obviously, exactly one bridge $\boldsymbol{E}_i$ in $\boldsymbol{G}_c$ corresponds to each bridge $\boldsymbol{H}_i$ of $\boldsymbol{G}$ ($\boldsymbol{E}_i$ consists of one vertex P_i and all edges (P_i, X), with X being boundary point of $\boldsymbol{H}_i$ in $\boldsymbol{G}$ with respect to $\boldsymbol{C}$), and vice versa.

In analogy to the incompatibility concept for edges with respect to a hamiltonian circuit, we say that two bridges $\boldsymbol{E}_i$ and $\boldsymbol{E}_j$ of $\boldsymbol{G}_c$ with respect to $\boldsymbol{C}$ are *incom-*

patible if in case of any (planar) embedding of $\boldsymbol{C}$ and $\boldsymbol{E}_i$ and $\boldsymbol{E}_j$ the two bridges $\boldsymbol{E}_i$ and $\boldsymbol{E}_j$ are embedded in different domains (for example, $\boldsymbol{E}_i$ in the interior domain of $\boldsymbol{C}$ and $\boldsymbol{E}_j$ in the exterior domain, or vice versa). It is not difficult to recognize already in $\boldsymbol{G}$ the compatibility or incompatibility of two bridges $\boldsymbol{E}_i$ and $\boldsymbol{E}_j$. Let $X_1, X_2, \ldots, X_s$ be the boundary points of $\boldsymbol{H}_i$ lying in cyclic order on $\boldsymbol{C}$, and let $\boldsymbol{W}_k$ be the path on $\boldsymbol{C}$ joining X_k to X_{k+1} (here let $X_{s+1} = X_1$ be true with $k = 1, \ldots, s$). If there is an index k such that all boundary points of $\boldsymbol{H}_j$ belong to $\boldsymbol{W}_k$, then $\boldsymbol{E}_i$ and $\boldsymbol{E}_j$ in $\boldsymbol{G}_c$ are *compatible*, otherwise they are *incompatible*. In the first case, we shall also call $\boldsymbol{H}_i$ and $\boldsymbol{H}_j$ in $\boldsymbol{G}$ compatible (independently of whether or not $\boldsymbol{H}_i$ and $\boldsymbol{H}_j$ are planar). Here again, we construct for $\boldsymbol{G}_c$ an auxiliary graph $\boldsymbol{F}_c$ by assigning a vertex R_i to each bridge $\boldsymbol{E}_i$ and by joining in $\boldsymbol{F}_c$ two vertices R_j and R_k by an edge if and only if $\boldsymbol{E}_j$ and $\boldsymbol{E}_k$ are incompatible. We can now also state that $\boldsymbol{G}_c$ is planar provided that $\boldsymbol{F}_c$ is a bipartite graph. Otherwise, planarity cannot be deduced.

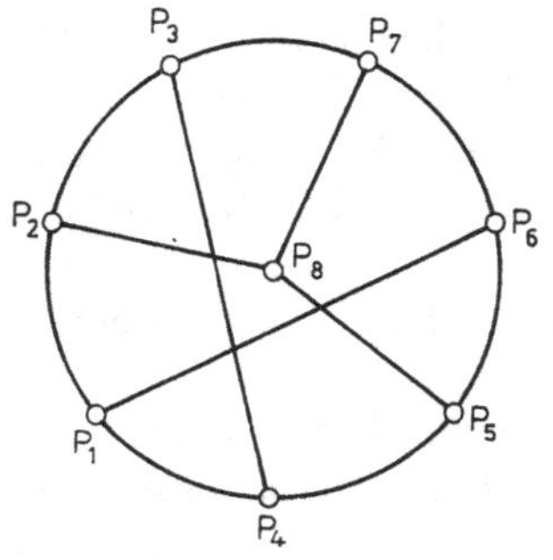

Fig. 10.13

In order to find out whether $\boldsymbol{G}$ is planar, we have to check up each bridge $\boldsymbol{H}_i$ separately for its planarity, that is, to be more precise, we have to investigate the planarity of the graph $\boldsymbol{G}(\boldsymbol{H}_i)$ for each i.

Let us take a little example. Let the graph $\boldsymbol{G}$ be given by its adjacency matrix

	1	2	3	4	5	6	7	8
1		1		1		1		
2	1		1					1
3		1		1			1	
4	1		1		1			
5				1		1		1
6	1				1		1	
7			1			1		1
8		1			1		1	

Fill in all empty cells of the matrix with a 0 (cf. Fig. 10.13). We seek a circuit, say

$$C_1 = (2, 3, 7, 6, 5, 4, 1, 2).$$

Obviously, it is not possible to enlarge the circuit in the manner described above. Thus, we set $C = C_1$ and consider all bridges. There are three to be found, namely

$$H_1 = \{(P_3, P_4)\},$$
$$H_2 = \{(P_1, P_6)\},$$
$$H_3 = \{(P_8, P_2), (P_8, P_5), (P_8, P_7)\}.$$

It is seen without difficulty that these three bridges are pairwise incompatible, i.e., the auxiliary graph F equals K_3, the complete graph with three vertices. This one, however, is not bipartite, and the graph indicated is therefore not planar.

It is clear that in the case of non-existence of a hamiltonian circuit (or when no hamiltonian circuit can be found although one possibly exists), the algorithm becomes useful only if the required circuit C generates at least two bridges. This is achieved by a successive extension of the circuit C generating the bridges so far as the initial graph is 3-connected (prove this statement!).

As was proved in [6], all operations required by the algorithm can be performed as matrix operations.

10.5. THE EMBEDDING ALGORITHM OF DEMOUCRON, MALGRANGE AND PERTUISET

The idea of this embedding algorithm is as follows: Let G be a simple graph which is at least 2-connected. The embedding of G in the plane is carried out step by step starting with a circuit $G_1 \subsetneqq G$ and a successively added path which decomposes a face into two parts.

Algorithm

(i) Let C_0 be a circuit of G. We set $G_1 = C_0$ and seek an embedding G_1' of G_1 in the plane.

(ii) Let G_k' be an embedding of G_k, and let H be any bridge of G with respect to G_k, and this bridge can be embedded in only one face of G_k'. Then, we choose in H a path W whose end-points (and only these) belong to G_k (and which are distinct from each other, which is assured because of the double connectivity), embed this path in the face found above and denote by G_{k+1} the union of G_k and W.

(iii) If (ii) cannot be performed, we seek a bridge H in G_k that can be embedded in two faces of G_k', and embed a path of H (which has besides its end-points no common vertex with G_k) in one of the two domains. Here again, we set $G_{k+1} = G_k \cup W$. If neither (ii) nor (iii) is applicable, the algorithm terminates.

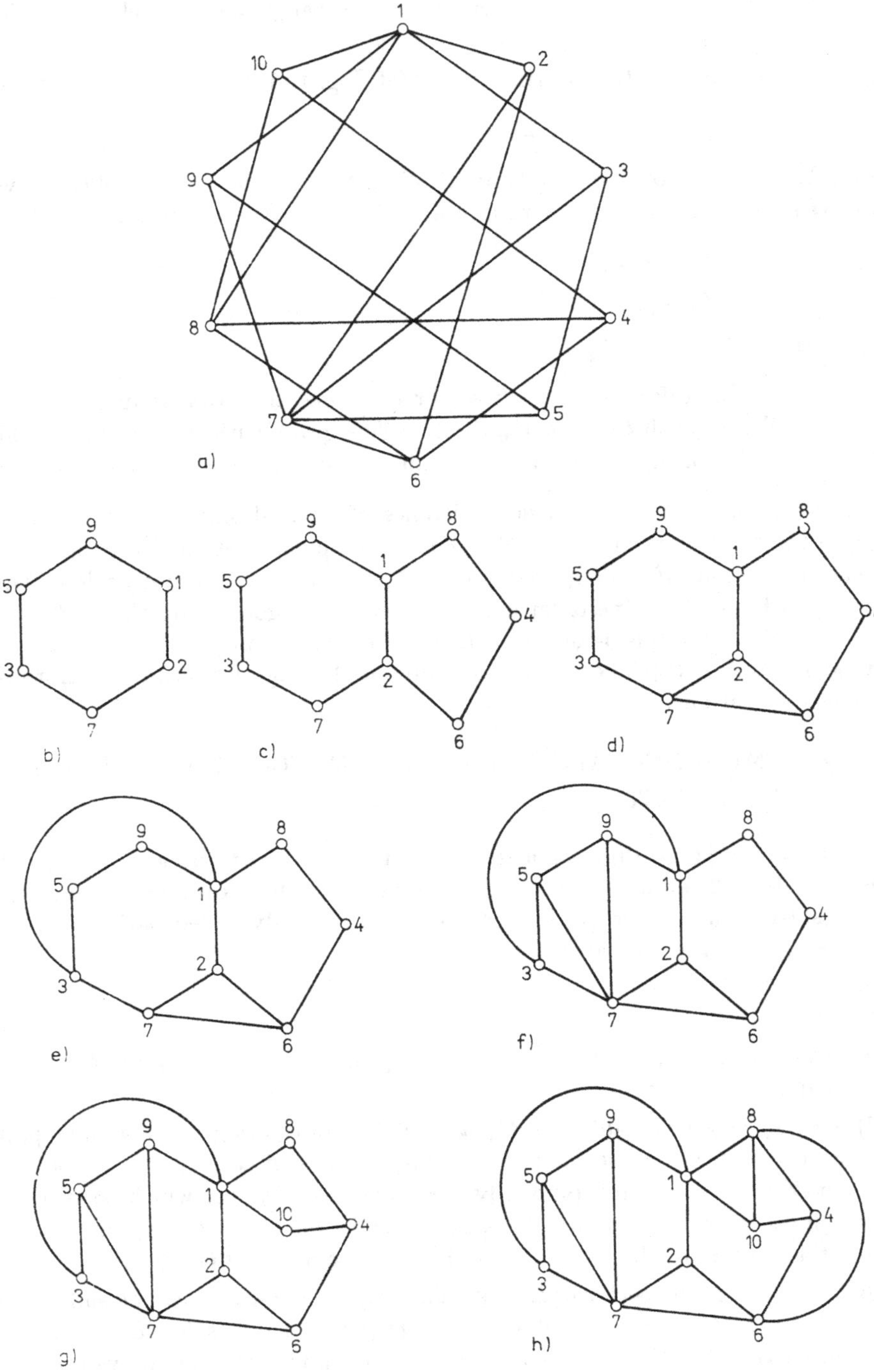

Fig. 10.14

The following theorem holds:

THEOREM 10.7. *Let G be a simple 2-connected graph with m edges and n vertices. G is planar if and only if the algorithm breaks off after the construction of* G_{m-n+1}.

Since on the one hand the graph G_k possesses exactly $k + 1$ faces after the construction of the sequence $\{G_i\}$, and since on the other hand, however, exactly $m - n + 2$ faces exist in G according to Euler's theorem, the algorithm terminates (in case of planarity) after the construction of G_{m-n+1}.

The proof that, in the case of a planar graph, G_{m+1-n} can always be constructed (by using the algorithm presented) will not be given here. The interested reader is referred to the original literature.

The example of Fig. 10.14 serves to illustrate the embedding of a given graph. An algorithm suggested by G. Hotz [12] is similar to that described above. Hotz does not use the concept of a bridge, but seeks only paths (as parts of bridges).

Investigating all bridges in the case of relatively large graphs requires a great deal of work, which makes the procedure described above rather complicated.

The embedding algorithm of G. Hotz is of great clarity and is described in detail by H. Sachs [21].

10.6. THE PLANARITY ALGORITHM OF TUTTE

The following is devoted to an algorithm of W. T. Tutte which is presumably the most ingenious idea of solving the embedding problem.

Let G be a primitive (that is, 3-connected) planar graph with at least three vertices.

We call a circuit C of G *peripheral* if C generates exactly one bridge in G.

Let $P_1, \ldots, P_r$ be the vertices lying on C in the order of the traversal of C, and let $P_{r+1}, \ldots, P_n$ be the remaining vertices of G. We imagine the graph $G(\mathfrak{X}, \mathfrak{U})$ to be embedded in the plane, and let P_i have the Cartesian co-ordinates (x_i, y_i).

The idea of the planarity algorithm of Tutte is as follows: We embed the peripheral circuit C in the plane such that the points $P_1, \ldots, P_r$ span a convex polygon (no three points are allowed to lie on one and the same straight line). These vertices are provided with the co-ordinates (x_i, y_i) when performing the embedding $(i = 1, \ldots, r)$. The remaining $n - r$ vertices P_j $(j = r + 1, \ldots, n)$ get as co-ordinates (x_j, y_j) such values that P_j just lies at the centroid of its neighbours (i.e., of those vertices to which P_j is adjacent).

How does the algorithm present itself now? Let A be the admittance matrix of the graph G, thus

$$a_{ij} = \begin{cases} -1 & \text{if } P_i \text{ and } P_j \text{ are adjacent,} \\ v(P_i) & \text{if } i = j, \\ 0 & \text{otherwise.} \end{cases}$$

Here, $v(P_i)$ stands for the degree of the vertex P_i, i.e., it indicates the number of the vertices lying in the neighbourhood of P_i. Then, the co-ordinates (x_i, y_i) of P_i $(i = r + 1, \ldots, n)$ result as the solutions of the two systems of equations

$$\sum_{j=1}^{n} a_{ij}x_j = 0, \qquad \sum_{j=1}^{n} a_{ij}y_j = 0 \qquad (i = r + 1, \ldots, n).$$

It is supposed that the co-ordinates (x_i, y_i) of the vertices P_i $(i = 1, \ldots, r)$ of the peripheral circuit $\boldsymbol{C}$ are known, like the co-ordinates of the convex r-gon. Thus, we have to solve the two systems of equations

$$\boldsymbol{Bx} = \boldsymbol{p}, \qquad \boldsymbol{By} = \boldsymbol{q}$$

where $\boldsymbol{x}$, $\boldsymbol{y}$, $\boldsymbol{p}$, $\boldsymbol{q}$ are column vectors of the dimension $n - r$, and $\boldsymbol{B}$ is a square matrix with $n - r$ rows, which results from the admittance matrix $\boldsymbol{A}$ by deleting the first r rows and the first r columns.

One can verify in the following manner that the matrix $\boldsymbol{B}$ is regular (that means that the two systems of equations admit a unique solution in case of given $\boldsymbol{p}$ and $\boldsymbol{q}$): The following matrix is defined by us as the admittance matrix $\boldsymbol{A} = (a_{ij})$ of a (connected) graph $\boldsymbol{G}$ (which is not necessarily without multiple edges):

$$a_{ij} = \begin{cases} v(P_i) & \text{for} \quad i = j, \\ -r_{ij} & \text{for} \quad i \neq j \end{cases}$$

(here, let r_{ij} be just the number of the edges joining P_i to P_j for the case $i \neq j$).

If we delete in the graph $\boldsymbol{G}$ all edges of the peripheral circuit $\boldsymbol{C}$ and if we contract the vertices of $\boldsymbol{C}$ to one point P, then the graph $\boldsymbol{S}$ results. Let $\boldsymbol{A}_s$ be the admittance matrix of $\boldsymbol{S}$. Here, we imagine the adjacency relations of P to be indicated in the first row and column. Since $\boldsymbol{S}$ is connected, it possesses a spanning tree. According to the matrix tree theorem of Kirchhoff and Trent (cf. [20]), the number of the spanning trees of $\boldsymbol{S}$ equals the determinant of the submatrix of $\boldsymbol{A}_s$ which results by deleting any (say the first) row and column. This submatrix $\boldsymbol{A}_s$, however, is obviously equal to $\boldsymbol{B}$. Since $\boldsymbol{S}$ has a positive number of spanning trees, the determinant of $\boldsymbol{B}$ is also positive. Thus, $\boldsymbol{B}$ is regular.

In order to be able to apply the algorithm (to solve the two systems of equations), we need a peripheral circuit. We think of a primitive graph (only such graphs are considered) embedded in the plane. Then, each circuit bounding a face is peripheral, because the existence of more than one bridge to such a circuit would contradict the triple connectivity of a primitive graph (task!). Thus, we know that each edge of a primitive graph lies in at least two peripheral circuits (and, as can easily be seen, also in two at the most), namely just in those two circuits bounding the face with which the edge is incident. However, this does not yet enable us to find a peripheral circuit. By utilizing the reflections made in paragraph 10.3. (planarity algorithm of Dambitis), we shall be in the position to solve the task, since we have seen that in case of a planar graph, each circuit basis has at least two circuits bounding a face. We had obtained them by seeking pendant rows in the circuit matrix. The circuits that correspond to the terminal edges are just peripheral.

Now, we can formulate the theorem of Tutte:

THEOREM 10.8. *Let G be a planar 3-connected graph, and let C be a peripheral circuit of G with $r \geqq 3$ vertices $P_1, \ldots, P_r$. If we fixe these r points as the corner points of a convex r-gon, then it holds: If each of the remaining vertices is put at the centroid of its neighbouring points* (neighbourhood = adjacency), *then an embedding of G is obtained by connecting adjacent vertices by a straight line* [22].

This theorem will not be proved by us either. The reader is referred to the original literature.

But what happens if the initial graph is not planar? The given system of equations also admits a unique solution in case of triple connectivity of the graph. Joining adjacent vertices by a straight line yields either edge crossing or coincidence of vertices (in case they have the same neighbouring points).

This latter embedding algorithm is fairly obvious. After finding a peripheral circuit, one has only to solve a system of equations (with two different right-hand sides). This algorithm is very suitable for being applied on a computing machine. If it is possible in addition to connect an X-Y-plotter to the calculator, then the planar (or also the non-planar) graph can even be plotted.

The following reference books are recommended to all those who wish to get a deeper insight into the theory of planar graphs in general and into the embedding of planar graphs in the plane in particular: [9—11, 13, 15, 16, 18, 23—26].

10.7. BIBLIOGRAPHY

[1] Auslander, L., and S. V. Parter: On imbedding graphs in the sphere, J. Math. Mech. **10** (1961), 517—523.

[2] Bader, W.: Das topologische Problem der gedruckten Schaltung und seine Lösung. Arch. Elektrotechnik **49** (1964), 2—12.

[3] Berge, C., und A. Ghouila-Houri: Programme, Spiele, Transportnetze, 2nd edn., Leipzig 1969. (Translated from French.)

[4] Dambitis, J. J.: Methode zur Konstruktion ebener Graphen (Russian), Latv. matem. ežegodnik **6** (1969), 41—63.

[5] Demoucron, G., Y. Malgrange et R. Pertuiset: Graphes planaires: Reconnaissance et construction de représentations planaires topologiques, Rev. Franç. Recherche Opérat. 8, Nr. 30 (1964), 33—47.

[6] Fischer, G. J., and O. Wing: An algorithm for testing planar graphs from the incidence matrix, in: Proc. 7th Midwest Symp. on Circuit Theory, Ann Arbor, Mich., May 1964.

[7] Fischer, G. J., and Wing, O: On correspondence between a class of planar graphs and bipartite graphs, IEEE Trans. Circuit Theory **CT-12** (1965), 266–267.
[8] Guillemin, E. A.: How to grow your trees from given cut-set or tie-set matrices, IRE Trans. Circuit Theory **CT-6**, Spec. Suppl. (1959), 110–126.
[9] Halin, R.: Bemerkungen über ebene Graphen, Math. Ann. **153** (1964), 38–46.
[10] Halin, R.: Über simpliziale Zerfällung n beliebiger (endlicher oder unendlicher) Graphen, Math. Ann. **156** (1964), 216–225.
[11] Harary, F., and W. T. Tutte: A dual form of Kuratowski's theorem, Canad. Math. Bull. **8** (1965), 17–20.
[12] Hotz, G.: Einbettung von Streckenkomplexen in die Ebene, Math. Ann. **167** (1966), 214–223.
[13] Jung, H. A.: Eine Verallgemeinerung des n-fachen Zusammenhanges für Graphen, Math. Ann. **187** (1970), 95–103.
[14] Kuratowski, C.: Sur le problème des courves gauches en Topologie, Fund. Math. **15** (1930), 271–283.
[15] Lempel, A., S. Even and I. Cederbaum: An algorithm for planarity testing of graphs, in: Théorie des Graphes, Paris/New York 1967, p. 215–232.
[16] Lin, P. M.: On methods of detecting planar graphs, in: Proc. 8th Midwest Symp. on Circuit Theory, Colorado State Univ., June 1965, p. 14–15.
[17] MacLane, S.: A structural characterisation of planar combinatorial graphs, Duke Math. J. **3** (1937), 340–472.
[18] MacLane, S.: A combinatorial condition for planar graphs, Fund. Math. **28** (1937), 22–32.
[19] Mayeda, W.: Necessary and sufficient conditions for realizability of cut-set matrices. IRE Trans. **CT-7** (1969), 79–81.
[20] Sachs, H.: Einführung in die Theorie der endlichen Graphen, Teil I, Leipzig 1970.
[21] Sachs, H.: Einführung in die Theorie der endlichen Graphen, Teil II, Leipzig 1972.
[22] Tutte, W. T.: How to draw a graph, Proc. London Math. Soc. **13** (1963), 743–767.
[23] Tutte, W. T.: A theorem on planar graphs, Trans. Amer. Math. Soc. 82 (1956), 99–116.
[24] Tutte, W. T.: Separation of vertices by a circuit, Department of Combinatorics and Optimization, Research Report CORR 74-18, University of Waterloo, Ontario, 1974.
[25] Wagner, K.: Über eine Eigenschaft der ebenen Komplexe, Math. Ann. **114** (1937), 570–590.
[26] Wagner, K.: Eine Klasse minimaler nichtplättbarer Graphen, Math. Ann. **187** (1970), 104–113.
[27] Whitney, H.: Non-separable and planar graphs, Trans. Amer. Math. Soc. **34** (1932), 339–362.
[28] Whitney, H.: Planar graphs, Fund. Math. **21** (1933), 73–84.

Algorithms

Author Index

Subject Index